Günter Wagner

Vertebrate Limb Regeneration

Vertebrate Limb Regeneration

H. WALLACE

Department of Genetics
University of Birmingham

A Wiley–Interscience Publication

JOHN WILEY & SONS
Chichester · New York · Brisbane · Toronto

British Library Cataloguing in Publication Data:

Wallace, H.
  Vertebrate limb regeneration.
  1. Extremities (Anatomy)—Regeneration
  2. Vertebrates—Physiology
  I. Title
  596'.03'1   QP90-2   80-40963

  ISBN 0 471 27877 7

Typeset by Preface Limited, Salisbury, Wiltshire
and printed in Great Britain by Page Bros. (Norwich) Ltd.

Acknowledgements

I am indebted to all the investigators whose results form the basis of this text. Whenever possible, I have constructed new figures or tables to summarize their published data. Other figures and a few tables are copied here from the original ones published elsewhere, by kind permission of the authors, publishers and copyright holders listed below.

The Company of Biologists Ltd
The Society for Experimental Biology and Medicine
The Wistar Institute
Academic Press, Inc.
Springer-Verlag
Birkhäuser-Verlag

I am especially grateful to Dr. C. M. Illingworth for lending me the original photograph which illustrated her article 'Trapped fingers and amputated finger tips in children' in the *Journal of Pediatric Surgery*, volume 9, pages 853–858 (1974). It is reproduced here by permission of the author and Grune & Stratton, Inc.

Contents

Preface

When I first became interested in this subject about 10 years ago, I badly needed guidance from a book such as this of what was known about limb regeneration and what remained to be discovered. Picking out discrepancies between the most recent reviews allowed me to identify some areas of uncertainty, but it proved much more difficult to distinguish the reliably established aspects of limb regeneration from those which depended more on tradition than evidence. The distinction seemed to be missed, commonly by oversight but also sometimes deliberately obscured in an attempt to avoid controversy, even to the extent of ignoring evidence which conflicted with the accepted tradition. It is this conservative tradition which is mainly reflected in more general surveys of regeneration and which is summarized in the better textbooks of embryology and developmental biology, gaining strength from mere repetition. Consulting the original literature, especially where my own research drove me to do so, gradually convinced me that this traditional dogma contained several gaps and inconsistencies.

So this monograph was written with a certain degree of missionary zeal. It should provide the casual reader with enough detail to appreciate the point of most relevant investigations. It summarizes and compares the results of these investigations in a manner calculated either to clarify and reinforce the traditional view or to expose deficiencies in our knowledge and interpretation of the subject. In particular, it discriminates between alternative interpretations by attempting to justify or refute them. I have not dealt thoroughly with the older histological descriptions of limb regeneration nor with the modern histochemical ones, preferring to concentrate on a few examples which provided most insight into the subject. Apart from that area and a few inevitable oversights, this monograph assesses the merits of all investigations published during the last 50 years and incorporates all the useful information I could extract from the older literature.

Challenging a tradition has inevitably involved explicit criticism of the conclusions reached by several investigators. I trust they will take such criticism in good part. Whether they accept it or not, it should be less infuriating than an account which blandly ignored their efforts. Many errors and some tactless comments have already been pruned from the text on the advice of several colleagues. I am particularly grateful to Malcolm Maden for helping this book through a long gestation, and to Peggy Egar for acting as an emergency midwife. I shall try to blame them for any fault which has escaped their scrutiny, but cannot pretend they entirely share the opinions expressed here. I do not expect the reader to accept all my opinions either, but I should be disappointed if this book does not provoke some attempts to settle the outstanding issues.

University of Birmingham   H. WALLACE

Chapter 1

Descriptive

Regeneration is the formation of a missing structure by the rest of an organism. Like any other physiological process, it can be studied for its own intrinsic interest and possible medical applications. The replacement of what has been lost often occurs by a process which at least superficially resembles the initial development of the same structure. Consequently, one of the main motivations for investigating regeneration has been the hope of gaining some insight into developmental mechanisms. Amphibian limb regeneration is well suited to this purpose, being a rather precise process which usually ensures that the original and regenerated portions are integrated into an anatomically and functionally normal unit. Sometimes, however, the regenerate shows minor imperfections, most often by a reduction in size or an abnormal number of digits. In extreme cases, particularly after experimental intervention, there only occurs an outgrowth without an identifiable hand or foot. Whether or not such growths are termed regenerates has depended on the discretion of the investigator, a discretion which has been exercised much too freely. Various descriptions contain terms such as 'initiation of regeneration' or 'regenerative process' when the structure formed could equally well represent an abortive attempt at growing a hand, foot, tail, or even antlers. Unless such growths bear a reasonable resemblance to the missing structure, they cannot fairly be called regenerates even though they may share some of the developmental processes typical of true regeneration.

Virtually perfect regeneration occurs routinely after amputation in any limb region of many larval and adult urodeles (newts and salamanders), and in the tadpole larvae of frogs and toads. Nothing approaching the structure of a normal limb has been found after such amputations in postmetamorphic frogs or toads, despite several claims of regenerative tendencies either in natural conditions or after treatments intended to enhance regeneration.

Any regeneration which began during larval life is inhibited at the onset of metamorphosis in anurans, and is noticeably retarded during and after metamorphosis in urodeles. Amputated limbs or tails of adult *Salamandra*

1

regenerate very slowly and imperfectly (Roguski, 1961; de Conninck *et al.*, 1955). Schwidefsky (1935) found most *Triturus alpestris* limbs regenerated abnormally, while Scadding (1977) has listed similar cases and recorded that *Necturus maculosus* is incapable of limb regeneration. Several investigators have also remarked on a seasonal influence, in that regeneration occurs relatively slowly during the winter, even under laboratory conditions, and in complete darkness (Maier and Singer, 1977). Such changes in regenerative ability suggest a hormonal influence which will be considered in a later chapter.

The inhibition or interference associated with metamorphosis has caused most studies of regeneration to be performed either on young urodele larvae, whose rapid regeneration can be completed before metamorphosis, or on adult urodeles. The most consistently popular choices have been adults of the european crested newt, *Triturus cristatus*, and the eastern spotted newt of North America, *Notophthalmus viridescens*. These share the virtues of being fairly common and adaptable to laboratory conditions, and seem remarkably tolerant of crude surgery. Perhaps only such relatively aquatic adult urodeles are capable of consistent limb regeneration. Most studies on larvae have involved various species of American mole salamanders, especially the spotted salamander, *Ambystoma maculatum*. The axolotl, *A. mexicanum*, offers a peculiar advantage in having suppressed metamorphosis to retain its larval form and aquatic habit throughout life. Consequently, axolotls have become world-wide curiosities and are maintained routinely as laboratory stocks. Both young axolotls and adults have served in regeneration studies for more than a century (Philipeaux, 1867). Young specimens will regenerate an arm in about three weeks but adults take considerably longer, resembling other adult urodeles in this respect. Regeneration has only rarely been studied in other urodele larvae, even of those species which are used frequently as adults. Larvae of the common European salamander, *Salamandra salamandra*, have been used occasionally and regenerate well, although adults do not do so consistently. The ribbed newt, *Pleurodeles waltl*, is becoming more popular now that several breeding colonies have been established. Larvae of this species also regenerate rapidly and perfectly, but the intervention of metamorphosis impedes the process and sometimes causes structural defects to appear in the regenerate. The few species mentioned above have been reclassified enough times to give them an identity crisis. To avoid confusion, I have consistently translated the names used in the literature to the currently accepted ones by means of the conversion table shown here (Table 1.1).

Most observations on tadpoles and attempts to induce regeneration after their metamorphosis have been conducted on whatever anuran species happened to be available. Limb stumps of adult clawed toads, *Xenopus laevis*, grow quite striking cartilage spurs or spikes which lack articulations (Beetschen, 1952). The contrast between perfect regeneration of larval legs and the spikes produced at and after metamorphosis in *Xenopus*, demonstrated by Dent (1962), dissuade me from considering the latter as

Table 1.1 Correct names of urodeles, and synonyms found in the literature.

Correct name*	Previous names	Distribution	Vernacular names
1. Salamandridae			Newts and salamanders
Triturus cristatus	*Triton, Molge*	Europe, Asia	Crested newt, warty newt
Triturus vulgaris	*T. taeniatus*	Europe, Asia	Common newt, smooth newt
Notophthalmus viridescens	*Triturus, Diemyctylus*	N. America	Red-spotted newt
Cynops pyrrhogaster	*Triturus*	Japan	Fire-bellied newt
Pleurodeles waltl	*P. waltlii, P. watlii*	Spain, Morocco	Ribbed or Iberian newt
Salamandra salamandra	*S. maculosa*	Europe, S.W. Asia	Fire salamander
2. Ambystomidae			Mole salamanders
Ambystoma maculatum	*Amblystoma punctatum*	N. America	Spotted salamander
Ambystoma tigrinum	*Amblystoma*	N. America	Tiger salamander, Colorado axolotl invented for persistent larvae
Ambystoma mexicanum	*Siredon pisciformis*	Mexico	Axolotl, or Mexican axolotl to avoid confusion with above

*According to opinions 635, 649 and 662 of the International Committee on Zoological Nomenclature (*Bull. Zool. Nomen.* vols 19–20, 1962–3) and Wake (1976).

genuine regeneration. Although commonly described as regeneration, only very limited growth has been recorded for limb stumps of late metamorphic stages of *Rana* and *Bufo* (e.g. Michael *et al.*, 1969, 1970, 1972) and of juvenile or adult specimens of other anurans (Goode, 1967). Considerable efforts to enhance such growth have not revealed any reliable means of causing frogs or toads to regenerate.

Many lizards regenerate tails after autotomy or amputation, to achieve a structurally abnormal but fairly satisfactory functional replacement. Despite that, lizards seem as incapable of limb regeneration as any other amniote. Perhaps the best authenticated case of perfect structural regeneration on a limited scale occurs in man (Illingworth, 1974). Children up to the age of eleven can regenerate most of the terminal phalanx of a finger, including a finger nail, after minimal treatment. There is surprisingly little evidence on which to decide if other amniotes share this ability, or if it can be enhanced or extended in humans. Newly metamorphosed reed frogs, however, also regenerate a terminal phalanx and digital pad much more successfully than a larger arm segment (Richards *et al.*, 1977). Treatments intended to improve regeneration have proved as unsuccessful with lizards and mammals as they were with frogs. Chapters 2 and 5 provide some detail of these treatments and the whole topic will be resumed in the final chapter. Our understanding of vertebrate limb regeneration is necessarily restricted at present to amphibians, and overwhelmingly obtained from half a dozen urodele species.

1.1 Normal regeneration

The sequence of events following simple amputation is essentially the same, whatever the position of amputation on either the arm or leg of larval or adult urodeles (see Figure 1.1). The same sequence has also been described for the legs of tadpoles provided they were able to complete regeneration before the climax of metamorphosis. The end of the limb stump is rapidly covered by a distal migration of the epidermis. This forms a complete transparent sheet over the wound in less than a day for larvae or within a few days for adults. Very little cell division occurs during the epidermal migration and the thickening of the new apical epidermis is achieved initially by continued distal migration without local mitosis. The mesodermal tissues exposed by amputation are soon protected from external conditions, therefore, either by an initial blood clot or by the new epidermal covering. The amputation has damaged local mesoderm cells, however, and degenerative changes occur there during the next few days. Phagocytes accumulate at the tip of the limb stump to remove cell debris and to attack the extracellular matrix of the skeleton and connective tissue. Some fluid often accumulates at the end of the stump, giving it an oedematous or distended appearance. Erosion of the matrix liberates other mesodermal cells which are quite viable. Several investigators have identified these cells as 'activated', usually by the criterion of their enlarged nuclei and nucleoli, and

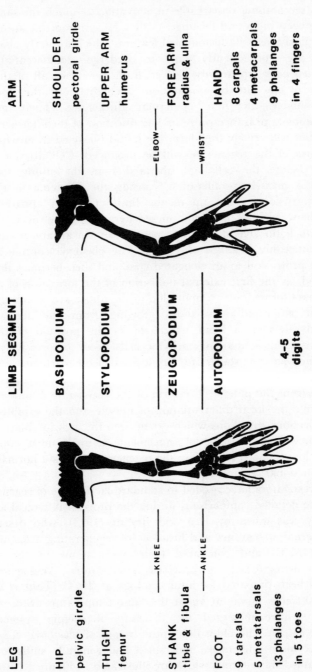

ARM

LEG

LIMB SEGMENT

SHOULDER
pectoral girdle

UPPER ARM
humerus

—ELBOW

FOREARM
radius & ulna

—WRIST

HAND
8 carpals
4 metacarpals
9 phalanges
in 4 fingers

BASIPODIUM

STYLOPODIUM

ZEUGOPODIUM

AUTOPODIUM

4-5
digits

HIP
pelvic girdle

THIGH
femur

—KNEE

SHANK
tibia & fibula

—ANKLE

FOOT
9 tarsals
5 metatarsals
13 phalanges
in 5 toes

Figure 1.1. Limb skeleton of an axolotl with the conventional terms used to describe different segments and skeletal elements.

defined activation as the first characteristic phase of regeneration. Other investigators see nothing remarkable in activation as much the same process can be recognized in wound healing. As they escape from the stump tissues, these cells quickly lose the histological features which formerly characterized them, and are consequently said to undergo dedifferentiation. The dedifferentiated cells, apparently derived from several or all of the internal tissues close to the site of amputation, accumulate at the apex of the limb-stump under the new epidermal thickening. Their generalized mesenchymatous appearance prevents identification of their former or future tissue type, but collectively they form the initial blastema from which all the internal tissues of the regenerate will be produced. Cell division probably begins even before the cells are liberated from the stump tissues and continues as a massive proliferation, causing the blastema to grow as a colourless protrusion at the tip of the limb. There is apparently some interaction between the blastemal mesenchyme and the overlying apical epidermis, which thickens during this period to form an apical cap without protruding noticeably (Tank et al., 1977). The blastema continues to grow from a slight protrusion to an elongated cone and then becomes flattened or palette-shaped, as the first external indication of the formation of a hand or foot. Blastemata formed after amputation in the upper arm reach this palette stage almost as rapidly as those growing from the fore-arm, and simultaneously develop a bend which marks the position of the elbow. Condensations of blastemal mesenchyme cells occur progressively between the cone and palette stages to produce skeletal cartilage. Additional cartilages are marked out later in the same way as successive digits protrude and elongate from the palette. The form of the regenerate is now essentially complete, while the local differentiation of muscle and the establishment of nervous connections (by axons which extend into the growing blastema) soon render the new limb capable of movement. Growth must continue for considerably longer, however, before the regenerate attains a normal size and becomes a functional replacement.

Several investigators have needed to standardize the rate of regeneration in order to make detailed comparisons during the process. A typical account of such a study was given by Iten and Bryant (1973), who described the changing external appearance and histology of regenerating arms of adult *N. viridescens* at 25°C, and compared their observations to the stages of regeneration distinguished by previous investigators. Analogous staging systems have been prepared for adult axolotls at 21 °C (Tank et al., 1976) and for *A. maculatum* larvae at about the same temperature (Stocum, 1979). These reports have increased the diversity of staging systems to an unreasonable extent. The older descriptive stages of blastema, cone, palette and notch are more universally understood and generally sufficient. I have generally retained these terms, which are illustrated in Figure 1.2 and related to other staging systems in Table 1.2. The rate of regeneration assessed by these recognizable stages is virtually the same whether an arm is amputated

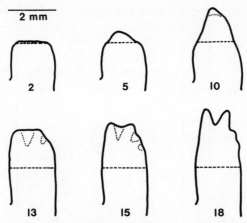

Figure 1.2. Successive stages of regeneration 2–18 days after amputating the arm of a 50 mm axolotl, to show the appearance of a healed stump, early blastema, cone, palette, notch and 2–3 digit regenerate.

through the humerus or through the radius and ulna. In the former case rapid growth continues for longer and thus compensates for the larger missing structure. That observation was first made by Spallanzani (1768) and supported by Iten and Bryant (1973) but may be only approximately true (Smith *et al.*, 1974). Table 1.2 shows that the rate of regeneration is quite sensitive to temperature. Schauble and Nentwig (1974) have investigated the effect of temperature on adult *N. viridescens*, finding a temperature optimum for normal regeneration at 20–25°C and that regeneration virtually ceases at 10°C. We can assume that the rate of limb regeneration shown in Table 1.2 for *N. viridescens* is typical of adult newts, as *T. cristatus* regenerates at a very similar speed (Schwidefsky, 1935; Smith *et al.*, 1974) although *T. vulgaris* and *T. alpestris* tend to regenerate more rapidly and abnormally. Limb regeneration occurs about twice as rapidly in larval urodeles, as shown here for young axolotls and reported in passing for several other species.

Table 1.2 only serves as a rough guide to the rate of regeneration, without predicting accurately the timing in different circumstances. Seasonal changes and conditions of illumination can cause an appreciable difference, as mentioned earlier, and so can experimental conditions. For instance, it is a common experimental practice to compare the regeneration of an operated arm to that of the contralateral 'control' arm. These control arms furnish excellent standards of normal regeneration for most purposes, being subject to the same general conditions as the experimental arms. Tweedle (1971), however, discovered that an arm stump regenerates more slowly if the contralateral arm has been amputated simultaneously. He traced this retardation to a nervous control of regeneration (see Chapter 2), for

Table 1.2 Descriptive stages of regeneration for urodele limbs.

Phase*	Stage*	Alternative description†	Adult *N. viridescens*		*A. mexicanum* at 20°C		
			At 25°C	At 20°C†	25 mm larva	50 mm larva	>150 nm adult§
Healing and dedifferentiation	Wound healing	—	0–5	0–2	0–1	0–1	0–2
	Early dedifferentiation	—	6–8	7	2	2–3	3–5
	Late dedifferentiation	Epidermal mound	8–11	13	3	4	6–7
Accumulation and growth	Moderate early bud	Accumulation	10–14	15	4	5	8–9
	Early bud	Bulb	12–17	19–21	5	6	10–12
	Medium bud	Blastema	14–20	21–25	6–7	7–8	13–15
Redifferentiation and morphogenesis	Late bud	Cone	18–24	25–30	8–9	9–11	16–20
	Palette	Paddle	22–28	31–35	10–12	12–14	21–25
	Early digit	2–3 digits (notch)	24–33	37–40	13–16	15–19	26–29
	Medium digit	4 digits	30–40	43–50	17–19	20–22	30–32
	Late digit	Complete	34–	48–	20–	23–	33–

Column group header: Time in days after bilateral amputation

*Phase, stage and times at 25°C from Iten and Bryant (1973).
†Compiled from Chalkley (1954) and others.
§cf. Tank *et al.* (1976).

amputation or merely cutting the nerve supply of one arm caused a mild reaction involving chromatolysis and some fibre degeneration of the neurons supplying the other arm. The degeneration of contralateral axons has also been observed by Maden (1977a). Most operations and all amputations inevitably cause considerable destruction of local nerve fibres, and must be presumed to produce this contralateral effect. The retardation caused by bilateral amputation, according to Tweedle (1971), is quite marked in larval axolotls, perhaps increasing the times shown in Table 1.2 by 25%, but much less evident in adult *N. viridescens*.

There are other minor differences between larvae and adults which will need to be considered later. The replacement of cartilage by bone and the thickening of the skin, both of which occur gradually in axolotls and more rapidly during metamorphosis of other urodels, as well as overall growth, probably all contribute to such differences. Hormonal changes certainly occur and can influence regeneration to a certain extent (see Chapter 3). Such differences, however, should not obscure the important generalization that the basic nature of limb regeneration is identical in larval and adult urodeles, as well as in anuran tadpoles. That is one reason why the preceding description of normal regeneration has been reduced to a single paragraph—apparently a travesty of the many painstaking histological studies devoted to it (reviewed by Schmidt, 1968). Even minute histological observations have rarely been able to discriminate between rival theories of regeneration, particularly concerning the origin and destiny of blastemal cells, or even to surmount the investigator's preconceptions. That provides a justification for considering only a few relatively specialized histological studies before examining in more detail the experimental analysis of regeneration.

The regeneration of amphibian tails shows the same sequence of events described here for limb regeneration: wound healing, local dedifferentiation, and the proliferation of blastemal cells occur in rapid succession. Growth of the apical blastema can continue for a long period while the basal regions redifferentiate, accentuating the proximo-distal sequence which can also be recognized in limb regenerates (Iten and Bryant, 1976a,b). Tadpoles regenerate a perfect replica of the amputated tail tip, sometimes thus delaying its resorption during metamorphosis. Larval urodeles show a curious modification in tail regeneration. The axial skeleton differentiates as a series of vertebrae whose centra replace the unregenerated notochord, anticipating a change which usually occurs at metamorphosis (Holtzer, 1956). The similarities are so great that we need not hesitate to generalize observations on regeneration of tails or limbs.

1.2 Quantitative histology

Chalkley (1954) conducted a meticulous study of normal limb regeneration in a homogeneous group of adult *N. viridescens* kept at 20.5°C and fed

10

sufficiently to maintain their size. He amputated arms 2 mm above the elbow and fixed samples at 6 day intervals to provide serial transverse sections of the regenerate and neighbouring portion of the arm-stump. He then counted the cells and mitotic figures in selected sections 160 μm apart, and assigned them as far as possible to particular tissues.

Although the epidermis covered the wound within a day as a layer four to five cells thick, and later thickened to 14–15 cells when forming an apical cap during the third week, Chalkley recorded that epidermal mitoses were most frequent proximal to the amputation site during this period. Subsequently, as the early blastema elongated to the cone and palette stages,

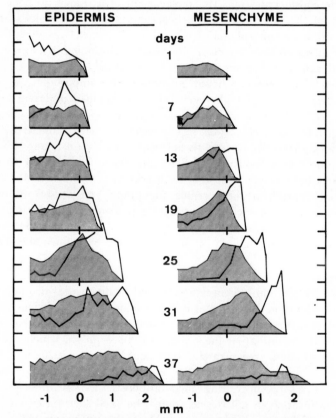

Figure 1.3. Distribution of cell number and mitotic index along the distal arm stump and growing regenerate of *N. viridescens* (modified from Chalkley, 1954). Vertical scale intervals are 500 cells/section (shaded areas) or 0.5% increments of mitotic index (thick lines). Cell division is stimulated in the distal 1.5 mm of the stump and continues there for at least 3 weeks, but becomes more pronounced at the apex of the growing blastema. Abscissa: 0 mm is the amputation plane. (Reproduced by permission of The Wistar Institute.)

the apical epidermis reverted to a normal thickness of three to four cells with a new basement membrane and renewed mitotic activity, most divisions occurring at the base of the regenerate (Figure 1.3). Chalkley concluded on this basis that the epidermis did not contribute any appreciable number of cells to the internal blastemal mesenchyme. Manner (1953) had reached the same conclusion after counting mitotic figures in the same species. Both these studies thus denied the possibility that epidermal cells were converted into mesenchyme, which had been proposed by Godlewski (1928) and supported by Rose (1948) on the basis of cell counts as well as experimental results to be considered later. Rose (1948) detected a sharp drop in epidermal cells counts during an early period in regeneration. Chalkley (1954), however, recorded a steady increase of epidermal cells over the regenerate with remarkably little fluctuation. It is worth remembering that adult newts cast the outer layers of their skin at unpredictable intervals and that specimens kept together under the same conditions may do so simultaneously. Such moulting affects the epidermis of the regenerate and could cause cell counts to fall dramatically, while only being coincidental to regeneration.

Chalkley recorded a steady increase of internal cells for 30 days after amputation, accompanied by distal growth, so that most mesenchyme cells and mitoses were near the stump apex for the initial 20 days but later occurred in the growing blastema. There was no abrupt transition between stump and regenerate for such counts (Figure 1.3) which thus concerned a uniform population of cells engaged in a continuous process. Chalkley assessed this process as the formation of a few thousand mesenchyme cells by dedifferentiation during the first 3–4 days, a later increase mostly by mitotic proliferation in the stump to over 60 000 cells, which migrate distally and divide once or twice more on average in the growing regenerate before redifferentiation sets in (Figure 1.4). These phases overlap, of course, and migration continues from the stump into the growing regenerate. Only initially could a reasonable proportion of the cells be identified as belonging to a particular stump tissue, but some dedifferentiating and dividing cells could be assigned to most tissues present—periosteum, muscle, and various connective tissue layers. All mesodermal tissues seemed to be involved in the production of blastemal cells, therefore, and roughly in proportion to the numbers of cells of each tissue in a cross-section of the stump apex. In other words, the predominance of connective tissue mitoses (85%) perhaps only reflected the ubiquity of connective tissue. This observation acts as a corrective for a considerable number of earlier studies which had described dedifferentiation in one tissue or another and often given the impression that one particular tissue might be the principal or even exclusive source of blastemal mesenchyme. Chalkley's data show that the mesenchyme could be derived entirely from the mesoderm of the stump apex, but no particular tissue could be identified as the origin on the basis of histological observation. Furthermore, as Chalkley (1959) emphasized, his own and

12

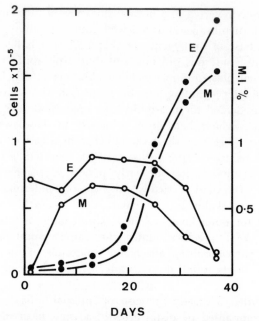

Figure 1.4. Total number of cells (solid circles) in the blastemal mesenchyme (M) and the epidermis (E) with the corresponding mitotic indices (open circles) during regeneration of *N. viridescens* arms (from data of Chalkley, 1954). Note how a declining mitotic index maintains the steady increase in cell number during the fourth and fifth weeks. (Reproduced by permission of The Wistar Institute.)

similar studies are quite compatible with the notion that blastemal cells remain determined to revert to their previous tissue-specificity as soon as circumstances permit or, alternatively, that an unidentified population of undifferentiated cells in the stump proliferates to form most of the blastema and perhaps all the regenerated mesoderm. Both of these concepts have been proposed repeatedly under the titles of modulation, reserve cells and neoblasts (e.g. Weiss, 1939, 1973; Holzer, in discussion of Hay, 1970) and both, I believe, have been invalidated by experimental means.

1.3 Autoradiography

Hay and Fischman (1961) found that they could distinguish between the movements of epidermal and internal tissues, during limb regeneration of adult *N. viridescens*, by injecting specimens with radioactive thymidine and later identifying labelled cells in autoradiographs of longitudinal sections. They administered four injections of 1.25 μCi tritiated thymidine at hourly

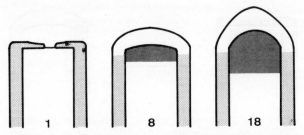

Figure 1.5. Newt arm stumps fixed after a 4 hour
exposure to tritiated thymidine to show the distribution
of S-phase cells, 1, 8, and 18 days after amputation.
Shading indicates the intensity of labelling by
autoradiograph silver grains in this and Figures 1.6 and
1.7, which are all modified from Hay and Fischman
(1961). Epidermal cells of the stump become labelled,
but not those of the wound epithelium and apical cap.
The only labelled internal cells are those which have
dedifferentiated to form a blastema.

intervals and fixed the arms either 1 hour later or after delays of up to 28
days. Most of the thymidine is incorporated into chromosomal DNA as it is
synthesized by cells preparing for division, and consequently their nuclei are
marked by silver grains after the autoradiographic processing. Prompt
fixation after the series of injections should identify those cells which were
undergoing DNA synthesis (cells in S-phase) while the radioactive tracer was
available, 4 hours in this case (Figure 1.5). A series of preparations fixed at
intervals after labelling can reveal mass cellular movement and the
occurrence of cell division, for the amount of radioactivity in each nucleus is
apportioned more or less equally to each daughter nucleus and is thus
halved at each division, provided the tracer is used up soon after being
injected.

In the absence of amputation, the only cells to become labelled in the
adult arm are those of the epidermis and some circulating blood cells. Cells
at the base of the epidermis divide routinely in order to replace the
outermost cells which become keratinized and are eventually sloughed off in
moulting. Although only a small proportion of the basal cells are actually
dividing at any one time, their extensive S-phase and the prolonged labelling
period conspire to label about 15% of them. This pattern of exclusively
epidermal labelling is maintained even after amputation by most of the arm
to within a few millimetres of the stump apex, corroborating earlier
observations and experimental findings that a regenerate is produced only by
cells close to the site of amputation.

Labelling shortly before amputation, Hay and Fischman traced the
epidermis as it migrated over the wound surface to form an apical cap
(Figure 1.6). Some of the cap cells divided 3–4 days after amputation, but
no further dilution of label was observed even when the epidermis expanded

14

Figure 1.6. Newt arms exposed to tritiated thymidine shortly before amputation. Only the epidermis becomes labelled, as shown by fixing 1 day later. The apical epidermis is still heavily labelled 15 days later (for little division has occurred there) but no blastemal cells are labelled.

and thickened over the initial blastema. Some apical epidermal cells remained strongly labelled until they were sloughed off 20–28 days later. The epidermis immediately proximal to the plane of amputation gradually lost its label by repeated divisions during this time and continued to migrate distally, augmenting and eventually replacing the original apical cap. Labelled leucocytes rapidly accumulated in the oedematous wounded region, persisted for about 2 weeks and sometimes apparently fused to produce multinucleate osteoclasts (Fischman and Hay, 1962). The mesenchymal cells which consolidated into a blastema during the second and third weeks, however, consisted exclusively of unlabelled cells—and thus almost certainly originated from the unlabelled internal tissues of the limb stump.

This labelled pattern could be completely reversed by injecting tritiated thymidine at various intervals after amputation (Figure 1.7) providing complementary evidence which reinforces the interpretation given in the last paragraph. The new apical epidermis showed a subnormal proportion of labelled cells, 5% initially and even less later, which demonstrated the virtual

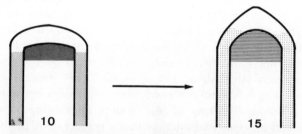

Figure 1.7. Newt arms exposed to tritiated thymidine 10 days after amputation to label the early blastema and stump epidermis but not the wound epithelium. Fixing 5 days later reveals a general dilution of label (by cell division) accompanied by blastemal growth, and a distal movement of the epidermis.

absence of DNA synthesis (and hence cell division) from 3 days after amputation until the blastema began to elongate into a cone some 3 weeks later. A zone of proliferation could be detected, however, in the epidermis near the end of the stump, showing an increased frequency of labelled cells (25–35%) between 5 and 10 days after amputation. Comparing a series of specimens fixed at increasing intervals after the injection revealed that these epidermal cells moved distally as their labelling intensity decreased. Presumably, the local proliferation compensated for and contributed to the distal migration of non-dividing apical epidermis. Mesodermal cells within 1 mm of the wound surface began DNA synthesis, and thus became labelled, as they dedifferentiated and escaped from the muscle, periosteum and layers of connective tissue, beginning as early as 4 days after amputation. The subsequent steady increase in labelled mesenchymal cells, either at later labelling times or after longer intervals from the same labelling time, indicated a continued proliferation and probably also some continued recruitment of dedifferentiated cells from the internal tissues adjacent to the end of the stump. By 20 days after amputation, the autoradiographs revealed labelled blastemal mesenchyme under a virtually unlabelled apical epidermis.

Many of the observations mentioned above are duplicated in a very similar study also conducted on adult *N. viridescens* by O'Steen and Walker (1961). By means of a larger dose of tritiated thymidine, a single injection of 15 µCi, they traced dedifferentiating cells from internal stump tissues and early blastemal mesenchyme to all tissues of the regenerate except for cartilage. They found no evidence of any interconversion of epidermal and internal cells during regeneration, but found it difficult to exclude that possibility because injections as much as 5 days prior to amputation still resulted in labelling some blastemal mesenchyme. Injections 10 days before amputation, however, did confine labelling to the epidermis both of the stump and of the regenerate. O'Steen and Walker concluded that there could be no substantial contribution of epidermal cells to the blastemal mesenchyme. That conclusion is a provisional one, for it depends upon the validity of their other conclusion to explain the discrepancy in their results: that a detectable fraction of the tracer persisted in the body for a period exceeding 5 days. There is no firm basis for disputing this contention, but it has been generally ignored and perhaps disbelieved. Hay and Fischman (1961) could detect no remaining radioactivity in blood plasma 3 hours after their final injection of thymidine, which is supported by the finding that as much injected thymidine is taken up by the skin of larval *A. tigrinum* in 2 as in 4 hours (Scheving and Chiakulus, 1965). These results concern relatively low doses of the tracer, however, and it is possible that some free thymidine may persist in intracellular pools until it can be incorporated in DNA during the next S-phase several days later. Callan and Taylor (1968) observed a marked persistence of thymidine which labelled cells during two successive S-phases following a single 20 µCi injection into adult *T. vulgaris*, and Steen

(1968) recorded that doses of 5–15 μCi/g persist for longer than 24 hours in axolotls.

The clarity and consistency of the two autoradiographic studies described here provide strong support for the earlier conclusion of Chalkley (1954) that many or all of the internal tissues of the stump provide dedifferentiated cells which proliferate sufficiently to form all the blastemal mesenchyme. The autoradiographic observations also constitute evidence that virtually no mesenchymal cells are derived from the epidermis. Rose and Rose (1965) believed they could trace the conversion of some basal epidermal cells into mesenchyme and later into new chondrocytes of the early regenerate by means of essentially the same labelling technique, a 3 μCi injection administered within 1 day before or after amputation. It is quite likely that the labelled mesenchyme cells recorded by Rose and Rose can be ascribed to a delayed incorporation of persisting tracer, as their observations merely confirmed the general labelling described by O'Steen and Walker (1961) whose interpretation has not been refuted.

1.4   Cell cycles

Labelling with tritiated thymidine can be used to obtain fairly precise estimates of the average time spent by cells between one division and the next, the mitotic cycle time (CT), as well as of the interphase periods (G$_1$ and G$_2$) before and after the S-phase when DNA replication occurs. These periods can be deduced from a graph of the proportion of division figures which are labelled in samples fixed at intervals after administration of a pulse of radioactive tracer. Although this technique was devised for injections

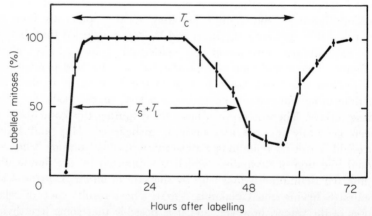

Figure 1.8. Estimation of the cell cycle in blastemal mesenchyme of young axolotls (simplified from Wallace and Maden, 1976b). Labelling was confined to a 2 hour pulse (T_l), terminated by a cold thymidine chase. The duration of an average cycle(T_c), of the S-phase (T_s) and of G2 can be read directly from 50% intercepts of the graph.

into whole animals (Quastler and Sherman, 1959), it has proved far more amenable to cells in tissue culture and plant root-tips which can be exposed to a brief and well-defined pulse of thymidine. The amounts of the tracer which need to be injected into whole animals, in order to achieve adequate labelling, are liable to persist for perhaps as long as the cell-cycle itself. Despite this difficulty, it is of considerable interest to apply the technique to regenerating limbs where an increased mitotic index in the blastema suggests that many cells are dividing regularly and perhaps more rapidly than usual.

Figures 1.8 and 1.9 illustrate an attempt to determine the length of the cell cycle in the limb blastema of young axolotls. A single injection of tritiated thymidine at about 1 μCi/g body weight caused all dividing cells to become labelled and must have persisted in the body at least 15 hours for that to happen. A chase of non-radioactive thymidine, at a concentration of 1000 times that of the tracer, successfully terminated the labelling period and thus achieved fairly reliable results, which are shown in Table 1.3. A few similar studies had been conducted previously on several tissues of *N. viridescens*, but always using the more laborious technique of identifying labelled division figures in sections after an undefined period of labelling. All the results obtained from *N. viridescens* reveal distortions of the labelled-mitosis curve which can be attributed to a prolonged labelling period. Unless compensated in some way (e.g. statistically, as Yamada *et al.* (1975) did), these results yield over-estimates of the S-phase and under-estimates of the total cycle. It seems clear that subtle and perhaps unconscious compensations have been applied to the other results. The estimates for blastemal cells and adjacent connective tissue of *N. viridescens* appear starkly different in Table 1.3, for instance, although the labelled-mitosis curves from which they are obtained could well be identical. The data in Table 1.3, therefore, only permit rough generalizations. Both the total cycle time and the S-phase are extremely protracted, considering the rapid blastemal growth which they support. That might be expected on the basis of the large amounts of nuclear DNA in urodele cells. The postulated relationship between DNA content, S-phase and cycle time documented in higher plants by Van't Hof (1965) and Evans and Rees (1971) predicts that these times should be perceptibly longer in *N. viridescens* than in the axolotl. Table 1.3 does not support that prediction, but probably does not rule it out either. Regeneration clearly involves an increase in the number of dividing cells, beyond those normally involved in the growth or maintenance of tissues. It may also involve an acceleration of the cell cycle to a degree which might be characteristic for different organs and tissues, or which might alter characteristically during the process of regeneration as seems likely for iris epithelium.

Contrary to these suppositions, labelling studies and estimates of the mitotic index depict the growing blastema as a homogeneous population of proliferating cells and generally confirm conclusions based on cell counts. Probably the earliest cell counts were those of Litwiller (1939), who followed arm regeneration in *Cynops pyrrhogaster* for 100 days at an undisclosed

Table 1.3 Duration in hours of cell-cycle phases in adult *N. viridescens* and larval *A. mexicanum*.

Tissue or region	Temperature (°C)	$G_1 + \frac{1}{2}M$	S	$G_2 + \frac{1}{2}M$	Total	References
Newt						
Intestinal epithelium	20	13.9	55	5.8	74.7	Grillo and Urso (1968)
Limb blastema	20	4.5	34–41	3	45	Grillo (1971)
Limb connective tissue	20	15	30–40	5	50–60	Grillo (1971)
Regenerating lens	22	44	19	2	65	Zalik and Yamada (1967)
Iris forming lens	21–2	11	27	8	46	Yamada et al. (1975)
Rest of iris	21–2	30	40	8	78	Yamada et al. (1975)
Axolotl						
Fore-arm blastema	20	10	38	5	53	Wallace and Maden (1976b)
Fore-arm blastema	20	11.5	30	6.5	48	Maden (1976)
Upper-arm blastema	20	13	28	7	48	Maden (1976)
Upper-arm blastema	21.5	2.4	32	5.6	40	McCullough and Tassava (1976)
Same but denervated	21.5	16.6	30.5	6.4	53.5	Tassava and McCullough (1978)

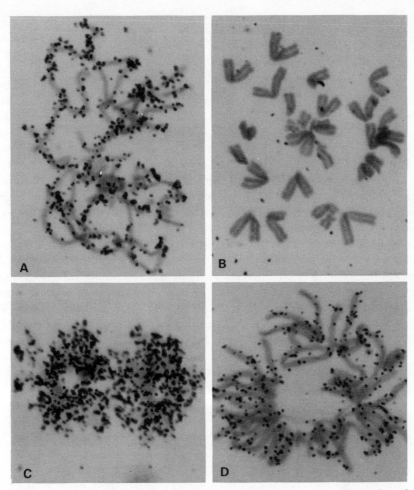

Figure 1.9. Examples of blastemal division figures in autoradiographs of squashed preparations (from Wallace and Maden, 1976b). A, B, labelled prophase and unlabelled metaphase from the same preparation at 6 hours; C, D, labelled anaphase fixed at 18 hours and a second metaphase fixed at 64 hours from the same specimen.

temperature. He estimated that dedifferentiation resulted in about 8000 mesenchymal cells before he could detect any cell division. Each of these cells must have divided 5–6 times on average to produce a population of 340 000 in the late regenerate. Chalkley's estimates (Figure 1.4) are quite similar: less than 5000 dedifferentiated cells proliferate to about 150 000 in adult *N. viridescens*, with each founder cell engaging in at least five successive divisions at 2–3 day intervals mainly during the third and fourth weeks after amputation. More precise information has been obtained from labelling studies on larval axolotls, where the proportion of cells in S-phase begins to increase 4 days after amputation and the mitotic index rises on the following day (Tassava *et al.*, 1974, 1976). Maden (1976) has confirmed that dedifferentiation predominates during the first 4 days after amputation and is confined to the apical 0.4 mm of the arm stump. This produces a small blastema of about 5000 cells. Dedifferentiation must continue throughout the first week, however, for the number of blastemal cells increases more rapidly than can be accounted for by cell division. The second week is devoted entirely to cell division which occurs at 2 day intervals throughout the blastema until basal cells stop dividing in order to differentiate. On this time scale, five successive divisions of the progenitor cells could occur before the forearm blastema reaches a palette stage of about 50 000 cells and skeletal primordia first appear. Slightly later redifferentiation in upper arm blastemata allows an extra cycle of cell division and commensurate extra growth at the palette stage, but blastemata on the upper arm and those on the forearm show virtually identical cell cycles (Table 1.3).

Consideration of these cells cycle studies amply confirms O'Steen and Walker's (1961) observation of marked persistence of injected thymidine. I advocate the use of a chase injection to dilute out remaining tracer and thus effectively terminate its incorporation, certainly for cell cycle analyses and probably for other studies which involve radioactive labelling. There is always a danger, of course, that an excessive dose of thymidine might inhibit DNA synthesis, thus perturb the cell cycle and eventually hamper regeneration. Urodeles seem quite tolerant of thymidine, however, as doses of 2.5 mg cause no harm to adult newts (Callan and Taylor, 1968) so that an injection calculated to produce less than 1 mM thymidine in the whole body volume will probably benefit subsequent analysis.

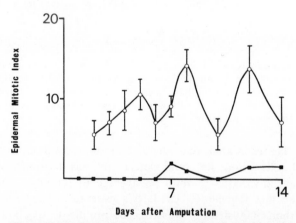

Days after Amputation

Figure 1.10. Peaks of cell division in the epidermis recorded 5, 8 and 12 days after amputation of young axolotl arm stumps (open circles). Irradiated arms (solid squares) show very few epidermal divisions. (From Maden and Wallace, 1976: reproduced by permission of The Wistar Institute.)

Cell division is apparently unsynchronized in the growing cone stage blastema, perhaps as a consequence of the gradual release of cells from dedifferentiating stump tissue as well as an inherent variability in the cell cycle itself. Some diurnal rhythmns of the mitotic index have been suggested (Litwiller, 1940), but have not been pursued in any detail. Epidermal divisions are more closely related to the original stimulus of amputation and seem initially to be synchronized to a considerable extent (Figure 1.10). Such synchrony may be entrained by external stimuli such as day-length and temperature fluctuations, but suggests the existence of a 3 day epidermal cell cycle.

1.5 Summary

Regeneration is defined as the formation of a structure which replaces, and must bear a reasonable resemblance to, a missing structure. Limb regeneration probably occurs universally among larval amphibians and commonly in adult urodeles, but is inhibited at metamorphosis in anurans. Studies of limb regeneration have thus been overwhelmingly restricted to a few species of newts and salamanders and less frequent observations on tadpoles. The normal course of limb (and tail) regeneration shows the same sequence of events in all these species. Wound healing occurs by a distal migration of epidermal cells which later accumulate as an apical cap. This may direct the subsequent regeneration of internal tissues but does not contribute any cells to them. All internal tissues at the apex of the limb stump dedifferentiate and liberate mesenchymal cells which proliferate and accumulate under the apical cap as a blastema. The blastema grows by continued cell division and a wave of redifferentiation moving distally from its base. The gradual elongation and disposition of new skeletal elements allow several stages of regeneration to be distinguished. These occur in a strict chronological sequence and so prove useful both for description and analysis.

Quantitative histology by means of cell counts and selective labelling of tissues with radioactive tracers substantiate and add precision to this general description of regeneration. Analyses of the cell cycle in proliferating blastemal cells have begun to provide details of a fundamental aspect of regeneration.

Chapter 2

Nervous Control

It has been suspected for 150 years that intact limb nerves are required for at least the initial stages of regeneration (Todd, 1823). That suspicion has gradually gained strength despite investigators who denied the existence of any nervous influence, advocated diverse mechanisms of nerve action, or ascribed a major importance to one type of nerve or another. Although such postulates are interesting and instructive in revealing the kinds of error which can creep into experimental design, their variety tends to confuse the issue. The following account is therefore restricted to a few notable examples from the older literature which are often overlooked, before considering in detail the numerous experiments performed since 1940.

Goldfarb (1909, 1911) conducted a series of operations designed to destroy either the sensory or the motor innervation of an amuptated hind-limb of *Notophthalmus viridescens*. He recorded a variable number of successful regenerates after each type of operation, although regeneration was often considerably delayed. He ascribed all delays and failures of regeneration to the trauma of the operation, which must have been considerable when he amputated both legs and the entire tail, and then reamed out the sacral spinal canal. His conclusion that nerves were not needed for regeneration depended more on argument than fact, for he only obtained two completely denervated legs, neither of which regenerated. All other cases retained a partial nerve supply to the experimental limb, which must often have been augmented by the growth of severed and adjacent nerves as the use of the limb was frequently regained. The more precise conclusion that no particular kind of nerve is exclusively required for regeneration is in fact buried in the text (Goldfarb, 1909), but did not provide the sweeping generalization required of the time.

Locatelli (1924, 1925) severed the lumbar sensory or motor nerves of adult *Triturus cristatus*, found that only the former operation regularly inhibited regeneration, and sensibly concluded that the regenerating leg normally requires a supply of sensory nerves. These results have been repeatedly confirmed, although her interpretation has been modified (Singer, 1946b).

22

On deviating brachial nerves from the arm to the shoulder region, Locatelli (1925, 1929) was able to induce the formation of accessory arms and ascribed a morphogenetic power to the nerves on that basis. These results have also been confirmed but the conclusion disputed on the grounds of similar and contemporary experiments which will be considered later (see chapter 8). By extirpating the spinal cord from the lumbar region, or alternatively by severing the central connections of the dorsal ganglia, Locatelli established that the only nervous requirement for regeneration lay in the integrity of the connections from the ganglia to the limb.

Schotté (1926a) reported the results of an enormous number of operations in which all the nerves were severed at varying intervals before and after amputation in the brachial plexus of *T. cristatus*. He also obtained enough similar results on the sciatic (or lumbo-sacral) plexus, besides using adult *Salamandra salamandra* and larvae of both species, to ensure his conclusions concerned a general property of limb regeneration in urodeles. Schotté's main conclusion was that urodele limbs will not regenerate while they remain devoid of nerves. His results firmly established this basic point for the first time, exposed the fallacy of Goldfarb's earlier arguments, and found support in the similar results obtained by Locatelli and Weiss (1925) during the same period. Schotté (1926a) clarified two aspects of the nervous influence which had previously caused much confusion. Firstly, transected nerves grow back into the limb quite quickly and completely. Consequently, if amputation is delayed for a few weeks, the limb-stump is liable to be re-innervated and to regenerate normally. Secondly, the converse experiment of delaying denervation for increasing periods after amputation revealed that regeneration is only dependent on nerves for a limited period: the palette and later stages continue to develop even after denervation (cf. Grim and Carlson, 1979). Schotté believed these results supported his earlier proposal that regeneration depended principally on the sympathetic nervous system, a conclusion questioned by Guyénot and Schotté (1926a,b), criticized by Locatelli (1929) and finally refuted much later by Singer (1942).

The general nervous control of early regeneration has since been confirmed and explored in more detail, by means of total denervation at varying times shortly before or after amputation on the arms of larval *Ambystoma maculatum* (Butler and Schotté, 1941; 1949; Schotté and Butler, 1941; 1944) and on the legs of frog tadpoles (Schotté and Harland, 1943b). The most striking novelty of these experiments was that amputated larval limbs rarely formed a persistent stump as is commonly found in adult newts after these operations; they usually continued to regress (be resorbed) as long as they remained denervated, and they began to regenerate whenever nerves returned to the limb apex. Larval limb regeneration could only be prevented by repeated resection of the nerves at weekly intervals. The generally accepted explanation of this difference is that the amputated stump of a denervated adult arm heals over and forms scar-tissue in such a way that regrowing nerves cannot fully penetrate to the apex. Even when re-innervated, such stumps rarely regenerate unless amputated once more. The denervated and amputated larval arm, on the other hand, regresses by a continuous erosion of the stump apex which remains unscarred with a perpetually replaced thin epidermal covering and which thus can be completely penetrated by regrowing nerves. Various commentators have suggested that these differences between larval and adult arms indicate that regeneration may require a re-innervation of the apical epidermis. Sidman and Singer (1960), however, devised a means of testing this proposition by extirpating the dorsal ganglia to permanently deprive the arms of sensory nerves, and transecting the ventral roots to allow eventual re-innervation by motor nerves which are not expected to enter the epidermis. Almost half these arms regenerated normally when amputated several months later, yet showed no sensory reactions and only occasional nerve fibres in the apical epidermis of specimens sampled during the course of regeneration and stained by Bodian's technique (which has limitations, see Egar and Singer, 1971).

This discriminating experiment was only one of the ingenious manipulations which became practicable as a result of Singer's lengthy series of denervations performed on adult *N. viridescens* (Singer, 1942; 1943; 1945; 1946a,b; 1947a,b; 1949; 1952). The principal operations involved severing virtually all combinations of 3rd, 4th and 5th spinal nerves, or either their dorsal or ventral roots before they merged in the brachial plexus (Figure 2.1). An operation on one side was followed by amputation of both arms above the elbow. Only the early phases of regeneration were recorded, and

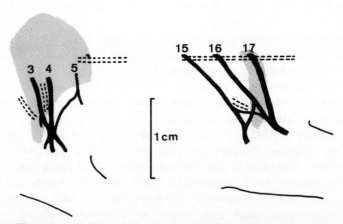

Figure 2.1. Left brachial plexus and sciatic plexus of an adult axolotl. Spinal nerves 3–5 can be exposed by lifting the back of the scapula (stippled). Spinal nerves 15–17 (or sometimes 16–18) are deeply embedded in muscle under the ilium (stippled). Additional landmarks are provided by the base of the limbs (in outline) and major blood vessels (broken lines).

an experimental arm was classified as inhibited if it did not begin to regenerate within a few days after the contralateral control arm. The limited observations seem justified as cases of delayed regeneration occurred only rarely and then more than a month later when severed nerves could have regrown into the limb. The standardized surgical procedures allow comparisons to be made between different series of experiments, and the results show an enviable degree of consistency.

Singer first re-examined the findings described in the preceding paragraphs. He found that severing both the sensory and motor components of all three brachial nerves consistently prevented regeneration of arms that were amputated immediately after denervation (Singer, 1942). This result confirmed the major conclusions of Schotté (1926a), but went further to show that regeneration was inhibited whether the sympathetic nerves were also severed or if they were left intact. Conversely, severing only the sympathetic nerves did not affect regeneration. Cutting the ventral roots leading to the brachial plexus rarely prevented regeneration, and some of the anomalous cases of inhibition could be discounted when later examination showed their dorsal ganglia had been unintentionally damaged (Singer, 1943). Extirpating the dorsal ganglia of all three spinal nerves without damaging the ventral roots yielded a uniform failure of regeneration of those arms which were amputated immediately afterwards (Singer, 1945). These two sets of results duplicated those of Locatelli (1929), and confirmed the basic conclusion that the normal sensory innervation is sufficient to support regeneration, but motor nerves are not. Singer perceived that this contrast need not discriminate between the nature and function of the two types of innervation, for the sensory nerve fibres are considerably more abundant than the motor fibres, while some experimental situations could be interpreted better in terms of the quantity (as opposed to the type) of innervation.

2.1 Quantitative hypothesis

A rather elaborate combination of observation and experiment was required to establish this point. Singer (1946b) severed spinal nerves or their ventral roots so as to reduce the brachial innervation to that of a single segment or just its sensory component. Allowing 23 days for isolated axons to degenerate, he then counted the remaining nerve fibres in sections taken at the usual amputation level, two-thirds down the upper arm. Such counts after the six different operations were corrected for a small amount of regrowth by severed nerves, and allowed the number of motor fibres to be estimated as the difference between total and sensory fibres. Amputation immediately after 23 types of partial denervation in series of 5–20 specimens allowed a rough assessment of the frequency of regeneration. The results showed a striking correlation between the number of intact nerve fibres and the frequency of regeneration, establishing Singer's quantitative hypothesis, which can be summarized as follows for adult *N. viridescens* amputated in the upper arm.

(1) Regeneration always occurs if more than 50% of the normal nerve supply is present, but never happens if less than 30% of nerve fibres remain intact.

(2) Within the intermediate 'threshold' range of 30% to 50%, any increase in the nerve supply results in an increased frequency of regenerating specimens.

(3) The ability to regenerate appears to be independent of whether the nerve supply is composed of sensory or motor fibres.

These conclusions can be illustrated from data in Table 2.1, which shows that each of the three spinal nerves has insufficient fibres to ensure regeneration. Regeneration does occur with merely the total sensory innervation. The normal motor innervation is insufficient by itself to support regeneration, but can reinforce a depleted sensory supply for that purpose. The complete 4th and 5th spinal nerves ensure regeneration, for instance, but their sensory elements sufficed in only 80% of the cases. Singer (1946a) had previously discovered an even more striking case, where regeneration occurred in arms deprived of all sensory innervation. The sensory nerves were permanently removed by extirpation of the three dorsal ganglia and the motor nerve roots were severed in the same operation. The motor nerves regrew, perhaps following all available pathways including those left vacant by degenerating sensory axons, and may have branched to become abnormally dense. When amputated 4–5 months later, about half of the specimens regenerated their insensate arms. Six of these regenerated arms were then re-amputated with resection of the motor nerve roots, and failed to regenerate now that their nerve supply was virtually eliminated. This experiment, repeated and confirmed by Sidman and Singer (1960), surely proves that the nervous control of regeneration can be exercised by motor fibres as well as by sensory ones. It does not really impinge on the

Table 2.1 Contribution of different spinal nerves to the number of major fibres in the upper arm of adult *N. viridescens* (data of Singer, 1946b).

Spinal nerve	Estimated number of fibres			Result of removing all other nerves	
	Sensory	Motor	Total	Nerves remaining (%)	Cases regenerate (%)
3	793	7	800	30	56
4	911	374	1285	50	75
5	304	194	498	20	0
Total	2008	575	2583	100	100

Complete sensory innervation alone (80% of all fibres) always allows regeneration.
Complete motor innervation alone (20% of all fibres) is inadequate for regeneration.

Table 2.2 Contribution of different spinal nerves to the number of major fibres in various regions of the arm of adult *N. viridescens* (data of Singer, 1947a, b).

Region	Number of fibres from each spinal nerve			Independent estimate of total ($\pm$ S.D.)	Density in fibres/ mm^2 of amputation surface	Median threshold value for regeneration in fibres/mm^2
	3	4	5			
Upper arm	800	1285	498	2357 $\pm$ 292	2380	960
Fore-arm	792	1161	309	2125 $\pm$ 238	1950	940
Hand	506	807	196	1580 $\pm$ 286	1380	320
2nd digit	177	170	34	333 $\pm$ 50	2000	860
3rd digit	192	233	61	458 $\pm$ 112	1910	880
4th digit	90	136	23	242 $\pm$ 71	1920	810

Note that the number of fibres decreases distally, but their density changes only slightly (except for the hand) and each spinal nerve contributes a fairly constant proportion even in the different digits.

quantitative hypothesis, however, as no nerve counts were performed and no controls involving delayed amputation exist.

The sympathetic nervous system, which Singer (1942) found had no detectable influence on regeneration, furnishes less than 5% of the total number of fibres. That is too small a fraction to make any appreciable difference to the estimated frequencies of regeneration. The sympathetic nerves have consequently been ignored in the preceding paragraph, although a minor effect on their part is allowed for in the quantitative hypothesis. While concluding that sensory and motor nerves reinforced each other in mutually supporting regeneration, Singer (1946b) was careful to avoid giving the impression that all fibres had an identical activity in this respect. He also pointed out that an absolute number of nerves must have different effects at other levels of amputation. The third finger, for instance, contains only about 450 fibres, of which 200–300 are required to ensure its regeneration. Singer (1947a,b) pursued this type of study at four level of amputation and found a general correlation between the frequency of regeneration and the density of severed fibres at the amputation surface (Table 2.2). The correlation could be improved somewhat by neglecting the skeleton and calculating density with respect to the surface area of the soft tissues permeated by the nerve fibres, although the significance of this is dubious and the nerve threshold in the hand remained anomalous.

2.2 Threshold density

It is important to examine the meaning of this correlation, usually epitomized in the phrase that a threshold density of 10 nerve fibres/$(100 \ \mu m)^2$ of amputation surface (or 15 fibres/0.01 mm^2 of soft tissue) is required for regeneration. The main justification for preferring the density

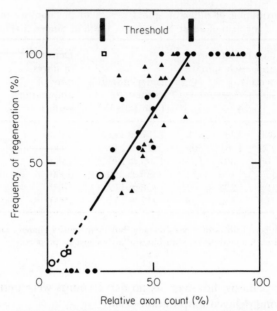

Figure 2.2. Nerve requirements for regeneration of the upper arm (solid circles), fore-arm, hand and individual fingers (triangles) of *N. viridescens*. The data of Singer (1946b, 1947a) have been converted to percentages of the normal innervation. A linear regression of the data between 20% and 70% innervation (the supposed threshold) extrapolates to zero and thus accommodates additional data from transplanted limbs (squares: Singer and Mutterperl, 1963) and supernumerary limbs (open circles: Bodemer, 1960).

of nerves to their absolute number here is that it allows the results from different levels of amputation to be roughly superimposed on the same coordinates. That can be achieved equally well by converting the number of nerves to a percentage of the normal innervation at each level of amputation, as shown in Figure 2.2. Although the density of nerve fibres may well have a physiological significance, later research suggests that the neurotrophic factor acts when nerves grow and branch among the mesenchymal cells of the early blastema. In that case density should be a function of the entire amputated surface which is then all soft tissue, while the number of growing points probably exceeds the fibre count in the stump. Figure 2.2 attempts to summarize these results impartially and suggests that the concept of a threshold is not particularly important to our understanding of the nervous control of regeneration. Although no regeneration occurred with less than 20% of the normal innervation and 70% was enough to ensure regeneration, the intermediate values yield a linear regression which extrapolates to the origin of the graph. The frequency of regeneration thus seems to be directly proportionaly to the degree of innervation, and 7% of the nerves should allow 10% of the arms to regenerate unless other factors intervene. One such factor is the tendency of mesenchyme to redifferentiate as scar tissue, which may occur when a poorly innervated blastema cannot maintain itself by cell proliferation; another is that markedly delayed regenerates were deliberately not scored. This analysis thus emphasizes the possibility that the action of nerves is simply to provide a growth factor, operating in terms of cell division, which is analogous or even identical to some of the hormones considered later. Growth measurements of Singer and Egloff (1949) support this concept.

Assuming that the threshold nerve fibre density is a significant parameter, subsequent research has cast doubt on the reliability of the original estimate of 1000/mm^2. Egar and Singer (1971) detected about twice as many fibres by electron microscopy, owing to the presence of fine unmyelinated axons which are usually overlooked in conventional Bodian preparations for the light microscope. These axons contributed little to the total cross-section of the nerve or exposed axoplasm which had perhaps only been underestimated by 3%. Peadon and Singer (1965) extended the latter's counts of nerve fibres in the upper arm of *N viridescens* by making similar counts on the aquatic larval and terrestrial red eft stages. They found that the total number of fibres increases with age, but less than the arm grows in thickness, so that the nerve density decreases as the newt approaches maturity. Furthermore, Eiland (1975) obtained nerve counts which were very similar to those of Singer in the fore-arm of much larger specimens. She derived a threshold density of 300/mm^2. Transplanted limbs will regenerate with considerably lower nerve densities in adult *N. viridescens* (Singer and Mutterperl, 1963) and larval *A. maculatum* (Thornton and Tassava, 1969). Figure 2.2 shows that

Table 2.3   Regeneration of partially innervated arms of 40–50 mm long *A. maculatum* larvae (data of Karczmar, 1946).

No. of cases	Intact nerves	Percentage of normal innervation	Frequency of regeneration (%)	Average delay (days)
Many*	3 + 4 + 5	100	100	~10
30	3 + 4	90	93	13
19	4 + 5	68	74	17
19	4	58	63	20
25	3 + 5	42	56	23
25	5	10	20	34
Many*	—	0	0	∞

*Normal regeneration occurs routinely, but is prevented as long as an arm remains totally denervated (Schotté and Butler, 1941).

only a fraction of the normal innervation is needed to support the growth of supernumerary arms (see section 2.6).

A less precise but general confirmation of the quantitative nervous control has been obtained in larval *A. maculatum*. Larvae are clearly less suitable than adults for these investigations as the nerves require repeated transection to prevent regeneration, and regression changes the level at which regeneration eventually begins. Karczmar (1946) estimated the relative contributions of the three segmental nerves which innervate the arm according to their cross-sectional areas in the brachial plexus; spinal nerves 3, 4 and 5 occupied 32%, 58% and 10% of the plexus area in these larvae. He then recorded the frequency of regeneration among arms retaining various combinations of these nerves when amputated above the elbow and redenervated at 7 day intervals. He obtained a good correlation between the frequency of regeneration and the estimated residual innervation (Table 2.3). The delay before regeneration and the rate of regression were correlated to the extent of denervation, and there was no sign of a threshold value below which regeneration could not occur (Figure 2.3). Deck (1961a) repeated this experiment using such small (18–23 mm) larvae of the same species that he felt obliged to use Karczmar's estimates of relative nerve contributions, and his results are also shown in the same figure. Once more, the frequency of regeneration was proportional to the available nerve supply and no lower threshold value was evident. An extrapolation of Deck's results suggests that more than 60% of these young arms might regenerate even after total

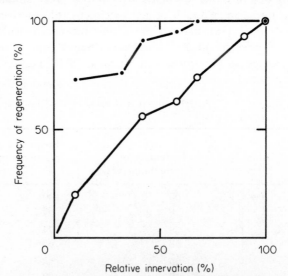

Figure 2.3. Nerve requirements for regeneration of the upper arm of larval *A. maculatum* (open circles: derived from Table 2.3) and from the fore-arms of extremely small larvae (solid circles: data of Deck, 1961a). No threshold limits are evident.

denervation, a control operation which Deck omitted in the belief that limbs cannot regenerate in the absence of nerves. That is a dubious assumption, for nervous connections are only being established in such young larvae and regeneration can occur in arms lacking a nerve supply in similarly sized larvae (Yntema, 1959a). Although these results showed as simple quantitative nerve-dependency, Deck attempted to reconcile them to the prevailing theory of a threshold requirement of nerve number or density. He pointed out that partial denervation might be rapidly counteracted by the formation of collateral branches from the ends of uninjured fibres. If that were true to an appreciable extent, it is surprising that the experiment yielded any meaningful results at all. Peadon and Singer (1965) also argued that these results could be explained on the basis of a constant threshold requirement which became a more stringent limit as the arm grew and its normal nerve density declined. Their argument contained too many assumptions to be taken seriously unless tested, and they did not test it on larval newts.

2.3 Anurans

A nervous control of regeneration has also been demonstrated for the hind limbs of anuran tadpoles. Schotté and Harland (1943b) found that the amputated legs regressed as long as complete denervation was maintained, but eventually regenerated after nerves were allowed to grow back into the stump. Innervation seemed to be a prerequisite for the thickening of the wound epithelium into an apical cap and for the accumulation of blastemal cells (Thornton 1956a), although subsequent growth of the blastema was suppressed by metamorphosis.

The same quantitative nervous control postulated for urodeles probably applies here, but has not been satisfactorily established. Van Stone (1955) counted the number of nerve fibres in the thigh and shank of advanced *Rana sylvatica* tadpoles, which regenerated to some extent when amputated in the latter region but not in the former. Although the density of nerve fibres relative to the surface area exposed by amputation was greater in the shank than in the thigh, this correlation to regenerative ability does not establish a causal connection. Van Stone (1964) later attempted to extend the correlation over a series of developmental stages as the ability to regenerate was progressively lost in the thigh and later in the shank. No correlation to nerve density appeared for the thigh and at most a slight one for the shank. These results are probably biased by the decision to define regeneration as the formation of a blastema, even when the blastema did not develop further. This is clearly different from urodeles, where the presence of nerves is required for both the formation and considerable growth of the blastema in larval *A. maculatum* (Schotté and Butler, 1944) and in adult *N. viridescens* (Singer and Egloff, 1949; Schotté and Liversage, 1959). Such a comparison emphasizes the likelihood that the gradual loss of regenerative ability during tadpole metamorphosis is quite unrelated to the nervous control of regeneration—a conclusion diametrically opposed to claims derived from the following studies on anurans.

The permanently aquatic African clawed toad, *Xenopus laevis*, is the only adult anuran which has been recorded as regularly 'regenerating' amputated limbs. This provided an opportunity to compare the nerve number or density between these limbs and those of *Rana sylvatica* adults which certainly do not regenerate. Such a comparison revealed no suitable difference in nerve number or density, but showed that *Xenopus* axons are relatively thick. Rzehak and Singer (1966; Singer, Rzehak and Maier, 1967) argued that the thick axons explained the latter's ability to regenerate. That is a considerable distortion of the quantitative hypothesis to accommodate one exceptional case. When it is realized that what is blandly termed a regenerate, or conceded to be a heteromorphic regenerate, of adult *Xenopus* limbs consists of a simple spiky outgrowth containing a cartilage axis but no muscle (Dent, 1962), it becomes apparent that the meaning of regeneration has also been distorted in this instance. We can dismiss this comparison as a triumph of enthusiasm over objectivity.

Singer (1952, 1965) has repeatedly expressed his opinion that the inability of adult anurans and higher vertebrates to regenerate amputated limbs might be caused by an insufficient number or density of nerve fibres. He devised an ingenious experiment to test this possibility on recently metamorphosed frogs (Singer, 1954). He freed the long sciatic nerve from a frog's left leg and deflected it forward under the skin so that it emerged from the amputated surface of the left upper arm. When the major branches of the sciatic nerve were rerouted in this manner, the total innervation of the arm was increased by about 50%. The virtually denervated leg was amputated in the thigh with a flap of skin sutured over the wound, and the incapacitated frogs survived surprisingly well. Singer stated that arm regeneration occurred in virtually all of the 32 extremely young frogs treated in this way, but in less than a half of a small sample of rather larger control specimens. His criterion of regeneration was an apical growth of the arm stump, between 1 and 7 mm in a period of over 3 months. These 'regenerates' showed most of their growth initially, with the external appearance of a small blastema; some produced a wavy surface 'anticipating digit formation' (which never occurred), or grew one or two 'finger-like projections' (meaning protrusions). In the two control series which Singer described, he identified some regeneration by up to 2.5 mm growth in about 15% of cases after amputation alone. Amputated arms containing an additional sciatic nerve which had no spinal connections grew as much as 3 mm in almost 30% of specimens from other *Rana* species.

There are several obvious objections to Singer's claim that he had induced regeneration by augmenting the nerve supply. First of all, it is unreasonable to identify any of these cases as regeneration. Perfect regeneration would include the reformation of the elbow, fore-arm, hand and digits with movable joints and coordinated function. Even defective regenerates should include some recognizable elements of these parts as a minimum criterion, yet none were evident. Secondly, the experiment was poorly conducted in that the controls and experimentals were not matched for identity of species, age or size. All these factors are known to contribute to variability in the regeneration of tadpole legs as regeneration becomes progressively inhibited during metamorphosis. This uncontrolled variability detracts from any significance one might attach to the more frequent and slightly larger growths on experimental arms. Thirdly, if the experimental arms did show an enhanced response relative to simple amputation, then an extra isolated nervous tract caused an intermediate response. Frogs grow quite rapidly after the completion of metamorphosis and normal growth at the tip of the healing arm stump could perhaps account for the occasional slight projections which Singer termed regenerates after amputation alone. The addition of sheath cells from the degenerating tips of deviated nerves might perceptibly increase the number of cells involved in the process of wound-healing and scarring, thus resulting in a modest increase of the frequency and size of projections. In short, Singer's results can be interpreted according to a completely different mechanism from the trophic effect of intact axons in urodeles, which formed the theoretical basis for his experiment. It is amazing how widely and uncritically Singer's conclusion has been accepted, until one realizes that other investigators have required equally spurious criteria of regeneration in order to claim some significance for their own results (see Table 2.4).

Konieczna-Marczynska and Skowron-Cendrzak (1958) repeated Singer's nerve deviation on adult *Xenopus laevis*. They re-routed the left sciatic nerve into the right hind-leg of 21 specimens aged 2–3 years and amputated the hyperinnervated legs 1 month later. The growth they detected during the next year consisted primarily of cartilage spikes which occasionally bifurcated. Two limbs with an augmented nerve supply seemed superficially to form three and four digits but the skeleton was a continuous piece of cartilage. The amount of growth seemed to be related to the degree of innervation, however, and complete denervation usually prevented growth. I have obtained similar growths on amputated legs of juvenile *Xenopus*, during a regime of prolactin injections which stimulated overall growth as much as that of the regrown spikes. Even under normal innervation the amount of growth achieved by *Xenopus* 'regenerates' is quite impressive (Beetschen, 1952) but enhanced growth does not alleviate their structural defects.

2.4 Mechanism of nervous control

Neurons which differentiate in the neural tube of vertebrates mostly remain within the confines of the brain and spinal cord, or migrate a short distance to form the dorsal ganglia which occur as pairs in most postcranial segments. The cell body containing the nucleus thus remains within the spinal cord or ganglion, while major cytoplasmic processes or axons grow out and make

34

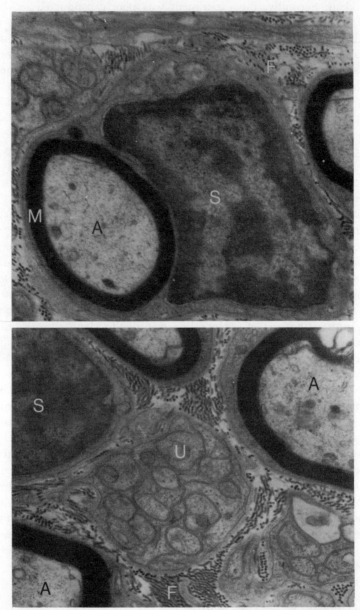

Figure 2.4. Fine structure of a nerve trunk from the fore-arm of *Pleurodeles waltl* shortly after metamorphosis. A, axon in myelin sheath (M) and associated Schwann cell nucleus (S). U, unmyelinated axons are also embedded in grooves of Schwann cells. F, collagen fibrils of the connective tissue sheath.

contact with all regions of the body, and smaller processes make synaptic connections with other neurons of the central nervous system. All the peripheral axons are surrounded by a neurilemma produced by a series of Schwann cells, each of which encloses a segment of axon. This complex is called the nerve fibre; large fibres are usually termed myelinated because the neurilemma is composed of multiple layers of Schwann cell membrane; segments of smaller axons may be grouped within the folds of a single Schwann cell to form unmyelinated fibres. We are concerned here with large myelinated fibres and an equal number of fine unmyelinated axons, both sensory fibres emanating from the dorsal ganglia and motor fibres coming from the ventral roots of the spinal cord which mingle with each other in the brachial or sacrolumbar plexus to form the mixed nervous trunks of the limbs (Figure 2.4). Throughout their length the nerve fibres are surrounded by connective tissue layers: the endoneurium surrounds each fibre, several fibres are enclosed in a common perineurium, and all the fibres of one visible nerve are surrounded by an epineurium. The connective tissue matrix is secreted by fibroblasts and supports both the nerves and associated blood vessels. Severing nerves, as typically performed for denervation of a limb, results in neuronal chromatolysis (Tweedle, 1971) and causes the anucleate distal parts of the axons to degenerate—a process known as Wallerian degeneration after its discoverer (Waller, 1850). The proximal nucleate portion of the axon may recede only to the nearest undamaged node of Ranvier and then grows out again after some delay. It frequently re-establishes its former peripheral connections unless prevented or diverted by scar tissue. A larval arm which has been completely denervated by cutting the brachial plexus, for instance, can recover to show normal movements in about 3 or 4 weeks. Even diverted motor nerves seem able to find their correct end-organs (Grimm, 1971; Cass and Mark, 1975), although sensory nerves do not (Johnston et al., 1975). A feature of Wallerian degeneration is that the neurilemma distal to the cut fragments into its constituent Schwann cells which lose their myelin content, become amoeboid and may transform into macrophages. These Schwann cells have been observed to migrate out from the severed ends of the nerve tract and apparently serve as guidelines for the regrowing axons, which they can follow and eventually surround with a new myelinated neurilemma. The connective tissue layers probably persist unchanged during Wallerian degeneration and axon growth, although the severed region must be repaired at some time perhaps involving a local movement of fibroblasts. Reviews of nerve repair give more details on this subject (see Young, 1942; Weiss and Hiscoe, 1948; Mark, 1969).

Transection of the brachial plexus isolates the distal parts of axons from the cell body and central nervous system. The Schwann cells and connective tissue of the nerve sheath persist undamaged at the site of amputation, yet regeneration is inhibited until axons have regrown. The control of regeneration, therefore, is clearly by an axonally transmitted factor and is not a property of the sheath cells. Severing the ventral spinal roots shows that the remaining intact sensory axons are sufficient to support regeneration, which still occurs when the dorsal ganglia are isolated from the spinal cord (Locatelli, 1925; Singer, 1943). Regeneration thus must require an adequate supply of axons which remain connected to their cell bodies, i.e. intact neurons, but does not demand a complete pathway to the central nervous system. The conclusion that regeneration depends on a neuronally transmitted factor invited the further assumption that the typical nerve function of electrical impulses may be involved. There is some reason for rejecting such an assumption, as sensory axons and motor axons typically propagate impulses in opposite directions, yet either can support regeneration. This is not an infallible reason, for an axon can propagate impulses in either direction, the barriers to antidromic conduction being at synapses. A few investigators have certainly ignored the argument anyway and attempted to demonstrate an electrical influence on regeneration. The application of constant low currents and potential differences to the arms of postmetamorphic frogs usually produced a barely perceptible response (see chapter 5). The action potentials of muscles can certainly be dispensed with during regeneration, which continues during prolonged paralysis induced by botulinum toxin (Drachman and Singer, 1971).

2.5   Neurotrophic theory

A far more likely explanation of the influence of nerves on regeneration stems from the 'trophic' effects of nerves in maintaining the functional integrity of the sensory or motor organs which they serve. A rather antiquated but suitable example is provided by the lateral line sense organs of the catfish, studied by Parker (1932; Parker and Paine, 1934). If the lateral line branch of the tenth cranial nerve is severed, the isolated distal parts of the axons and the associated sense organs degenerate about 2 weeks later in a cephalo-caudal sequence. When the axons begin to grow back along the lateral line, the sense organs are reformed in the same sequence. The same reversible trophic effect can be identified in other sense organs (Zelena, 1964) and in the atrophy of muscles which have lost their innervation (Guth, 1968). The requirement of intact nerves for limb regeneration clearly resembles these examples of neurotrophic effects, particularly where larval limbs or young blastemata of adult urodeles not only stop regenerating after denervation but are resorbed (Schotté and Liversage, 1959; Powell, 1969). A common explanation of all these phenomena would be that some substance which stimulates maintenance or growth is synthesized in the cell body of the neuron, transported down the axon, and eventually diffuses out to the end-organ. This should be a feature of both sensory and motor nerves, irrespective of the direction in which they normally conduct electrical impulses. It is now well established that the static appearance of axons masks a considerable turnover of substances and organelles. Most proteins and lipoproteins are replaced by a slow distal transport from the cell body at about 1–2 mm/day, but fast transport systems supply materials needed for synaptic transmission, such as acetylcholinesterase at rates of up to 400 mm/day in mammals, and apparently proximal transport systems co-exist with them (Ochs, 1972; 1974).

Singer (1943) quickly adopted this trophic effect as the most likely explanation of how nerves support limb regeneration, and developed his trophic theory with several modifications to accommodate other experimental results which will be discussed later. The essential features of the trophic theory have met with general acceptance, but it can only have the status of a working hypothesis until an extract or chemical is found to duplicate the effects of nerves. The search for such a substance has proved a disappointing adventure so far (Singer, 1960; 1974; Singer et al., 1976).

Table 2.4 lists some of the grafts and injections which have been performed in the hope of identifying a neurotrophic factor. The effects recorded in the table should be interpreted cautiously, for some of the references are to enthusiastic preliminary reports to which I can find no sequel. If a trophic factor had been found which reliably elicited even a modest degree of regeneration, it would certainly have been well publicized. On the other hand, the limited success or complete failure of any substance to affect regeneration may only mean that it was not applied correctly. Normal nerves presumably dispense their postulated trophic factor continuously, while even the best infusion techniques only introduce a solution for about 6 hours each day (Deck, 1971).

The most direct way to test potential trophic substances or the tissues which may produce them is by application to amputated denervated limbs. Adult urodeles are most suitable for this test as their limbs show very little regression, but precautions against the eventual regrowth of nerves need to be exercised. A more rapid and sensitive test involves attempting to counteract the regression of denervated early blastemata. All nervous tissue seems to elicit this trophic response. Transplanted limbs can gain a direct innervation from the spinal cord which is sufficient for regeneration (Thorton, 1956b). Spinal ganglia implanted into denervated blastemata allow regeneration to continue (Kamrin and Singer, 1959). Their additional finding that implanted ganglia which had been killed by heating or freezing could not halt regression is rather more interesting: it probably implies that the trophic factor must be dispensed over a prolonged period by living tissue. Perhaps that explains why none of the injections or infusions tried so far have yielded reliable cases of regeneration (Table 2.4). Extracts of blastemal tissue may at best increase the incidence of arrested early blastemata, and probably do so by causing local damage and thus enhancing dedifferentiation. The complete regeneration claimed by Burnett et al. (1971) occurred after a sufficient delay for normal reinnervation. No purified substance gives any promise of being or mimicking the trophic factor, but

Table 2.4 Tissue grafts and injections used in attempts to influence limb regeneration or identify a neurotrophic factor.

Graft (G) or injection (I(	Species	Effect on regeneration	Reference
(A) On denervated urodele limbs			
G Spinal ganglia	*N. viridescens*	Permits blastemal growth	Kamrin and Singer (1959)
I Nerve extract	*N. viridescens*	Arrested early blastema	Deck (1971)
I Nerve extract	*N. viridescens*	Reduces decline in protein synthesis	Lebowitz and Singer (1970)
I Brain extract	*N. viridescens*	Restores protein synthesis	Singer *et al.* (1976)
		Restores DNA synthesis	Jabailly and Singer (1977)
I Blastemal extract	*N. viridescens*	Arrested early blastema	Deck and Futch (1969)
I Blastemal supernatant	*N. viridescens*	Regeneration claimed	Burnett *et al.* (1971)
I Acetylcholine	*N. viridescens*	No effect	Singer (1960)
I Cyclic AMP	*N. viridescens*	Marginal increase of mitosis; supports DNA and protein synthesis	Foret (1973); Foret and Babich (1973); Babich and Foret (1973)
(B) On innervated urodele limbs			
I Colchicine	*N. viridescens*	Inhibits	Singer *et al.* (1956)
I Atropine, eserine, acetyl-choline	*N. viridescens*	No effect	Singer (1960)

I	Botulinum toxin	*N. viridescens*	No effect	Drachman and Singer (1971)
I	Cyclic AMP, theophylline	*N. viridescens*	No effect initially	Sicard (1975a, b)
I	Nerve growth factor	*A. maculatum*	Accelerated early phases	Weis and Weis (1970)
I	Hemicholinium-3	*A. tigrinum*	Retards growth	Hui and Smith (1970)
(C) On innervated anuran limbs after metamorphosis				
G	Diverted sciatic nerve	Rana species	Regeneration claimed	Singer (1954)
G	Diverted sciatic nerve	*X. laevis*	Improved growth	Konieczna-Marczynska *et al.* (1958)
G	Embryonic tissue	Rana species	Regeneration claimed	Malinin and Deck (1958)
G	Larval epidermis	Rana species	Regeneration claimed	Gidge and Rose (1944)
I	Newt blastemal extract	*R. pipiens*	Blastema	Burnett *et al.* (1971)
I	Nerve growth factor	*R. catesbiana*	More arrested blastemata	Weis (1972)
G	Salt irritation	Rana species	Regeneration claimed	Rose (1944, 1945)
G	Salt irritation	Rana species	Same but nerve dependent	Singer *et al.* (1957)
G	Repeated wounding	*R. temporaria*	Regeneration claimed	Polezhaev (1972)
(D) On innervated limbs of lizards and a marsupial				
G	Deviated sciatic nerve	*Anolis*	Regeneration claimed	Singer (1961)
G	Deviated sciatic nerve	*Lygosoma*	Regeneration claimed	Simpson (1961)
G	Brain implant	*Didelphis*	Regeneration claimed	Mizell (1968)

extracts of nervous tissues can partly restore the rates of DNA and protein synthesis in denervated blastemata.

The simpler test of injecting solutions into normally regenerating limbs in the hope that excessive regeneration might reveal a trophic factor or that inhibition might identify a specific antagonist has not been particularly helpful either. Colchicine disrupts microtubules and thus probably inhibits axonal fast transport, but also prevents cell division and so directly arrests blastemal growth. Atropine is used as a routine antagonist of cholinergic nerve action and inhibits regeneration when applied at generally toxic doses, yet acetylcholine, which is the transmitter produced by these motor nerves, has no apparent effect on regeneration and does not compensate for denervation. Botulinum toxin prevents the release of acetylcholine from nerve endings while regeneration continues unscathed. Injections of eserine, which inhibits cholinesterase and thus should preserve acetylcholine levels, seems equally ineffective in these tests. On this basis, Singer (1960) presented a convincing argument that the elusive trophic factor could not be acetylcholine. Nerve growth factor presumably causes a temporary acceleration of blastemal growth because it increases the density of nerve fibres.

Similar grafts and injections have been performed on anuran limbs after metamorphosis and these are included in Table 2.4 for the sake of completeness. When these treatments were intended to demonstrate a trophic action they depended on the assumption that the limbs would regenerate given adequate innervation. The results provide no support for that assumption and in fact tend to contradict it. Nerve growth factor has the same effect on the limbs of metamorphosing bullfrog tadpoles as on larval salamanders: it increases the innervation and accelerates the formation of a blastema, but the tadpole blastemata are still arrested at metamorphosis. Increasing the local innervation by surgical means has had no greater effect in adult anurans, lizards or mammals.

The use of embryonic tissues and blastemal extracts both for frogs and denervated newt limbs tests another questionable assumption which will be discussed later: that tissues other than nerves produce the trophic factor, but in smaller amounts. Embryos and blastemata have been considered relatively rich sources of trophic substances, but Table 2.4 provides little justification for the idea unless an incipient blastema is held to be a suitable criterion for regeneration. Polezhaev's (1946, 1958) claim to have induced regeneration in frogs by repeated injections of homogenates is not only tainted by this theoretical background but is unconvincing in that a remarkable number of controls regenerated equally well. The effect of salt solutions is considered in more detail in the next chapter, for contralateral limb stumps show the same modest growth exhibited by the treated ones, indicating a systemic and possibly hormonal response.

Assuming that a trophic substance exists and is characteristic of living neurons, it might be more readily identified by considering the constituents which are known to be transported down axons. The list might even be restricted somewhat if the trophic factor could be associated with either the fast or slow transport systems. This approach can only be attempted at present by piercing together heterogeneous data and comparing them to the different velocities of transmission along an axon. Slow transport or bulk axoplasmic flow occurs at a rate of 1–2 mm/day; fast transport is usually recorded at 100–400 mm/day according to temperature and the means of measurement; nerve impulses travel at rates around 1–100 m/s depending mainly on axon diameter and myelination. The most pertinent evidence here is the fast transport recorded in the sciatic nerve of bullfrogs as 150 mm/day at 25°C (Abe *et al.*, 1973) and the changes in blastemal metabolism following denervation in newt arms at the same temperature (Singer and Caston, 1972). Taking the length of isolated nerve in the upper arm as 15 mm, cutting the brachial plexus should only affect the blastema after a considerable delay of at least 5 days if only slow transport is involved, but perhaps within 3 hours via fast transport and almost instantaneously via action potentials. RNA synthesis had already increased when assayed 2 hours after denervation of these newt arms (Figure 2.5), but that is not the kind of response expected to follow deprivation of a trophic factor or to cause later degeneration and resorption of the blastema. Allowing for a time lag while RNA accumulated to detectable levels, it probably resulted from the repetitive firing of the damaged axons. The more appropriate response of subnormal synthesis was recorded after 12 hours but could well have occurred within 6 hours for RNA, as shown earlier by Dresden (1969). The postulated trophic factor had thus become depleted in a period which is only compatible with its fast transport. Since the decline of protein synthesis can be alleviated by infusing a brain or nerve extract into the denervated blastema, except when the extract had been inactivated by heat or trypsin, a soluble protein seems implicated as the trophic factor (Lebowitz and Singer, 1970; Singer *et al.*, 1976). Contemporary models of fast transport suggest it should occur in a particulate fraction of the axoplasm which includes newly synthesized proteins, cholinesterase, phosphatidylcholine and probably all neurotransmitter substances (Lasek, 1970; Ochs, 1974). Acetylcholine and cholinesterase seem to have been eliminated as candidates (Table 2.4). Another neurotransmitter, norepinephrine, is associated with sympathetic nerve endings, and regeneration can occur without them.

Older observations on the destruction of nerves and sense organs in the lateral line and barbels of catfish (Parker and Paine, 1934; Torrey, 1934, 1936) provide some support for this interpretation. A wave of degeneration spreads distally in each case at a rate of 20–30 mm/day at 18°C, which is more easily explained as attenuated transport than as an accelerated bulk flow. Harris and Thesleff (1972) have provided a modern example of a neurotrophic factor affecting rat skeletal muscle which is transported at about 120 mm/day.

Although denervation curtails the synthesis of RNA, DNA and protein in

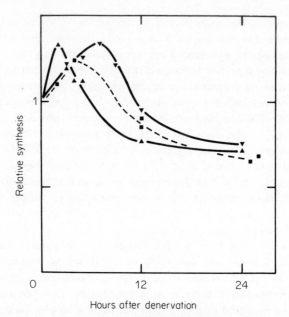

Figure 2.5. Synthesis of RNA (▲), DNA (▼) and protein (■) by denervated blastemata expressed as a proportion of the synthesis in control blastemata on the contralateral arms of the same specimens of *N. viridescens*. The data, obtained by Lebowitz and Singer (1970) and Singer and Caston (1972), are specific activities (cpm/μg protein) following 2–3 hours incorporation of a radioactive precursor). (Reproduced by permission of the Company of Biologists Limited.)

a young blastema, it does not abolish these metabolic activities. The pattern of an initial burst of synthesis followed by a gradual decline (Figure 2.5) is also characteristic of the blastemal mitotic index. This increases 1–3 days after denervation and then decreases markedly, but some cell divisions occur during the following four weeks (Singer and Craven, 1948). It thus seems quite likely that infusing a brain extract which partially restores DNA and protein synthesis in denervated blastemata (Table 2.4) would also help to correct the decline in cell division and thus improve blastemal growth, but perhaps only to a marginal extent. Since a variety of enzyme activities seem to be only mildly reduced by denervation (Manson *et al.*, 1976), the available evidence suggests the trophic factor controls the metabolic rate of blastemal cells in a rather general way (cf. Bast *et al.*, 1979). The reduced rates of DNA synthesis and cell division are best interpreted as a general retardation of the cell cycle, as Maden (1979a) has recorded a progressive accumulation of presynthetic G1 blastemal cells over a 4 day period following denervation. Extending the G1 period is a common means of prolonging cell cycles.

The impact of denervation on freshly amputated limb stumps has been examined by Tassava and his colleagues, mainly by autoradiography on young axolotls and adult newts. The latter material should provide the most discriminating results because their arms have virtually stopped growing and so have particularly low baseline values for DNA synthesis and cell division of internal tissues. Simultaneously denervated and amputated newt arms did not differ from innervated control stumps for the first week after amputation. Normal wound healing and apical dedifferentiation occurred in both and about 15% of the apical cells were stimulated to synthesize DNA. The denervated arms appeared much the same during the second week, when control arms contained increasing numbers of S-phase cells and some dividing cells (Mescher and Tassava, 1975). Having obtained essentially similar results from young axolotls, Tassava (Tassava et al., 1974; Tassava and Mescher, 1975; 1976; Tassava and McCullough, 1978) concluded that denervation must prevent mitosis by blocking dedifferentiated cells in G2. They had to postulate that the blocked cells were selectively destroyed, because they could not detect an accumulation of G2 cells. These seem rather dubious conclusions, for the residual effects of severed axons persist for several days and cell division continues for much longer in denervated blastemata. Maden (1978) reduced the residual effect of axons by comparing axolotl arms which had been denervated 6 days before amputation to simultaneously denervated and normal innervated arm stumps. Wound healing and apical dedifferentiation occurred uniformly in all three series, and so seem genuinely independent of the trophic factor. All three series also showed an initial increase of S-phase cells 2–4 days after amputation and of dividing cells a day or so later, but the increases could be related to the continued delivery of trophic factor in control arms and to a residual supply of the factor in denervated arms (Figure 2.6). The good correspondence between the labelling index of S-phase cells and mitotic index curves for each series on this figure shows that the cells were not blocked in G2, even though the supply of trophic factor seemed to be exhausted. The results are more compatible with an extension of G1 to slow down the cell cycle, as suggested in the bottom line of Table 1.3, and an accumulation of G1 cells (Figure 2.7). The trophic factor therefore merely acts as a stimulant of the cell cycle, encouraging blastemal growth by ensuring that the rate of proliferation exceeds the combined rates of destruction and redifferentiation. Since wound healing and dedifferentiation occur in the absence of nerves, as does redifferentiation when late cone stages are denervated, blastemal growth is the only phase of regeneration which is dependent on the trophic factor. Geraudie and Singer (1978) measured DNA synthesis in the isolated epithelium and mesenchyme of denervated blastemata, finding that synthesis is reduced in both components although less so in the epithelium. They argued the epithelium must be independently sensitive to the trophic factor, a conclusion in keeping with the observation that nerve grafts stimulate cell division in the overlying epidermis (Overton, 1950). On the other hand, the

44

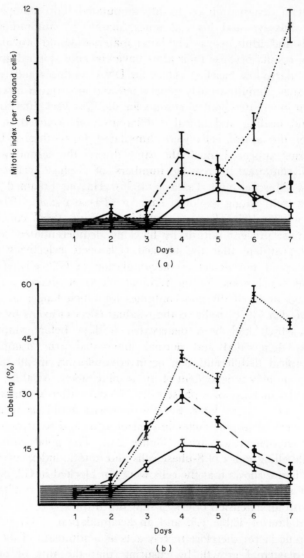

Figure 2.6. Mitotic index (*a*) and labelling index (*b*) in the mesenchyme of 50 mm axolotl arms, recorded in the first week after amputation by Maden (1978). The horizontal lines indicate the background frequencies of dividing or S-phase cells found in unamputated arms. ×, normal arm stumps; ■, denervated at the time of amputation; O, denervated 6 days earlier. (Reproduced by permission of the Company of Biologists Limited.)

epidermis could be more directly dependent on the growth of the underlying mesenchyme, and thus indirectly dependent on the trophic factor. The stimulus of amputation is apparently enough to set off a partly synchronized series of epidermal divisions to accommodate a similar stretch associated

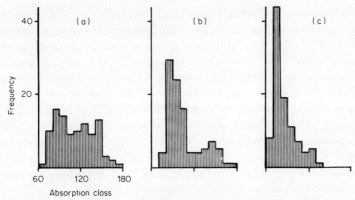

Figure 2.7. Frequency distribution of nuclei from (a) normal cone blastema, (b) 2 days after denervation; (c) 4 days after denervation (from Maden, 1979a). Prereplicative Gl nuclei constituting absorption classes 60–90 accumulate at the expense of S-phase and G2 nuclei. The absorption classes are measures of DNA content. (Reproduced by permission of the Company of Biologists Limited.)

with wound healing (Figure 1.10), and much the same wave of epidermal divisions has been recorded in denervated limb stumps (Tassava and Mescher, 1976).

2.6 Sparsely innervated limbs

Our inability to substantiate the neurotrophic theory by identifying a trophic factor serves as a reminder that the quantitative and threshold aspects of the theory are only based on correlations and can be contested on the basis of examples of regeneration when innervation is reduced or absent. Polezhaev (1946d) mentioned that transplanted limbs of tadpoles and newts sometimes regenerate so quickly that they must begin to do so before local nerves have penetrated them. Pietsch and Webber (1965) described similar cases after transplanting arms of very young *A. maculatum* larvae into the orbit. When amputated 2 weeks later the isolated hands contained very few nerve fibres, yet these arms regenerated at a normal speed and then contained less than a quarter of the normal nerve supply. Pre-amputated arm segments transplanted to the orbit occasionally also regenerated promptly, provided they obtained a good blood circulation. Singer and Mutterperl (1963) transplanted upper arm segments to the flank of adult *N. viridescens*, finding that rather few of the grafts regenerated and always after some delay. The arms which had regenerated contained on average less than one-third of the normal density of nerve fibres and some examples probably started to regenerate with less than one-tenth of the normal innervation. A similar fraction of the total nerve supply can induce the formation of supernumerary limbs when diverted to a surface wound (Bodemer, 1960). Thornton and

46

Tassava (1969) also recorded regeneration in orthotopically transplanted arms of *A. maculatum* larvae which were kept sparsely innervated by repeated denervations. Although the significance of these results has been questioned on the grounds that they arose after exceptional injury, they provide ample reason for discarding the threshold concept whose shaky basis was exposed earlier. They do not seriously conflict with the rest of the quantitative neurotrophic theory, however, for regeneration usually occurred at a reduced frequency and after sufficient delay for some re-innervation to anticipate blastemal growth. That may even be true for grafts pressed against the well-innervated orbital muscles: the normal period of wound healing and dedifferentiation at the apex of the graft could have been long enough for a few nerve fibres to reach the incipient blastema.

The most celebrated exception to the whole neurotrophic theory is furnished by the regeneration of aneurogenic limbs of *A. maculatum* larvae. If the hind brain and trunk nerve cord are excised when they first form as a neural tube, the defective embryo can survive and develop fairly normally but is incapable of normal movements or feeding. The arms develop quite well with either very few or no detectable nerve fibres and can reach the four digit stage in this species. The relatively late arm development of *A. mexicanum* is only completed after feeding has started, and consequently this operation only yields arms with two or three digits by the time the yolk reserves become exhausted. Defective larvae of either species can be kept alive indefinitely when joined in parabiosis like siamese twins, sharing a common circulation with a normal larva. Unfortunately, nerves from the normal nurse larva tend to grow into the defective one so the originally aneurogenic arms quite soon gain a sparse innervation. Amputation of both sparsely innervated and genuinely aneurogenic arms results in typical regeneration which is only marginally slower than normal (Yntema, 1959a, b). If an aneurogenic arm is transplanted orthotopically to replace the arm of a normal larva, the brachial nerves of the host grow out according to their usual pattern and probably find the correct end-organs, for the grafted arm becomes functional (Piatt, 1942). Aneurogenic arms which have been transplanted in this way or heterotopically to the flank frequently regenerate as duplicated arms, whether or not they have been penetrated by host nerves (Yntema, 1962; Thornton and Tassava, 1969). The regenerative ability of aneurogenic arms has been provisionally assigned to a property of the skin. Regeneration still occurs if the aneurogenic skin is removed and replaced before amputation, but when the aneurogenic skin is replaced by normal skin the arms do not regenerate (except in cases where regression may have eliminated the graft); radioactively labelled mesoderm from normal limbs, packed into the excavated aneurogenic limb-stumps, can be traced as a constituent of the regeneration blastema produced in the virtual absence of nerves (Steen and Thornton, 1963). The converse experiment of transplanting aneurogenic limb skin on to recently denervated and amputated arms has been performed, but only described as an experiment in progress which had not at that time yielded results (Thornton, 1965). Thornton and Thornton (1970) transplanted aneurogenic limbs orthotopically onto normal hosts and followed the response to their delayed innervation by periodically denervating and amputating them. They found that the limb regeneration became dependent upon nerves within 13 days of transplantation. Limbs left up to 33 days after transplantation in order to become well innervated, and then kept surgically denervated for a further 30 days, recovered their ability to regenerate in the absence of nerves in half the cases tested.

If normal larval limbs do not regenerate during prolonged denervation but sometimes can regenerate if they have been transplanted (Thornton and Tassava, 1969), it seems reasonable to suspect the transplantation of influencing the results of these aneurogenic limbs. Thornton and Thornton (1970) rightly considered their results as providing an overdue test of the trophic theory in its later manifestations, and clearly approved the ideas of transplantation trauma and addiction to innervation which are considered next.

2.7 Modified trophic theory

The quantitative theory and the trophic theory developed from it, have both proved fruitful explanations of how nerves influence limb regeneration. It would be absurd, therefore, to abandon them readily because of the anomalous behaviour of transplanted and aneurogenic limbs. The obvious alternative is to modify the theories in order to accommodate the known exceptions. It is unfortunate if the theories become more complicated as a result, but perhaps they explain a complex phenomenon. It is only disastrous if the modified theories are so vague as to lose their predictive ability and are thus no longer amenable to test. Although criticisms of this nature have been levelled (e.g. Pohley discussing Singer, 1965), the theories can be tested to some extent as described previously.

The first modification which has been proposed is that the trauma of transplantation permits limbs to regenerate when only a subnormal number of nerves are present. The trauma could then be thought to elicit an extra supply of the trophic factor from non-nervous tissue, presumably in the wounded region. Alternatively, the trauma might increase the sensitivity of the limb—by lowering its postulated threshold demand—so that it could respond to subnormal amounts of trophic factor (Singer and Mutterperl, 1963). These seem to be equally valid interpretations in terms of the trophic theory. No test has yet been devised to distinguish between them, nor, indeed, is it possible to really test such a deliberately vague term as trauma which has been invoked on other occasions besides this to obscure mechanisms which might conceivably intervene in regeneration. Savage incisions inflicted on denervated limbs in situ only lead to excessive regression, although intended to be even more traumatic than transplantation (Thornton and Tassava, 1969).

The second modification, employed to account for the regeneration of aneurogenic limbs, is that limb tissues produce their own trophic substance originally (or do not require one) but, once innervated, they become dependent upon the presence of nerves and the neurotrophic factor. Either limb tissues cease to produce trophic factor when it is supplied to them from nerve endings or the tissues' threshold requirement increases once acclimatized to the presence of neurotrophic substance. The requirement or dependency here is normally intended to mean for regeneration, but very probably also applies to normal growth which is curtailed by denervation. The increased dependency on a trophic factor does seem to result from a period of innervation (Thornton and Thornton 1970), although this could equally well involve an addictive raising of a threshold requirement or the compensatory reduction in synthesis of a trophic factor by innervated tissues.

The modified trophic theory as expressed by Singer (1965) thus retains its essential feature of limb regeneration being quantitatively dependent on a substance released from axons, but admits that other limb tissues may also produce a trophic substance and that variation either in the production of that trophic factor or in the sensitivity to it of limb tissues can obviate the normal threshold requirements either partially or absolutely. The theory is now entangled in unnecessary postulates: there is no evidence that embryonic, aneurogenic or transplanted limbs actually produce any trophic factor, and no reliable evidence of changing sensitivity. Abandoning the threshold concept allows us to dispense with these modifications, for there is then no need to invoke transplantation trauma nor to postulate extraneous production of the trophic factor in exceptional circumstances. The only remaining problem is to account for the regeneration of aneurogenic arms. These arms develop perfectly at first but then their growth becomes retarded and their muscles degenerate (Egar et al., 1973; Popiela, 1976). They regenerate well during the period of normal development and perhaps for some time after, but there might be a transition to nerve-dependency both for developmental and regenerative growth at about the time the larva begins to feed. Even when supported in parabiosis, aneurogenic axolotl arms only form three fingers and then virtually stop growing until they are invaded by the partner's nerves. The timing of growth retardation suggests that embryos might be provided with a limited reserve of a trophic factor which normally persists until well after sufficient innervation has occurred to support further growth and regeneration of the arm. If that were so, the requirement for the neurotrophic factor would increase as the intrinsic supply became exhausted, without any change in sensitivity of limb tissues to the factor. A postulated embryonic reserve thus provides an example of a single age-dependent modification of the neurotrophic theory which could explain the regeneration of larval aneurogenic arms. Judging from the way embryo extracts encourage the proliferation of tissue cultures, an embryonic trophic factor very probably exists but it is not capable of explaining some of the results analysed in the next section.

2.8 Testing the trophic theory

The results obtained from chronically denervated limbs especially those from transplanted aneurogenic limbs (Thornton and Thornton, 1970) neither support nor exclude the previously considered embryonic trophic substance. Transplanted aneurogenic arms remained effectively free of nerves for another week prior to amputation, and then regenerated perfectly while maintained virtually without nerves (Table 2.5, line 1). Host brachial nerves take 7 days to reach the transplanted fore-arm and the transplant subsequently resumes growth, becomes capable of movement and can regenerate; 10 days after transplantation, in fact, it has become identical to a normal arm, for all these features are now dependent upon the continued presence of nerves. The ability to regenerate after simultaneous amputation and denervation is gradually lost during the first week of re-innervation (lines 1–5). That loss cannot be merely a function of age, as similar transplants can regain the ability to regenerate without nerves when 2 months older (line 6). These not only resembled aneurogenic arms in that almost half of them regenerated but also in the limited amount of regression shown by the remainder. The striking difference from simultaneously amputated and denervated arms (contrast lines 6 and 9) could be attributed to any of the five features listed in Table 2.5 which need to be varied systematically in control series. Thornton and Thornton (1970) argued that the relatively brief experience of innervation, 12–26 days, as well as the prolonged subsequent denervation, allowed the regeneration in line 6. They thus concluded that a brief addiction to neurotrophic substance was reversible, in agreement with the second modification of the trophic theory. Thornton and Tassava (1969) had already supported the first modification—transplant trauma abolishes the normal nerve requirement—by obtaining the results shown in lines 8 and 9. There is a fallacy in their argument, however, for either transplantation or the period of denervation which preceded amputation could be responsible for the regeneration and limited regression found in line 8. Now, lines 6 and 8 show virtually identical results implying that they only differed in trivial experimental conditions: those of line 6 (including the use of aneurogenic arms) may be needlessly elaborate. A further comparison to the hypothetical example given in line 7 suggests the same results could be obtained without even transplanting the larval arms. The simplest explanation of the data in Table 2.5, therefore, is that a good proportion of young larval arms might regenerate in the absence of nerves provided they were denervated several days before amputation. Thornton and Thornton (1970) believed that could not be true but neglected to cite any evidence to support their belief. Admittedly, a delay of amputation for up to 3 days does not alter the rapid

Table 2.5   Regeneration of larval A. *maculatum* arms during a continued state of denervation.

Larval length (mm)	Development	Transplant (denervated)	Reinnervated	Denervated	No. of cases	Percentage of arms (all kept denervated after amputation)		
						Regenerate	Static	Regress
1. 20–23	Aneurogenic	7	0	0	13	100	0	0
2. 20–23	Aneurogenic	7	3	0	19	74	0	26
3. 20–23	Aneurogenic	7	4	0	9	44	0	56
4. 20–23	Aneurogenic	7	5	0	16	12	0	88
5. 20–23	Aneurogenic	7	6	0	12	0	0	100
6. 20–23	Aneurogenic	7	12–26	30	33	49	39	12
7. 18–23	Innervated	*in situ*	0	7	0	~60	~40	0
8. 30	Innervated	7	0	0	37	46	54	0
9. 18–23	Innervated	*in situ*	0	0	15	0	0	100

History of arm in days before amputation*

*Given in correct sequence. Transplantations were all orthotopic to normal host larvae and involve a 7 day period of denervation before host brachial nerves each the fore-arm level of amputation. The original data have been corrected to take this into account.
Lines 1–6 are from Thornton and Thornton (1970).
Line 7 is derived by extrapolation from Figure 2.3 (data of Deck, 1961a, b).
Lines 8–9 are from Thornton and Tassava (1969).

regression of a chronically denervated larval arm (Schotté and Butler, 1941) but a 10 day delay causes a markedly reduced rate of regression (Karczmar, 1946). Curiously enough, Singer (1946a) obtained a 50% frequency of regeneration in adult newt arms amputated 4 months after denervation. Singer established that these arms then lacked sensory nerves but had regained a motor innervation, which he surmised must have hypertrophied to reach some threshold density. In the absence of a merely de-afferented control series, he did not consider the possibility of delayed amputation being responsible for the regeneration. Just as larval limbs become acclimatized to an increased innervation (Table 2.5, line 1–5), however, they might gradually adapt to a diminished supply of the trophic factor. If that were true, it would allow a radical simplification of the current trophic theory.

This rather tortuous argument thus leads to a perfectly simple but neglected question: can an arm regenerate in the absence of nerves if it has been kept denervated for some period prior to amputation? That is certainly true of aneurogenic arms and Table 2.5 shows it is true for transplanted arms. We have no reason to consider them as exceptions until it has been proved false for normal arms. Table 2.6 shows that normal arms can indeed regenerate while chronically denervated from 3 or 4 weeks before amputation. It would be difficult to execute a more protracted test because the brachial plexus becomes progressively obscured by scar tissue at each weekly denervation. The success of the operation was assessed, however, by the immobility of the arm. Routine histological procedures suggested the final cone blastemata were virtually or completely devoid of nerves. These results clearly imply a slow adaptation to deprivation of the neurotrophic agent, complementing the more rapid addiction to nerves seen in Table 2.5. Regeneration apparently requires only the amount of trophic factors to which the arm has become accustomed, without any fixed threshold limit. A sudden withdrawal of the neurotrophic agent causes a temporary inability to regenerate followed by a recovery, with no evidence that trauma has any influence on the process.

Table 2.6 Regeneration of chronically denervated arms (unpublished data of Alan Watson, using 9 cm axolotls).

Weeks of denervation prior to amputation	Number of arms, scored 3 weeks after amputation			
	Total	Regenerating[*]	Static	Regressed
0	12	0	1	11
1	12	0	0	12
2	12	0	0	12
3	12	9	0	3
4	11	11	0	0

[*]Only these cases were examined for nerves; some were sparsely innervated and others were apparently devoid of nerves.

2.9 Conclusions

The normal dependence of limb regeneration upon the presence of intact nerve fibres has been adequately established for larval amphibians and for adult urodeles. The frequency of regeneration is related to the proportion of axons remaining in partially denervated limbs. Sensory and motor nerves have equivalent and additive properties in this respect. The quantitative aspect of nervous control thus commands support. The widely advocated idea of a minimum threshold number or density of axons being a prerequisite for regeneration can now be abandoned, however, partly because estimates of the threshold level have differed so widely, but mainly because regeneration can occur without any nerve supply.

Limb regeneration is almost certainly sensitive to a substance which is produced in all neurons and delivered to all nerve endings by the axonal fast transport system. This may well be the same factor responsible for the maintenance of sense organs and muscles, and for the general growth of the larval limb. It stimulates the proliferation of mesenchymal cells and thus accelerates blastemal growth, but is not required for other aspects of regeneration. The trophic factor has not been identified and there is no evidence that other tissues apart from neurons can produce it, or even of a reservoir of trophic material in the egg to be used up during embryonic growth. Anomalies such as the regeneration of aneurogenic limbs, which seemed to demand some modification of the trophic theory, can now be accomodated by postulating a cellular capacity to adapt to increased or decreased innervation. Whether or not the quantitative aspects of innervation can be related to the inhibition of limb regeneration at metamorphosis in anurans, or its failure to occur in amniotes, is still completely unknown. Tests based on this premise have yielded only temporary or trivial indications of regenerative activity.

Chapter 3

Hormonal Influence

For a long time, the only widely acknowledged example of hormonal intervention in limb regeneration concerned the action of the pituitary gland in adult urodeles. Partly because the pituitary controls other secretions and partly because of more recent claims concerning other hormones, we shall be compelled to consider several of the major endocrine glands of amphibia. It is not completely safe to assume that these glands show the same functional relationships or produce identical hormones to the better known endocrine system of mammals, although many similarities are known. The severity of the operations or treatments employed to demonstrate the action of hormones also causes some difficulty in distinguishing between the requirement of a hormone for survival and a specific requirement for regeneration. It follows, as a general principle, that replacement therapy should provide more secure evidence than mere extirpation of an endocrine gland. Unfortunately, it has always been too difficult or expensive to obtain preparations of mammalian hormones pure enough to ensure that active contaminants could be entirely disregarded. That and the possible difference from native hormones cast some doubt even on the results of replacement therapy.

3.1 Pituitary

The pituitary matures gradually during larval life to initiate metamorphosis in most amphibians, and itself changes during metamorphosis (Copeland, 1943; Kerr, 1966). Extirpation of the pituitary (hypophysectomy) has no appreciable effect on the survival of most larvae or on their ability to regenerate, at least up to the normal time of metamorphosis. Adult newts rarely survive more than 30 days after hypophysectomy. They soon show a variety of symptoms related to the lack of different hormones: torpidity and refusal to eat, attributed to the lack of prolactin or of growth hormone; pale colour from the absence of melanocyte-stimulating hormone (MSH); a rough, dry skin indicating the absence of moulting and slime production,

53

caused by the lack of thyroid-stimulating hormone (TSH) and consequent deficiency of thyroxine (Dent, 1966). This syndrome and the mortality can be prevented by repeated injections with either growth hormone (Wilkerson, 1963) or prolactin (Connelly et al., 1968), but each of the commercial preparations used by these investigators was contaminated by the other hormone as well as by TSH.

Schotté (1926b) first reported that hypophysectomy prevented limb regeneration in adult *Triturus cristatus* and later returned to the subject to formulate a theory that a pituitary-controlled adrenal cortex hormone constituted an essential requirement for regeneration (review, Schotté, 1961). This theory will be considered in more detail in section 3. A standard feature of the experiments performed by Schotté and his associates over a 10 year period, however, was that adult *Notophthalmus viridescens* were completely unable to regenerate a limb if the pituitary had been removed at the time of amputation or earlier. They consistently described a rapid healing of the limb stump by a thick epidermal layer with a basement membrane and a variable amount of scar tissue over the end of the skeleton. Hall and Schotté (1951) noted an occasional late accumulation of dedifferentiated cells under the apical epidermis but, by analogy to the similar appearance of frog limb stumps, they insisted that such cells were imprisoned in scar tissue and could not form a blastema. In contrast to that, Tassava (1969a) has provided convincing evidence of the initial phases of regeneration in *N. viridescens* limbs amputated 5 days after hypophysectomy. He obtained examples of distinct blastemata by the simple expedient of feeding the newts well beforehand to prolong their lives after they stopped eating. Only the initial phases occurred and these became progressively retarded, as is to be expected in moribund specimens.

Similar indications of belated regeneration had been recorded previously for the tails of hypophysectomized *T. alpestris* and *T. vulgaris* (De Conninck et al., 1955), and have since been confirmed for limbs by temporary hypophysectomy in *N. viridescens* (Liversage and Liivamagi, 1971). Consequently, there is no adequate evidence that any pituitary hormone is really essential for regeneration in adult urodeles. Prolactin, and growth hormone, certainly affect the rate of regeneration in much the same way that they influence other metabolic processes, but they are only indispensible in terms of survival. It is convenient to consider attempts at replacement therapy for these two hormones first, and to postpone the other trophic hormones to later sections dealing with the endocrine glands which they control.

Hypophysectomized adult newts are found to survive much longer and show greatly improved limb regeneration if they carry ectopic grafts of pituitaries. They may survive indefinitely with pituitary autografts, and 6 months or more before homografts (Dent, 1967) or xenografts from axolotls (Tassava, 1969b) are suicidally rejected. In either event, the survival period and state of health are ample for complete regeneration (Schotté and Tallon,

1960). Such ectopic pituitaries should be largely independent of the normal regulation by hypothalamic nerves and neurosecretions. They continue to secrete normal amounts of TSH as judged by thyroid activity (Dent, 1966), besides sufficient growth hormone and prolactin for normal requirements. The evolutionary conservatism of these hormones is illustrated by the diverse sources of hormone which can be used for replacement therapy in *N. viridescens*. Richardson (1945) noted improved regeneration after injecting a crude extract of mammalian pituitaries known as Antituitrin G. Pituitary extracts obtained from frogs (Liversage and Scadding, 1969) or chicks (Liversage and Fisher, 1974) also improve the survival and hence the regeneration of hypophysectomized newts. Wilkerson (1963) obtained clear cases of regeneration in hypophysectomized newts injected with a relatively pure preparation of growth hormone. Connolly *et al.* (1968) and Tassava (1969a) confirmed these results but found that prolactin injections worked equally well. They argued persuasively that the amount of prolactin present as a contaminant in Wilkerson's growth hormone was sufficient to compensate for the missing pituitary. Any controversy over this point has been abolished by the later finding of Bromley and Thornton (1974) that a normal rate of regeneration can be supported in hypophysectomized newts

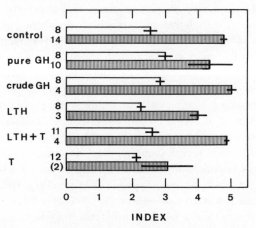

INDEX

Figure 3.1. Replacement therapy for the pituitary in *N. viridescens* (from data of Bromley and Thornton, 1974). Each histogram shows the mean and standard error of a histological index of regeneration, for the number of arms shown on the left scored at 24 days and 32 days (cross-hatched) after amputation. Index of regeneration numbers 2–5 correspond roughly to blastema, cone, palette and digit stages. Control: normal regeneration of untreated newts with pituitaries; all other results are from newts hypophysectomized 6 days before amputation, then injected on alternate days with about 0.03 USP units of growth hormone (GH), or 0.02 i.u. of prolactin (LTH), or kept in 0.1 μM thyroxine (T).

Crude GH containing about 0.01 i.u. LTH stimulated a normal rate of regeneration, being superior to pure GH. Thyroxine enhanced the effect of LTH. Some regeneration occurred even in the arms of two newts which survived 28 days in thyroxine.

either by prolactin supplemented with thyroxine or by an essentially pure preparation of growth hormone (Figure 3.1). Prolactin and growth hormone also stimulate the growth of limb regenerates in normal newts still in possession of their pituitary glands (Niwelinski, 1958; Schauble and Nentwig, 1974).

The similar effects of growth hormone and prolactin seem less surprising now that more is known of their composition and action. They show considerable homology in peptide sequence and secondary structure (Bewley and Li, 1971). Their spheres of influence overlap, for both of them act as trophic hormones to elicit the release of somatomedin from the liver and kidney, and thus indirectly stimulate skeletal growth (Francis and Hill, 1975). The release of these hormones from the pituitary is regulated by two distinct oligopeptides emanating from the hypothalamus, somatostatin which inhibits the release of growth hormone, and a tripeptide which stimulates the release of both prolactin and TSH. These recent discoveries only apply with certainty to mammals and few comparative studies have extended to amphibians. Several amphibians possess separable growth hormone and prolactin, however, which show distinct physiological properties but which are both immunochemically related to rat growth hormone, and thus to each other. *Ambystoma tigrinum* (and probably *A. mexicanum*) seems peculiar in possessing a prolactin active on pigeon crops and a protein which is partly homologous to rat growth hormone but which does not share its ability to stimulate growth in toads (Nicoll and Licht, 1971; Hayashida *et al.*, 1973). Since newts respond to mammalian growth hormone and prolactin, however, there is some hope of distinguishing between the corresponding newt hormones by treating newts with the specific hypothalamic control factors. Somatostatin reduces or delays limb regeneration in newts (Vethamany-Globus *et al.*, 1977). Unfortunately, somatostatin also inhibits the release of insulin and glucagon from the pancreas (Koerker *et al.*, 1974; Dubois *et al.*, 1975) with severe complications described in section 4, so it is impossible to draw a conclusion from this result. A drug which inhibits the release of prolactin, such as ergocryptine (Varga *et al.*, 1973), would serve the same purpose if it has no side-effects.

Considering growth hormone and prolactin together, it is evident that their absence following hypophysectomy retards regeneration but does not prevent its early phases from wound healing to the formation of a small blastema. Later stages are not prevented either, provided the specimen survives long enough. A blastema which is already present at the time of removing the pituitary, for instance, continues to differentiate and form digits, although further growth is curtailed (Schotté and Hall, 1952; Tassava, 1969a). If such a blastema is re-amputated it can still form a fairly normal small hand, perhaps by a morphallactic rearrangement, but re-amputated palette stages are not able to recover in the same way or to the same extent (Schotté and Hilfer, 1957). These pituitary hormones appear to accelerate cell metabolism and encourage proliferation, therefore, without exerting any

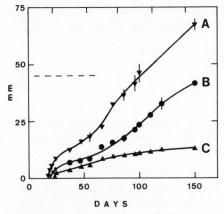

Figure 3.2. Growth of tail regenerates in adult axolotls (from data of Horak, 1939). Vertical lines show the range from duplicate specimens; the broken line indicates the approximate length of tail cut off initially. A, with three extra pituitaries; B, normal controls; C, hypophysectomized immediately before amputation.

specific influence on the process of regeneration. This conclusion is well illustrated by the results of Horak (1939) on the rate of tail regeneration in adult axolotls bearing different numbers of pituitary glands. Axolotls survive indefinitely after hypophysectomy but cannot tolerate the implantation of more than three extra pituitaries. As shown in Figure 3.2, hypophysectomized axolotls regenerate tails much more slowly than do normal control specimens, while those with extra pituitaries regenerate more quickly—and probably show considerable overall growth as well if their regenerates are not hypermorphic. The axolotl pituitary clearly produces hormones which enhance growth, but they are not essential for regeneration.

The effects of growth hormone and prolactin seem remarkably similar to those of the neurotrophic factor, sufficiently so to suggest they are at least functionally identical to it. Defazio (1968) examined this proposition by testing the ability of denervated newt limbs to regenerate during a 30 day regime of growth hormone injections. His failure to detect any regeneration at the maximum dosage employed, 0.024 USP units injected on alternate days, is open to two interpretations. Either the hormone is radically different from the neurotrophic factor and so cannot replace it, or the treatment was inadequate. According to data shown in Figure 3.1, the dosage was roughly equivalent to the intermittent presence of one extra pituitary, perhaps less as the injection probably reduced the amount of growth hormone secreted by the host pituitary. Either larger doses of both growth hormone and prolactin, or the permanent implantation of extra pituitaries should provide a more effective test of this intriguing idea.

3.2 Thyroid

The thyroid gland becomes active relatively late in larval life when, under the control of pituitary TSH, it releases increasing amounts of thyroxine and tri-iodothyronine which stimulate metamorphosis. Thyroid activity then declines to maintain a fairly constant concentration of thyroxine in blood plasma. The dramatic events of metamorphosis have promoted two opposing views of the influence exerted by thyroxine on regeneration. Since anuran tadpoles lose their former capacity for limb regeneration even before they lose their tails (section 6), regeneration is obviously subject to some form of pituitary–thyroid control. Yet these metamorphic changes are clearly organ-specific responses to the common signal of increasing thyroxine. Most experimental evidence also emphasizes the indirect nature of the thyroid's influence on regeneration. Guyénot (1926, 1927a) first demonstrated that regeneration was largely an autonomous property of the limb. He grafted legs from old *Bufo* tadpoles onto young salamander larvae, and recorded that none of them recovered the ability to regenerate when amputated 6–37 days later. Three leg buds from young tadpoles grafted in the same way developed normally but failed to regenerate when amputated 45 days later. He also grafted tadpole tails and found them capable of continued growth and regeneration on salamander hosts. The reciprocal grafts were quickly destroyed by their tadpole hosts. These results have been confirmed by the more obvious experiment of exchanging legs between young and old tadpoles of *Rana temporaria* (Liosner, 1931; Polezhaev and Ginsburg, 1939). A protracted exposure to increasing thyroxine concentration as experienced by older legs evidently causes some permanent change in their tissues, so that they cannot regenerate even in the presence of less thyroxine after metamorphosis or when placed on young tadpoles whose legs can regenerate. Schotté and Harland (1943c) produced a further argument against a direct hormonal control of regeneration. At a time when amputation through the thigh caused no regeneration in old *Rana clamitans* tadpoles, blastemata sometimes formed on amputated shanks and always formed on more distal leg stumps severed through the ankle or metatarsals. Similar results had been obtained earlier by Marcucci (1916) on *R. esculenta* and *Bufo* tadpoles. Identical amounts of thyroxine must have been available at each level of amputation, yet they responded differently. The extent of the response remains in some doubt, for the blastemata often stopped growing or were resorbed.

These arguments need to be modified if one questions whether frogs and toads are genuinely incapable of limb regeneration. The most extended investigations here are undoubtedly those of Polezhaev (1946a–d, 1972a) who forced limbs to regenerate by mechanical or chemical damage, using tadpoles which usually formed healed stumps after simple amputation. He applied similar treatments to fore-arm stumps of adult *R. temporaria* and *Bombina bombina* and recorded that they enhanced regeneration—enhanced because untreated controls also showed some regeneration in the spring. All these regenerates were so small and defective that they would be excluded by the definition advocated in chapter 1, yet they showed typical features of the process: a persistent wound epithelium, internal dedifferentiation, a little growth and redifferentiation of rather chaotic cartilages. On the basis of these features, one may doubt if the regenerative capacity of tadpole legs is entirely lost at metamorphosis, although it is certainly curtailed drastically. According to Polezhaev (1946b), either a vestigeal capacity is maintained throughout life or, if it is lost at any stage, it must reappear later.

The opposite point of view, that regeneration is essentially independent of thyroid activity, stems from the widespread use of a few species of urodeles (*T. cristatus*, *N. viridescens*, *A. tigrinum*) where perfect limb regeneration is normally encountered in both larvae and adults and where thyroxine concentration remains constant throughout regeneration (Liversage and Korneluk, 1978). That is a particularly superficial comparison, for limb regeneration occurs much more rapidly in larvae than adults even in these species. Furthermore, other species of *Triturus* and the related *Salamandra* whose larvae exhibit perfect limb regeneration are only capable of inconsistent or abnormal limb regeneration when adult. There is remarkably little information on what happens to regeneration in urodeles undergoing metamorphosis. Belkin (1934) immersed young axolotls in a 50 ppm thyroid

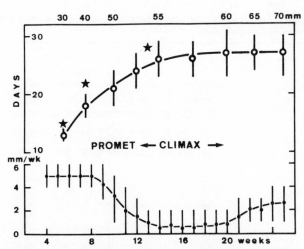

Figure 3.3. Growth rate of *Pleurodeles* during metamorphosis at 20°C (●), and the time required to form a notch regenerate (○) or a notch supernumerary arm (★); all plotted against age and total length. All three features are retarded early in metamorphosis.

extract which induced metamorphosis 29–39 days later. He stated that limb regeneration was unaffected by this treatment or by similar treatments beginning before or after amputation in large axolotls. Repeating the experiment much later, Belkin (1950) claimed to have detected an acceleration of regeneration during the period of induced metamorphosis. That is quite a surprise, as newts virtually stop growing for several weeks during spontaneous metamorphosis and their limbs regenerate more slowly then. Goodwin (1946) documented a retardation of regeneration in *N. viridescens* and *A. maculatum*, and I have noticed the same thing in *A. opacum* and *P. waltl*. Curiously, limb regeneration remains slow even when overall growth is resumed (Figure 3.3), but the regenerates produced during metamorphosis are often extremely defective (Table 3.1) This is clear evidence that some metamorphic event, perhaps a thyroxine overdose, exerts a permanent effect on the regenerative capacity of urodele limbs. It allows the speculation that thyroxine triggers the same changes in both main groups of amphibians: the virtually complete inhibition of regeneration in anurans being a drastic version of the defective regeneration seen in some urodeles and the retardation found in others. If this is not an idle speculation, it should be consistent with the effects of thyroxine deduced from experiments on adult amphibians described below.

On the basis of hypophysectomy, thyroidectomy and replacement therapy, Richardson (1940, 1945) concluded that thyroxine could not substitute for missing pituitary hormones but might be required in addition to them, particularly to ensure good skeletal differentiation in adult *N. viridescens*.

Table 3.1 Defective regeneration during metamorphosis of *Pleurodeles waltl* (unpublished data).

Specimen size (mm)*	No. of arms	Regenerates (%) classifed by no. of digits			
		1	2	3	4
30	20	0	0	0	100
40	12	0	0	0	100
50	12	0	0	25	75
55	12	0	0	8	92
60	12	8	33	33	25
65	10	60	40	0	0
70	10	20	0	50	30
80	10	0	10	30	60
90	12	0	17	25	58
100	12	0	0	33	67

*Length (±2 mm) at the time of amputation through both fore-arms. The specimens were selected for consistency of developmental stage: larval (30–40 mm), metamorphic phases (50–70 mm) and postmetamorphic juveniles (80–100 mm).

Later experiments generally support this conclusion (Connelly *et al.*, 1968; Tassava, 1969a). Schotté and Washburn (1954) also reported that thyroidectomized newts often grew regenerates which were retarded and stunted with skeletal defects. Yet a carefully documented and extensive study of thyroidectomized newts provides convincing evidence that regeneration can occur normally (Schmidt, 1958a, b). Schmidt's results did not differ particularly from previous ones when he had removed the thyroid at about the same time as amputating a limb. He noticed a precociously formed apical epidermal cap, perhaps related to the inhibition of moulting, and an early cartilage differentiation which could explain the stunting and skeletal defects described by Schotté and Washburn. When thyroidectomy preceded amputation by 20–30 days, however, regeneration occurred more rapidly than normal without exhibiting any conspicuous defects. This surprising result prompts the suspicion that the immediate effects of thyroidectomy include temporary fluctuations of secretion, hormonal hiccups, by the pituitary and possibly other glands. These settle down eventually and normal regeneration can occur at what is believed to be a very low concentration of thyroxine. There are some indications that thyroxine and prolactin are mutually antagonistic (see Li and Bern, 1976), and they may be controlled by the same common hypothalamic release factor even if not the one known in mammals (Vale *et al.*, 1973; Taurog *et al.*, 1974). If a reduced thyroxine level enhances prolactin activity by either of these means, then it should lead to accelerated regeneration. Schmidt (1968) has provided a more detailed appraisal of the pitfalls attending both surgical and pharmacological thyroidectomy. Accessory thyroid follicles or alternative sources of thyroid hormones have been recognized in amphibians, but perhaps not established with certainty. Thyroidectomy is followed by compensatory changes in other hormones, particularly those of the pituitary but perhaps also of parathyroid hormone and calcitonin. Furthermore, the specificity of thyroid antagonists is still open to question. Two such antagonists have been applied to late tadpoles, in which they seem to enhance dedifferentiation but impede blastemal growth of limb stumps. Thiourea caused little or no improvement of limb regeneration in *Rana pipiens* at and after metamorphosis (Peadon, 1953). Perchlorate stimulated limb regeneration slightly when applied to *Bufo regularis* tadpoles just before metamorphic climax (Michael and Aziz, 1976).

The chronic effect of excess thyroxine and of lethal amounts applied for shorter periods on limb regeneration have been investigated by Hay (1956) using adult *N. viridescens*. She found regeneration to be defective in these conditions: either inhibited cellular proliferation or precocious differentiation of cartilage tended to stunt growth. Stunted growth following either a gross excess or a deficiency of thyroxine is compatible with the general physiological conclusion that moderate concentrations of thyroxine stimulate diverse aspects of cellular metabolism (Lynn and Wachnowski, 1951). This general metabolic effect is expressed in an organ-specific manner, as illustrated by the deformities of tadpoles kept in too strong thyroxine solutions. One organ-specific change, apparently an irreversible one indirectly observed by amputating limbs at any subsequent time, is probably caused by the increase of thyroxine just before metamorphosis. Limbs then regenerate more slowly in urodeles, and almost imperceptibly in anurans. Although it still needs to be substantiated, this generalization is interesting enough to be considered further in Chapter 10.

3.3 Adrenal cortex

Any large wound such as that caused by amputating a limb is certain to cause considerable stress. Schotté and Hall (1952) postulated that the mammalian stress response, involving the pituitary ACTH and adrenal cortex hormones, might well apply to amphibians and provide a complete explanation of the failure to regenerate both in hypophysectomized adult newts and in adult anurans. The analogy needs to be formulated more precisely, for the initial rapid response to stress mediated by an ACTH-stimulated release of cortisone actually delays wound-healing in mammals, whereas the glucocorticosteroids which promote healing are released later. It is not at all clear how the initial healing of an amphibian limb stump affects later events. Schotté (1953, 1961) envisaged the normal level of cortisone as preventing a premature invasion of the wound by dermal fibroblasts which he considered to be inimical to the formation of a blastema. It is quite easy to produce a blastema which is entirely composed of dermal fibroblasts, however, and such blastemata grow into reasonably normal regenerates (Umanski, 1937; Namenwirth, 1974). Retaining Schotté's analogy to mammals, the wound must be perceived locally and transmitted by sensory nerves to the conscious centres of the brain and, perhaps independently, to the hypothalamus which causes a release of ACTH from the pituitary. The following examples are of breaks in this chain of command. Newt limbs regenerate perfectly when their sensory ganglia have been detached from the spinal cord, and when the sympathetic nerves have been severed, well before amputation. Liversage (1959, 1962) interrupted all conceivable afferent nervous pathways between the limb and brain of adult *N. viridescens* and larval *Ambystoma*, by transplanting limbs and by extirpating sections of the spinal cord. Weeks or months later when the original operational stress should have subsided, he amputated the isolated limbs and recorded that the specimen seemed unaware of the loss but the limb regenerated in due course. Liversage and Price (1973) have adapted methods for assaying total corticosteroid hormones, cortisol and corticosterone in the blood plasma of *N. viridescens*. They found these hormones were reduced to less than half the normal amount 10 days after hypophysectomy, which amounts to a formal demonstration that newts possess ACTH. The adrenal cortex showed considerable independence from pituitary control, however, and its secretions did not alter appreciably in response to limb amputation.

Schotté and Lindberg (1954) claimed that a hypophysectomized newt could recover its ability to regenerate after an adrenal gland from a frog had been implanted into it. Their best example used to illustrate this result was amputated at the time of hypophysectomy, and the adrenal was implanted 16 days later. This specimen survived remarkably well, for the limb was preserved 39 days after amputation and then showed a blastema containing some condensed skeletal primordia—about equivalent to the best regenerates obtained by Tassava (1969a) after hypophysectomy alone. The claimed adrenal effect, then, is simply a case of delayed regeneration which could have been achieved fortuitously by using well-fed newts in one experimental series which seems to have lacked controls. The optimistic conclusion drawn from this experiment was probably influenced by the contemporary successful replacement therapy for hypophysectomy, of daily injections with ACTH (Schotté and Chamberlain, 1955). Not only did these treated newts exhibit almost perfect limb regeneration after a slight delay, but they also survived 50 days or more and appeared to remain healthy for much of that period. These features indicate that the commercial ACTH preparation employed was a virtually complete pituitary extract. Pure ACTH has no effect on survival or limb regeneration when administered under the same conditions (Tassava, 1969a). Schotté and Bierman (1956) attempted to compensate for the lack of a pituitary by injections of cortisone, desoxycortisone and crude extracts of the adrenal cortex. These preparations would not be contaminated by pituitary hormones, of course, and in fact did not relieve any of the symptoms of hypophysectomy. Using doses which ranged from massive to lethal (but no uninjected controls), Schotté and Bierman detected some early blastemata among the few specimens which survived for 20–30 days. Once more using Tassava's (1969a) results as the most appropriate control, we may conclude that these cortical hormones did not enhance limb regeneration at all. Schotté and Bierman, however, were convinced their treatments had been effective and justified their belief by the following statement: 'It will be shown in a forthcoming publication that hypophysectomized and amputated newts kept over 20 days in a weak cortisone acetate solution regenerate in all cases without exception'. The nearest I have come to tracing that publication is the one missing from a symposium (Schotté, 1959). Schotté and Christiansen (1957) elaborated on the concept of a cortical control of regeneration by recording that amphenone B, a drug which suppresses the synthesis of hormones by the adrenal cortex and thyroid in mammals, prevents limb regeneration in adult newts for the duration of treatment at 1 mg/day. The newts survived this regime to recover and regenerate afterwards. Furthermore, the simultaneous administration of cortisone permitted normal regeneration, even in hypophysectomized newts, but ACTH did not relieve the inhibition caused by amphenone. It is impossible to evaluate these experiments which, like several of the preceding ones, were only published in the form of an abstract. The statement that cortisone can replace the pituitary gland is undoubtedly false, however, and this casts doubt on the criteria and reliability of the other results.

Newts with intact pituitaries react to cortisone treatment in much the same way that mammals do: the process of wound-healing and epidermal proliferation is hampered, so that regeneration is delayed at first but subsequently recovers (Manner, 1955; Williams, 1959). Schotté (1961) finally described some ingenious experiments intended to depress secretion from the adrenal cortex by imposing a regime of corticosteroid or ACTH overdose which terminated just before limb amputation. The experiments were mainly performed on larval urodeles and are considered in that context (section 5). It might as well be mentioned now, however, that Schotté could not demonstrate any interference with adrenal secretion because the hormone levels were not monitored and his results have since been contradicted. Neither these experiments nor any mentioned earlier in this section provide any reliable evidence for supposing the adrenal cortex has more than a temporary and trivial influence on regeneration.

3.4 Islets of langerhans

Besides advocating the idea of some adrenal cortex involvement, Liversage and his associates suggested that insulin is intimately concerned in limb regeneration (Liversage and Price, 1973; Vethamany-Globus and Liversage, 1973a, b). They found that organ cultures of tail blastemata from adult *N. viridescens* required insulin in order to show any growth or differentiation. The cultures responded best to a mixture of insulin, thyroxine, hydrocortisone, and growth hormone. Electrocautery of increasing amounts of the pancreas to eliminate islets of Langerhans enabled them to obtain results which could be related to the amounts of sugar in the blood. Total pancreotomy caused death in 4 days; partial pancreotomy resulted in a considerable delay of limb regeneration until the newt recovered from its diabetic condition, when the pancreas and its islets were found to have regenerated. More specific destruction of the pancreatic islet cells by injections of alloxan produced a similar spectrum of results including delayed limb regeneration and spontaneous recovery. Hypophysectomized newts eventually showed an atrophy of both pancreatic acinar cells and islet cells, perhaps reflecting the dual action of somatostatin mentioned earlier. The simplest interpretation of all these results, albeit not one favoured by the investigators, is much the same as reached previously for other hormones. Acute illness, whether diabetic or caused by electrolyte imbalance, places a restriction on general metabolism and thus indirectly delays or retards limb regeneration.

3.5 Larval urodeles

Schotté (1926b) first reported that larval urodeles survive quite well without a pituitary gland and are certainly capable of perfect limb regeneration. This discovery for 16–65 mm long larvae of *T. cristatus* and *T. alpestris* has been confirmed since for *N. viridescens* (Schotté, 1961) and for several species of *Ambystoma* (Liversage, 1967; Tassava *et al.*, 1968). Any hormones thought to be implicated in limb regeneration, therefore, cannot be produced by the larval pituitary or be subservient to it. The relative inactivity of the larval pituitary has been elegantly demonstrated by Schotté and Droin (1965) using *N. viridescens*. The aquatic larva of this species first changes into a prometamorphic 'brown eft' which transforms after a few weeks into the vivid red eft terrestrial stage. The latter commonly remains on land for 3 years before returning to a pond and undergoing a final metamorphosis into the green adult newt. This final migration or 'water-drive' is apparently induced by prolactin, and thus taken as a sign that the pituitary has finally matured. Only the red eft and adult require a pituitary for survival and limb regeneration; and only their pituitaries are capable of supporting these activities. A larval or brown eft pituitary is either inactive or insufficiently active to replace the normal pituitary of a red eft or adult, making it just as well that the larval stages can dispense with the gland. Table 3.2 is compiled from this and earlier studies to substantiate Schotté's (1961) remark that hypophysectomized adult newts could survive longer when they carried ectopic larval pituitaries from the same species or from *A. talpoideum* but still could not regenerate. If such carriers had survived much longer than 30–40 days, this study might have finally dissociated the pituitary's influence on survival from its speculative control over regeneration. In a misconceived attempt to refute this demonstration, Tassava (1969b) grafted pituitaries from axolotls into hypophysectomized adult *N. viridescens*, which then survived 6 months or more and retained the ability to regenerate their limbs. The fallacy of Tassava's argument lay in equating neotenous axolotls with

Table 3.2 Limb regeneration in larval, juvenile and adult *N. viridescens* when lacking a pituitary or carrying a transplanted one.

Hypophysectomized host (length in mm)	Pituitary donor (No.)	No. of cases[*]	Regeneration blastema		Estimated survival (weeks)
			Present	Absent	
1. Larva	—	2	2	0	8
2. Brown eft (32–40)	—	13	13	0	4
3. Red eft (40–75)	—	15	0	15	4
4. Adult	—	16	0	16	4–5
5. Adult	Larva (1–4)	57	0	57	5–6
6. Adult	Brown eft	11	0	11	4
7. Adult	Red eft	49	49	0	5–6
8. Adult	Adult	55	45	10	100

[*]After eliminating faulty operations which allowed the retention of host pituitary fragments, or the loss of a graft.
Data from Hall and Schotté (1951) (line 4); Schotté and Tallon (1960) (line 8); Schotté (1961) (lines 1 and 5); Schotté and Droin (1965) (lines 2, 3, 6 and 7).

typical urodele larvae. He actually obtained the pituitaries from 10–16 cm axolotls, which are equivalent to postmetamorphic juveniles in many respects. Ossification of the skeleton and production of the adult form of haemoglobin have already occurred (Keller, 1946; Dulcibella, 1974a, b). The pituitary is already active in producing MSH (Vorontsova, 1929; Dalton and Krassner, 1956) and either growth hormone or prolactin according to the results shown in Figure 3.2.

While still convinced of an ACTH–adrenal-corticosteroid synergism controlling limb regeneration, Schotté (1961) claimed that he could 'endocrinate' larvae to become dependent on this hormonal system. The indoctrination consisted of subjecting *N. viridescens* and various *Ambystoma* larvae, usually after hypophysectomy, to one of the following three treatments: either implantation of an adult pituitary for 15–47 days; or a series of 4–7 injections on alternate days of 0.25–0.5 i.u. ACTH or 0.25 mg cortisone. It may be recalled that similar injections often proved fatal to adult newts, and Schotté mentioned that hundreds of deaths occurred among the injected larvae but the survivors were capable of limb regeneration during the treatment period. Amputation performed within 5 days of removing the grafted pituitary or of terminating the injections, however, resulted in a marked failure of regeneration. Only 6 out of 32 larvae regenerated limbs following the pituitary regime; 15 out of 26 regenerated after deprivation of the accustomed ACTH; and cortisone deprivation was stated to yield similar results. After a further delay of 3–6 weeks, reamputation led to normal regeneration of all survivors. This experiment might be considered striking enough to warrant interpretation without any measurement of hormonal levels, or criticized as too elaborate to devise adequate controls and so rambling as to virtually preclude repetition. Fortunately, most of the experiment has been thoroughly repeated, with results which completely contradict those presented above. Tassava *et al.* (1968) recorded normal regeneration in larvae of four species of *Ambystoma* whether or not they had been hypophysectomized, both during and shortly after a 10–14 day regime of daily injections with 0.5 i.u. ACTH (which did not cause much mortality) or a 10–21 day subjection to an implanted adult newt pituitary. This surely leaves no doubt that something was seriously amiss with the experiment described by Schotté, but his concept of adaptation to particular hormone levels is an interesting one which presumably served as a model for the addictive aspect of the neurotrophic theory (Singer, 1965) considered in Chapter 2.

One further anomaly concerning larval urodeles deserves attention. In his original note on the effects of hypophysectomy, Schotté (1926b) recorded that 15 larvae of *Salamandra salamandra* responded to the operation like adult *Triturus*, showing an inhibition or retardation of regeneration and at least some mortality. These observations have never been repeated and may well be accurate. *S. salamandra* is ovoviviparous, giving birth to small but relatively old larvae whose pituitary glands might already be active.

3.6 Anurans

Tadpoles survive well in the absence of either pituitary or thyroid glands, continuing to grow in the absence of metamorphosis. Tail regeneration occurs normally (Allen, 1918; Wright and Plumb, 1960) but the legs grow so poorly that nobody has bothered to test their ability to regenerate (Allen, 1925). Warren and Bower (1939) induced precocious metamorphosis in *R. sylvatica* tadpoles by exposing them to iodotyrosine. Legs amputated during this treatment failed to regenerate but those amputated earlier regenerated more rapidly than usual. High concentrations of thyroxine will even inhibit tail regeneration, according to Li and Bern (1976) who also noted that growth hormone stimulated the process but prolactin had no effect on it. Hydrocortisone injections do not significantly alter the metamorphic suppression of limb regeneration in the same species, *Rana catesbiana* (Weis and Blier, 1973). This characteristic suppression certainly exhibits a variable timing or even degree in different species and at different positions along the leg. The diversity of staging systems employed to describe metamorphosis and of criteria used to define regeneration successfully mask any precise relationship between the suppression and metamorphic progress. The gradual decline of regenerative ability in *Xenopus* tadpoles (Figure 3.4) and the more abrupt loss implied in *Rana* tadpoles (Forsyth, 1946), however, both occur as thyroxine levels are increasing (Saxen *et al.*, 1957; Just, 1972). It seems unlikely that the thyroxine concentration ever subsides to premetamorphic levels, allowing the naive deduction that regeneration might still occur in hypothyroid frogs. Evidence cited in section 2 concerning the organ-specific and permanent changes are generally accepted as flatly contradicting any such notion.

Two separate claims to have stimulated limb regeneration in postmetamorphic frogs need to be examined in the context of hormones. Firstly, Schotté and Wilber (1958) successively transplanted 2–5 extra adrenal glands from *R. pipiens* into other frogs, mainly *R. clamitans*, while the hosts were growing juveniles despite being described as adult. This procedure caused the largest and most deformed outgrowths that have ever been recorded on *Rana* limb stumps (Figure 3.5), besides frequently resulting in less pronounced 'regeneration' in hypophysectomized frogs. Since the outgrowths were first detected 40–50 days after amputation and untreated controls were only observed up to 50 days, this experiment seems to have lacked adequate controls. Singer (1954) recorded that about 15% of rather smaller frogs regenerate spontaneously by the same criteria. Perhaps extra adrenals hasten the normal response to amputation, but it would take a better experimental design to establish that. Secondly, Rose (1944, 1945) produced smaller outgrowths of frog limb stumps by repeatedly dipping them in a strong salt solution. Singer (1954) confirmed that this response could be obtained routinely immediately after metamorphosis by dipping an amputated fore-arm into saturated salt. Upper arm stumps and denervated

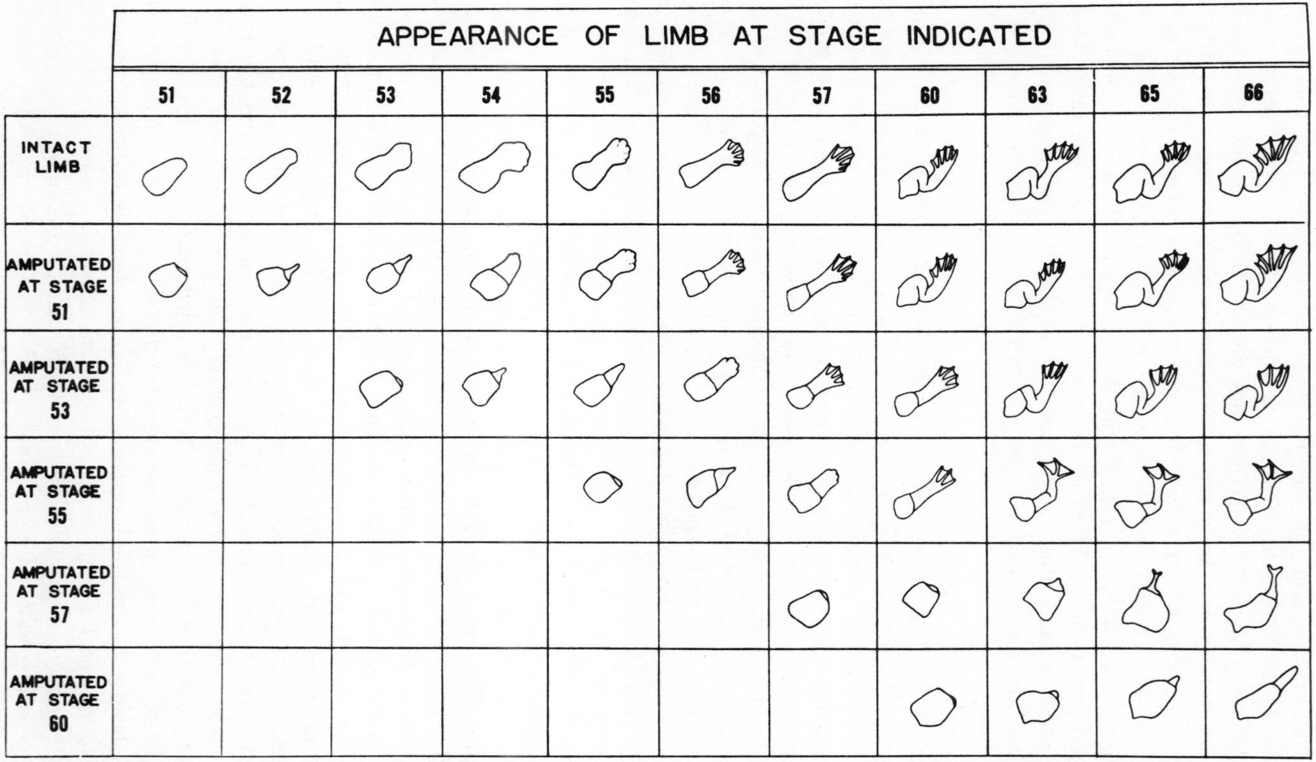

Figure 3.4. Growth of the leg during metamorphosis in *Xenopus laevis* and typical examples illustrating how later amputations produce progressively more defective regenerates. (From Dent, 1962: reproduced by permission of The Wistar Institute.)

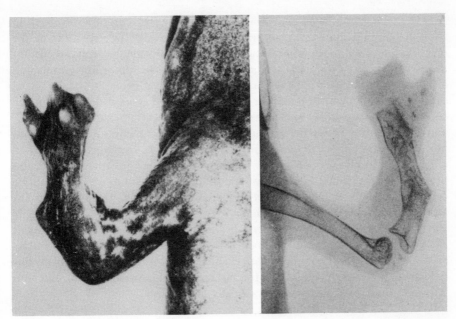

Figure 3.5. Extra adrenals were transplanted into the lower jaw of this *Rana clamitans* froglet, 1 and 2 weeks after its right arm had been amputated. The photograph and x-ray picture show the regenerated arm and its fused skeletal bone about 8 months later. (From Schotté and Wilber, 1958: reproduced by permission of the Company of Biologists Limited.)

stumps do not respond to this treatment (Singer *et al.*, 1957). Rose originally attributed his results to a local effect: osmotic shock and delayed wound-healing which allowed enough dedifferentiation to establish a small blastema. Later experiments incorporating controls, however, made him retract this explanation (Rose, 1964). If both arms are amputated but only one of them is repeatedly dipped in salt, then the other arm is also liable to produce an outgrowth. This indicates a systemic response, perhaps mediated by hormones. A still better explanation of all these results from young frogs is that normal growth of the limb stump provides enough cells which have no alternative to differentiating at the stump apex to yield a bizarre excrescence with no obvious relevance to regeneration.

3.7 Conclusions

Hormones are intermeshed in such a complex homeostatic system that changing any one hormone concentration may have unforeseen consequences, leading to unpredictable results in terms of regeneration. Taking many of the studies reviewed here at their own valuation, it is easy to sympathize with the belief of Liversage and Price (1973): 'It has been shown recently that ACTH-stimulated adrenal corticosteroids, somatotrophin

(STH), TSH-stimulated thyroxine, insulin and prolactin are involved in adult newt limb and tail regeneration and survival'. Indeed, most of the investigators cited here were almost as uncritical of other experiments as they were of their own. It is still possible that less drastic or abrupt changes of hormonal concentration than have been employed generally might reveal some perturbation of regeneration in the absence of general debility, but the only demonstrated effects of hormones have concerned health and survival, and growth rate. There is certainly no evidence of hormone fluctuations during regeneration or that such fluctuations could control successive phases of the process. The enhanced growth rate caused by growth hormone and prolactin seems to be merely a manifestation of their ability to stimulate cellular metabolism. Cells apparently adapt to prevailing hormone levels, and perhaps chronic changes deserve further investigation. The chronic effects of thyroxine, for instance, by inducing organ-specific permanent changes at metamorphosis indirectly make a distinction between urodeles and anurans in terms of regenerative ability. It is possible to reconcile this difference between the two groups: both show a reduced growth rate but reduced mildly in urodeles and drastically in anurans.

Chapter 4

Effects of Irradiation

One of the earliest discovered biological effects of X-rays was that they tended to prevent growth and regeneration. While X-rays have become a favourite experimental device for inhibiting regeneration, so that most of this chapter will be devoted to them, a few studies have shown that other ionizing radiations act in much the same way and even ultra-violet light can be considered in analogous terms. Radiomimetic chemicals and radio-protective substances have been employed so rarely that they need not be considered in this context, although some are mentioned in the following chapter.

Before the novelty of Roentgen rays and radium emanations had fully worn off and before their dangers had been fully appreciated, studies involving the growth and development of diverse organisms and the regeneration of planaria (Bardeen and Baetjer, 1904) and newts (Schaper, 1904) were epitomized in the law of Bergonié and Tribondeau (1906):

'Les rayons X agissent avec d'autant plus d'intensité sur les cellules que l'activité reproductrice de ces cellules est plus grande, que leur devenir karyokinétique est plus long, que leur morphologie et leur fonctions sont moins définitivement fixés'.

This may seem nothing more than an elegant way of expressing the observation that proliferating cells are more sensitive to radiation than are non-dividing ones, yet it can be held to encompass long-term effects of X-rays which were unknown at the time. Irradiated non-dividing tissues may survive well and function normally until their cells are provoked into division, at which time a considerable mortality occurs. Research into radiation biology expanded enormously following the stimulus of Muller's (1928) discovery that irradiation caused mutations, and the later developments of nuclear fission and fusion. Much of that research is outside the scope of the present discussion, but it would be helpful here to know what are the principal effects of irradiation on cells, if they are common to all cells or show a specific effect on different tissues, and whether cells are affected directly or indirectly. The

71

following summary convinces me that all living cells are vulnerable to the direct effects of radiation, although their sensitivity varies according to the law expressed previously.

Partial shielding of large cells such as amoebae, insect eggs or plant spores show that a portion containing the nucleus is much more sensitive than a purely cytoplasmic region to the lethal effects of ionizing radiation. The development of ultra-violet microbeams has allowed the same conclusion to be drawn for ultra-violet light and cells in tissue culture. The radiosensitivity of a wide range of organism (embracing viruses, bacteria, animals and plants) is proportional to their nuclear size and DNA content. Furthermore, incorporation of bromo-deoxyuridine, an analogue of thymidine, into their DNA increases the sensitivity of both bacterial and mammalian cells to X-rays and ultra-violet light.

All ionizing radiations (X-rays, γ-rays, α-particles or helium nuclei, β-particles or electrons, protons and neutrons) readily break DNA chains. A considerable proportion of these breaks may well be repaired perfectly, but some are repaired incorrectly and others are not repaired at all. The latter two categories produce both point mutations and a variety of visible chromosome aberrations, including inverted and translocated chromosome segments, acentric fragments, dicentric and ring chromosomes. None of these aberrations need hamper cell metabolism but all of them lead to a spectrum of disturbances in meiosis which reduce fertility, and several of them also perturb mitosis, so that at least one daughter cell is genetically deficient and may be inviable. The frequency of these chromosomal aberrations is a function of the radiation dose (approximately proportional to the square of the dose for X-rays), and the failure of proliferation in cell cultures shows a similar proportionality to the dose. Such cells which divide repeatedly are generally agreed to be most sensitive to irradiation during early prophase and G2, with sensitivity decreasing back through the cell cycle. Allowing for other complicating factors, it may be deduced that a considerable amount of radiation damage can be repaired provided enough time elapses before the next mitosis. Similarly, an intense burst of X-rays has a more pronounced effect than the same dose spread over a prolonged period. There is ample evidence for the existence of a repair system which corrects X-ray damage, although its mechanism is not so well understood as the analagous systems for repairing ultra-violet damage in bacteria. Most of these points are discussed in more detail by Bacq and Alexander (1961) and by Davies and Evans (1966). The latter review stresses the following pertinent aspect of the dosage effect on cells in tissue culture. Sufficiently high doses of X-rays will kill cells completely. Lower doses allow considerable survival but prevent cell division (which is often used as a measure of survival in normally proliferating cell cultures). Still lower doses only temporarily delay cell division, to a degree which becomes progressively smaller as the dose is reduced. A most intriguing feature of this graduated response is that doses permitting about 10% survival allow each cell to divide twice more on average, while a further reduction of the dose permits the cells to divide a few more times before most of the daughter cells either cease dividing or die.

These salient features of the cellular reaction provide a convincing argument that nuclear DNA is the critical target of radiation and that genetic damage is the major contributing factor to cell lethality. The investigations on which this conclusion is based include very few cases of regeneration, yet we shall see that the behaviour of irradiated amputated limbs matches the preceding account in so many details as to be beyond the realm of coincidence.

Early studies on the effects of irradiation on limb regeneration have been ably reviewed by Curtis (1936) who pointed out that they were mainly devoted to establishing the dose required, and to distinguishing between systemic injury caused by whole-body irradiation and a more direct effect on the limb itself. For example, Butler (1933) found that recently hatched larvae of *A. maculatum* could survive low doses of X-rays repeated at daily intervals, which prevented the development or regeneration of their limbs.

Butler noticed a time lag in the response of developing limbs which allowed the completion of one more digit before growth was arrested. This result has since been duplicated for regenerates exposed to a single dose of X-rays (Brunst, 1950a). Curtis (1936) perceptively interpreted this effect as indicating that irradiation inhibits cellular proliferation, for the continued differentiation of digital cartilage would make the digit protrude. Puckett (1936) extended Butler's observations on young *A. maculatum*, finding that small daily doses of X-rays or a single irradiation of the entire larva could suppress both limb development and regeneration. He concluded that undifferentiated mesenchymal cells were very susceptible to irradiation, showing a reduced ability to divide and differentiate or often being destroyed.

4.1 Characteristics of X-rays

The effects of X-rays have continued to be disputed up to the present, as has the dose of X-rays needed to inhibit regeneration. Both will be dealt with later in this chapter. All doses will be stated in rad (100 erg/g), which is the appropriate measure of absorbed radiation, although they were usually expressed in roengten units, R. The two values are so nearly identical for limbs containing soft tissue and skeleton as to be within the errors of measurement, so the numbers given in original articles remain the same. The dose of radiation can only be specified adequately by considering the penetrative power of X-rays, some indication of which is provided by the voltage setting during irradiation. The spectrum emitted from an X-ray tube shifts to more energetic higher frequencies (shorter wavelengths) in direct proportion to an increase of voltage (Figure 4.1). High energy 'hard' X-rays traverse living tissue, soft (~ 50 pm) rays are considerably attenuated after a few millimetres, whereas very soft rays (0.1–1 nm) cause mainly superficial damage. Equal doses under different conditions can thus give a misleading

74

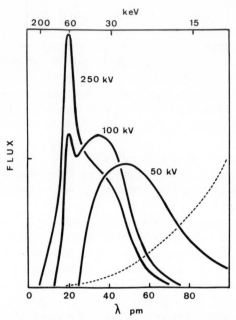

Figure 4.1. Relation of X-ray emission spectra to voltage, obtained from different X-ray tubes but normalized to represent equal emission. The characteristic K line of the tungsten target at 20 pm boosts the average photon energy at 80–200 kV, while inherent filtration reduces emission more at high wavelengths (the 50 kV spectrum is from a thin-window tube). Both features tend to skew spectra toward low wavelengths. Similarly, extra filters can virtually eliminate high wavelength soft X-rays, as shown by the absorption spectrum of copper (dashed curve).

impression of being equivalent treatments. Lazard (1968) provided an example by subjecting axolotl legs to 5 krad from two different unfiltered sources: regeneration occurred much more frequently and normally after exposure to 15 kV than to 60 kV rays (17% and 2% respectively showed at least some signs of regeneration). In both cases the total dose grossly

Table 4.1 Illustrative example of the changes in dose rate (rad/min/mA) produced by applying different potentials and filters to the same X-ray machine, measured 20 cm from the shutter.

| Potential (kV) | Filtration and nominal Half Value Thickness (HVT)* | | |
	Inherent HVT 0.9 mm	0.17 mm Cu HVT 1.5 mm	1.15 mm Cu HVT 3.0 mm
50	0.7	0.2	0.0
100	3.3	1.8	0.3
200	13.8	9.8	4.4
300	31.0	23.4	14.0

*HVT is the additional thickness of copper required to reduce the dose rate at 300 kV by a further 50%.

overstated the amount of penetrative X-rays, which are only a minor part of the spectrum at 50 kV (Figure 4.1). The proportion of soft X-rays can be reduced by means a thin copper or aluminium filter. This decreases the dose rate, considerably so at low voltage (Table 4.1), whereas high dose rates are more convenient and theoretically more effective. Since the dose rate is proportional to the square of the operating voltage, it is obviously better practice to employ the maximum voltage of which the machine is capable and to reduce the low frequency end of the spectrum by filtration, in order to obtain a more uniformly effective dose. Table 4.2 suggests a widespread ignorance of such precaustions.

4.2 Localized irradiation

Realizing that chronic whole-body irradiation was an unsatisfactory experimental condition, Butler (1935) adapted and improved the technique of Brunst and Scheremetieva (1933) whereby only part of the specimen was exposed to a single dose of irradiation prior to amputation, while the rest of the body was protected under a lead shield. This modified technique and Butler's introduction of the anaesthetic MS222 (ethyl *m*-aminobenzoate methanesulphonate, best neutralized with KOH) have become the standard procedure for irradiation. Butler's own results now demonstrated for the first time that X-rays have a direct local effect on the limb. When only one arm of an *A. maculatum* larva was exposed to 900 rad and then amputated, it did not regenerate (in fact it was gradually resorbed) although the contralateral shielded arm regenerated perfectly. Irradiation of an entire larva during a single period at the lower dose of 450 rad inhibited growth and regeneration of the limbs, but an unirradiated arm transplanted on to such an irradiated larva retained its capacity for normal regeneration. These results furnish the important proof that those cells which form a regenerate must originate in the limb stump; they cannot migrate from other regions of the body, as some descriptive studies had suggested. Concurrent experiments by Brunst and Scheremetieva (1936) confirmed the localized inhibitory effect of X-rays on adult axolotls and *T. cristatus*. Litschko (1934) had anticipated these conclusions, claiming that low doses of local irradiation prevented regeneration of axolotl limbs by causing a temporary stimulation of callus formation from connective tissue. This inhibitory mechanism has not been confirmed and is best ignored.

The following two experiments are conventionally cited as establishing most precisely the origin of blastemal cells, and thus deserve closer scrutiny. Scheremetieva and Brunst (1938) redesigned their shielding device to provide two adjacent apertures, 8 mm in diameter and 8 mm apart, positioned over the leg of a newt so that the thigh and ankle could be irradiated while the rest of the body, including the knee (which was covered by an additional lead plate), remained shielded. The main lead shield stood 50 mm above the bench, leaving a considerable space above the newt's leg, so each orifice was composed of a 'localizer tube' or lead cylinder 17 mm long intended to direct X-rays to the prescribed region of the leg. The effectiveness of these tubes has never been established. Almost certainly they reduced the dose to nominally irradiated areas, and this partly explains Brunst's conviction that extremely high doses of X-rays were needed to prevent regeneration. Active newts were clamped into this cumbersome lead coffin for the half-hour or so required to administer 5–7 krad. Understandably, some of them wriggled enough to spoil the localized exposure. Scheremetieva and Brunst assessed the correctness of their procedure by the results; stating that sometimes the results were quite conclusive (i.e. fitted their predictions) but other cases did not give clear-cut results, although whether that implies regeneration or not is left to the imagination. They supported their conclusions with one illustrative case which after successive amputations, failed to regenerate from the ankle then did regenerate from the shielded knee and finally did not regenerate from the upper thigh (Figure 4.2). The totally shielded contralateral leg was subjected to identical amputations and regenerated on all three occasions.

Butler and O'Brien (1942) performed the converse experiment of shielding the thigh and ankle while exposing the knee to 2.5–5 krad. An initial

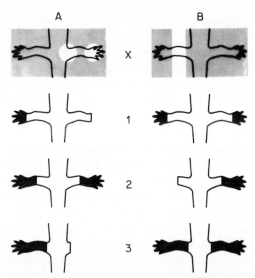

Figure 4.2. Complementary demonstrations of the local action of X-rays. Column A after Scheremetieva and Brunst (1938), column B after Butler and O'Brien (1942). Top diagrams show the arrangement of lead shields (hatched) to expose only part of one leg to irradiation (X). Successive amputations (1–3) of both legs allow regeneration (shown in solid black) only from shielded wound surfaces.

amputation of the foot led to regeneration in 3 weeks. Subsequent amputation at the knee showed that 5-krad inhibited regeneration for 9 weeks but lower doses allowed some cases of partial or delayed regeneration. A final amputation in the upper thigh resulted in normal regeneration within 4 weeks (Figure 4.2). The simplified procedure of placing lead plates over anaesthetized specimens certainly provided more consistent results than anyone had achieved earlier. Perhaps it is fortunate that the experiment was performed on larvae of the small brook salamander, *Eurycea bislineata*, which has seldom served as material for other regeneration studies. It has an unusually long larval life, reaching metamorphosis after 1–3 years (Wilder, 1924). Since they required large specimens to accommodate the lead shields, Butler and O'Brien almost certainly used 2-year-old larvae which had already undergone some initial metamorphic changes and whose irradiated limb stumps were not resorbed to the extent found in young larvae. The experiment thus fortuitously avoided a common complication of partly irradiated larval limbs, which often regress considerably to reach a shielded region and then regenerate.

Strictly speaking, neither of these experiments demonstrated a permanent inhibition of regeneration, but only that local irradiation caused a pronounced delay. They fit our modern interpretation but also permit alternative explanations, such as an eventual recovery of the irradiated cells. Brunst (1950) presented ample evidence that localized irradiation permanently inhibits limb regeneration. Both of these experiments were held by their authors to demonstrate that the blastema originates from cells in an extremely circumscribed area next to the site of amputation and, as an implicit corollary, competent cells must permanently occupy all regions of the limb. These conclusions are almost universally accepted, especially as they conform to repeated histological descriptions of dedifferentiation being limited to the most distal millimetre or so of the stump. The real vindication of the conclusions, however, surely lies in the many successful experiments based on them. Experiments involving quite intricate shielding patterns and tissue grafts (e.g. Goss 1957b; Conn *et al.*, 1971; Wallace, 1972; Maden, 1979c) depend upon the premise of a local origin of blastemal cells, and their results justify it. The main experiment designed to test this dual conclusion has failed to gain much attention, but is described in some detail below.

Wolff and Wey-Schué (1952) exposed the lower legs of adult *T. cristatus* to 4–8 krad, which should inhibit regeneration, while shielding the rest of the body. They arranged the edge of the shield slightly above the knee, and amputated different specimens between 1 and 5 mm away from the shield. All 10 specimens regenerated after amputation 2 mm away from the shield, but noticeably delayed regeneration occurred in fewer specimens amputated beyond that region—five out of 11 at 3 mm, and one out of 17 at 4–5 mm distance from the shield. Similar results have been obtained with axolotls (Lazard, 1968). Wolff and Wey-Schué did not detect any regression of the limb stump before regeneration and concluded that unirradiated cells must have migrated distally to reach the wound surface, a migration which they compared to that of the supposed neoblasts in planarians. This was quite an extravagant analogy to draw from such a modest displacement, for the results can easily be reconciled with the conventional notion of a localized origin of blastemal cells. It is a common observation, for instance, that the muscles and skin retract immediately after amputation so that the naked skeleton projects and thus distorts the plane of amputation and shielding by a millimetre or so. The skeleton itself is typically shortened by about 0.5 mm even in mildly irradiated adults of the same species (Smith *et al.*, 1974). It is also well authenticated that some dermal cells can accompany the epidermis as it extends distally to seal the wound (David, 1932), and such distal movement continues even when regeneration is delayed. In a later histological study, Wey-Schué (1953) identified connective tissue cells adjacent to the skeleton of the most distal shielded region which gave the impression of migrating through the eroded and necrotic irradiated tissue to form an apical blastema. We may take these results as a qualified confirmation of the previous conclusion: blastemal cells originate in a small local region which cannot be defined precisely, partly because minor tissue movements occur, perhaps partly because amputation always leaves a jagged surface and, no doubt, because dedifferentiated cells are capable of minor displacements.

F. and S. Rose (1974) may have stumbled across a situation where an appreciable cell migration occurs in the limb, although they interpreted their results in a completely different and improbable way. They irradiated the arms of very young *Ambystoma* larvae, while shielding most other parts of the body. The shielding pattern could not be maintained with any precision relative to such small arms, especially as many of the specimens were parabiotic twins. Those specimens which had completed the development of four digits were unable to regenerate their hands after 3–7 krad of local irradiation, but younger specimens possessing two to three digits often regenerated quite readily when amputated in the irradiated forearm. Although some intervening regression of the irradiated stump must be expected, it seems likely that shielded cells from the shoulder managed to migrate well into the upper arm in order to establish a blastema. There are two reasons for accepting this as a plausible explanation of the results. Firstly, cells are still entering the growing limb, notably Schwann cells which follow the growing axons, but also the epidermis and perhaps associated cells. Secondly, the required migration appears quite trivial in absolute distance, for the arm stumps of such young larvae are only about 2 mm long.

It is clear that neither of the last two experiments is rigorous enough to compel a re-assessment of the earlier conclusion that blastemal cells normally originate within a few millimetres of the site of amputation and, therefore, that regeneration can be inhibited locally. Inhibitory doses of X-rays are always harmful when applied as whole-body irradiation and may be lethal. Consequently, it is necessary to shield most of the body with lead during irradiation. The effectiveness of the shield depends upon the energy spectrum of the radiation, but 3 mm thick lead will usually absorb all but 5% of the incident dose. It is still important to limit the dose of radiation, for 500 rad affects the immunity system, probably by killing spleen cells (Cohen, 1966c), so 10 krad is a dangerous dose even using lead shielding.

4.3. Effective dose of X-rays

A survey of the doses actually found to cause a local inhibition of limb regeneration should allow an empirical determination of the minimum effective dose in most practical conditions. Table 4.2 suggests that 2 krad of fairly high voltage X-rays ensures inhibition for adult newts and most larval urodeles, but perhaps large juvenile and adult axolotls require 3–4 krad. There is little doubt that the dosages employed have been largely a matter of convention and habit, however, and there has been considerable argument about inadequate or excessive irradiation. Brunst (1950, 1960) has insisted that 5–6 krad is needed to prevent regeneration by adult axolotls, an opinion supported to some extent by Rose (1962; cf. Rose et al., 1955) for newts. Polezhaev (Polezhaev and Ermakova, 1960; Polezhaev et al., 1961–63; Polezhaev, 1972) was unable to prevent regeneration completely by doses of 7–10 krad. Lazard (1968) examined this problem in more detail by irradiating the legs of 10–18 cm axolotls, and found in effect that no satisfactory dose existed. She detected regeneration by minimal criteria of some growth which could be considered as a digit in about 1% of the specimens subjected to 4.2 and 5 krad. A dose of 7 krad abolished even that slight growth but also caused drastic necrosis leading to regression of the limb stump. All the investigators mentioned above also reported signs of damage, such as skin ulcers and depigmentation, inflammation or a temporary cessation of the blood circulation locally, after doses in the 7–10 krad range. Several of them regularly delayed amputation to allow a recovery period of several weeks after irradiation, assuming that once these symptoms had subsided they would not interfere with the rest of the experiment. That was an unwarranted assumption, for neither the symptoms nor the recovery have been investigated apart from a description by Brunst (1960). Fortunately, this is no longer a matter of much concern as experience over the last 20 years shows that regeneration can be prevented by much lower doses which do not cause severe radiation burns. When the doses are related to the conditions under which they were dispensed (Table 4.2), it becomes apparent that low doses generated at high voltage are just as effective inhibitors as the high doses of low voltage X-rays, while only the latter cause excessive superficial damage. Even soft X-rays penetrate some distance, often enough to inhibit the regeneration of thin larval limbs. Butler and his colleagues successfully used low doses of 60 kV X-rays on extremely young larvae of A. maculatum but had to use higher doses for the larger larvae of Eurycea bislineata. As Lazard demonstrated, the thicker limbs of large axolotls are even more resistant to unfiltered 60 kV X-rays. Relatively modest doses of hard X-rays, however, inhibit regeneration in axolotls and adult newts.

Table 4.2 Inhibitory dosages of X-rays under different conditions, classified according to specimen size as a measure of limb thickness.

Dose (krad)	Potential (kV)	Current (mA)	Filter	Dose rate (rad/min)	Distance (cm)	Shielding (mm lead)	Species	Reference
(A) Young larvae, 20–30 mm long								
0.9	60	6	—	30	25	6	A. maculatum	Butler (1935)
0.45	60	6	—	30	25	—	A. maculatum	Puckett (1936)
3	110	3	—	335	28	3	A. maculatum	F. and S. Rose (1974)
2	300	12	—	600	16.5	3	A. mexicanum	Wallace and Maden (1976)
0.66	180	6	3 mm Al	55	40	?	S. salamandra	Luther (1938, 1948)
(B) Older larvae, 30–80 mm long								
2.5*–5	60	6	—	85	25	1.5	E. bislineata	Butler and O'Brien (1942)
0.8*	60	4	?	?	10	?	A. mexicanum	Trampusch (1951)
1	50	2.5		150	10	?	A. mexicanum	Trampusch (1958, 1959)
2	250	12	1 mm Al	300	18	3	A. mexicanum	Wallace (1972)
2	300	12		600	16.5	3	A. mexicanum	Maden and Wallace (1975, 1976)
2	110	12	3 mm Al	160	17	3	A. mexicanum	Wallace et al. (1971)
2	110	12	3 mm Al	160	17	3	A. opacum	Conn et al. (1971)
2	80	2.5	—	160	5.5	3	P. waltl	Lheureux (1975b)
(C) Adult newts, 60–150 mm long								
5–7	43	3	—	?	17	?	T. cristatus	Scheremetieva and Brunst (1938)
10*–15	43	3	—	?	?	3–4	T. cristatus	Brunst (1950, 1952)
4–8	60?	?	?	?	?	?	T. cristatus	Wolff and Wey-Schué (1952)

2	80	12	1 mm Al	270	25	?	*T. cristatus*	Rahmani and Kiortsis (1961)
1.5	90	5	?	?	?	?	*T. cristatus*	Desselle (1968, 1974)
5	192	20	?	2500	10.6	?	*N. viridescens*	Thornton (1942)
2*–3.5	150	8	?	745	11.5	?	*N. viridescens*	F. Rose et al. (1955)
2	100	5	—	336	?	6	*N. viridescens*	F. Rose et al. (1965)
2	100	14	?	600	18	2	*N. viridescens*	Goss (1957b)
1*–2	180	30	—	1000	23.5	?	*N. viridescens*	Stinson (1964a–c)
1.6–2	193	30	0.25 mm Al	410	40	6	*N. viridescens*	Eggert (1966)
2	104	4	—	160	20	6	*N. viridescens*	Oberpriller (1968)
2	100	4	—	183	20	6	*N. viridescens*	Desha (1974)
2	100	5	—	187	15	2	*N. viridescens*	Wertz et al. (1976–80)

(D) *Juvenile and adult axolotls, 100–200 mm long*

0.33	40–160	7–10	HVT 1 mm Al	5.5	25	?	*A. mexicanum*	Litschko (1934)
5	100	3	?	?	24	?	*A. mexicanum*	Umanski (1937, 1938, 1951)
5–6	43–55	2–5	?	?	?	3–4	*A. mexicanum*	Brunst (1950, 1952)
4–6	100	30	1 mm Al	410	18	4	*A. mexicanum*	Brunst (1960)
3–5							*A. mexicanum*	Roguski (1961); Lagan (1961)
7*–10*	210	15	—	636	25	4	*A. mexicanum*	Polezhaev et al., 1960–64
5*–7	60	15	?	1400	10	2	*A. mexicanum*	Lazard (1968)
3	80	6	0.25 mm Al	600	?	?	*A. mexicanum*	Namenwirth (1974)
4	150	15	0.5 mm Cu + 1 mm Al	376	27	3	*A. mexicanum*	Carlson (1974)

*Doses causing less than complete inhibition.

The effective dose of irradiation can be determined on a less anecdotal basis, as outlined below, but the only applicable data concern the relatively resistant limbs of large axolotls and newts. According to the rather crude estimates available at present, the frequency of regeneration by irradiated limb stumps falls exponentially with increasing dose—as predicted by radiobiological theory. A logarithmic plot such as Figure 4.3 then indicates a family of straight lines whose gradients are related to the operating voltage, and thus to the predominance of hard X-rays. The dose which reduces the incidence of regeneration by one logarithmic unit or 37% (D_{37}) is about 200 rad for high voltage X-rays, and this could also be achieved probably at lower voltages by using filters. Consequently, 2 krad under these conditions yields an expected frequency of regeneration considerably less than 0.1%. One misleading result among a thousand specimens would be a quite acceptable hazard for regeneration experiments. The shoulder present in Figure 4.3 is commonly found in dose–response curves and its significance has been debated for years. It may be quantified by extrapolating the exponential part of the curve back to the ordinate, to yield an 'extrapolation number' of about two in this case. The shoulder or number derived from it may indicate that several sites must be damaged in order to inactivate a cell, or the operation of

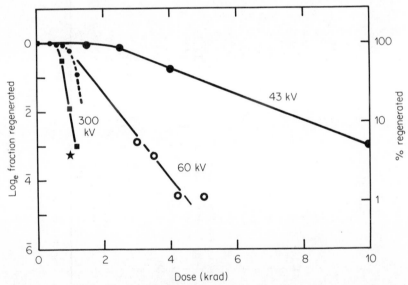

Figure 4.3. Dose–response curves relating the frequency of regeneration to local irradiation with unfiltered X-rays. D_{37}—the dose causing 37% decrements in regeneration frequency—in large axolotls is about 200 rad at 300 kV (see Figure 4.4), but 800 rad at 60 kV (data of Lazard, 1968), or 2700 rad at 43 kV for *T. cristatus* (according to data of Brunst, 1950a). The star marks a single estimate for 180 kV X-rays on *N. viridescens* arms (Stinson, 1964a). Probably the soft X-rays which predominate at low voltage only cause superficial damage.

a repair mechanism, or simply reflect the minimum number of cells which must remain capable of proliferation in order to establish a viable blastema. According to target theory which forms the basis for such dose–response curves, a D_{37} of 200 rad represents the dose which inactivates the average cell in terms of regenerative ability, for most exposed cells survive indefinitely after such a dose. Adapting a conventional argument explained by Dertinger and Jung (1970; cf. Maden and Wallace, 1976), this dose corresponds to about 12×10^{15} eV/g and is capable of causing 0.2×10^{15} primary ionizations (or hits) in each gram of exposed tissue. The most sensitive target is undoubtedly nuclear DNA, comprising 50 pg or more in diploid urodele cells. So 200 rad should produce about 10^4 hits in the DNA of each exposed nucleus. That may seem excessive but should not lead one to suppose the existence of more restricted or particularly vulnerable targets. Two or more adjacent hits might be required (to fracture chromosomes, for instance) and most radiation damage is probably repaired quickly and efficiently, so that only a minority class of irreparable or misrepaired lesions remain to cause the long-term inhibition of regeneration.

A second factor in the dispute about what dose inhibits regeneration is the subjective criterion of regeneration itself. Rose *et al.* (1955) noted that doses of

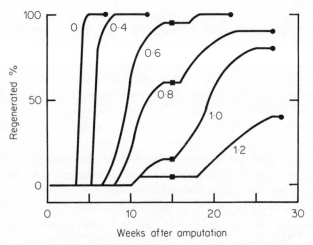

Figure 4.4. Cumulative incidence of regeneration related to the dose of local irradiation by unfiltered 300 kV X-rays; original data from 20 legs of 15–20 cm axolotls at each dose (0–1.2 krad), applied just before amputation. Regenerates scored up to the plateaux at 15 weeks were all fairly large with 3–5 digits; all the additional ones recorded later were either short spikes or minute hypomorphic structures containing little apart from cartilage. The symbols identify results obtained according to stringent or relaxed criteria of regeneration entered in Figure 4.3.

up to 2 krad did not always prevent *N. viridescens* limb stumps from growing a simple spike. Such spikes evidently correspond to the single digits which Lazard (1968) and others classed as hypomorphic regenerates. Others only recognize as genuine a regenerate with at least three distinct digits. Such arbitrary definitions are fairly satisfactory in a particular experiment, but must confuse comparisons between the results of different investigators. After attempting to compensate for different criteria of regeneration and the various X-ray spectra employed, I can detect no residual conflict among the results considered here (Table 4.2). Either at the low dose rates originally employed by Butler or those exceeding 500 rad/min attainable nowadays, a 2 krad local dose of hard X-rays will surely prevent limb regeneration in any urodele apart from a few giant species which have never been used for this purpose. A limb region which has been exposed to 200 rad or less is effectively unirradiated and should regenerate promptly. Intermediate doses, of course, give rise to variable results but regeneration is progressively delayed as the dose increases above 200 rad. Greatly delayed regenerates are usually reduced in size and hypomorphic, so that incompletely inhibited limbs only produce miniature regenerates or small spikes 3–6 months after amputation (Figure 4.4).

4.4 Histology of X-rayed limb stumps

The irradiated limb stump of an adult newt heals normally by means of an epidermal migration but never forms a blastema. The events described below are pieced together from various specialized studies, for no definitive description exists equivalent to Butler's (1933) account of irradiated larval limbs. Internal tissues at the end of the adult stump degenerate to some extent, eventually leaving a smooth layer of connective tissue or fibrous scar tissue over the end of the skeleton. The overlying epidermis gains a basement membrane and becomes pigmented. The scar tissue could be formed by a gradual advance of the dermis or, more probably, by a limited dedifferentiation and rapid redifferentiation of the internal cells (Desselle, 1974). Only very limited amounts of dedifferentiation have been detected, however, and several early investigators did not notice any. Both DNA synthesis and cell division occur, but in much fewer cells than found in regenerating control limbs (Desselle, 1974; Wertz *et al.*, 1976; Wertz and Donaldson, 1980). Rose and Rose (1965) recorded fairly normal amounts of epidermal cells undergoing DNA synthesis in irradiated limbs. The significance of these observations is questioned in the following section, but they have all been duplicated on large axolotls (Tuchkova, 1964–9).

Larval urodele limbs and tadpole legs show an apparently different and more spectacular response to irradiation and amputation, yet one that is fundamentally similar to that of the adult. Wound healing occurs in exactly the same way and then internal tissues at the tip of the stump begin to degenerate and are progressively eroded. The limb gradually shortens, being said to regress or be resorbed. The skeleton resists erosion longer than other tissues, so that the limb apex acquires a conical shape which can easily be confused with a blastema. Some of the first descriptions, in fact, described it as a regression blastema or pseudoblastema (Butler, 1933; Puckett, 1936; Horn, 1942). Regression can be quite rapid in very young larvae but is usually a slow process which may halt, in larger larvae, to form a stable stump like that found in adults. If regression continues so as to remove all the irradiated tissue, however, it is usual for the limb to regenerate eventually from the most distal shielded region. This superficial difference between adult and larval arms can probably be attributed to the different nature of the skeleton, although the dermal layer of the skin thickens considerably at metamorphosis and could also influence regression. The cartilage of young larval limbs is eroded relatively easily, while the partially calcified cartilage of older larvae is more resistant and adult skeletal bone resists erosion sufficiently to limit the loss of surrounding tissue. This inference drawn from descriptive studies is supported by some experimental evidence. If the skeleton of the arm is removed before amputation, considerable retraction and regression occur as a prelude to regeneration both in adults (Bischler, 1926) and larvae (Thornton, 1938b). Such regression continues unabated in irradiated adult arms lacking a skeleton (Schotté, unpublished). The continued erosion of apical tissue underneath an intact epidermis is most simply interpreted as the successive dedifferentiation and liberation of cells which fail to accumulate as a blastema, because they cannot proliferate or survive as a consequence of radiation damage. Early regenerates are also resorbed soon after being

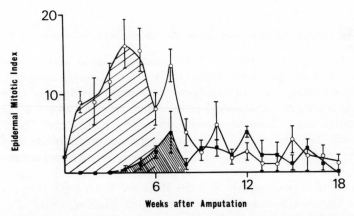

Figure 4.5. Epidermal mitotic index of normal (O) and irradiated (■) arm stumps of 50 mm axolotls (from Maden and Wallace, 1976). Hatched areas indicate a rough comparison of the inhibition, discounting the obvious delay caused by irradiation. Alternatively, shielded epidermal cells may have migrated down the experimental arms to supplant irradiated epidermis and provide most of the recorded divisions. (Reproduced by permission of The Wistar Institute).

irradiated, but palette and later stages usually continue to redifferentiate and grow slightly (cf. Pietsch and Bruce, 1965). Allen and Ewell (1959) have recorded a similar continuation of differentiation in irradiated tadpole limb-buds.

Presumably because they are still growing, larval arms show effects of irradiation which have not been reported for adult arms. Micronuclei appear in cartilage cells as they divide abnormally and the nuclei of other tissues become vacuolated (Maden and Wallace, 1976). Few dividing cells are seen in the pseudoblastema and epidermal divisions are drastically reduced for several weeks (Figure 4.5).

4.5 The inhibitory mechanism

There can be little doubt now that a sufficient dose (about 2 krad) of ionizing radiation causes a permanent local inhibition of regeneration when the irradiated limb is amputated. A similar dose applied during regeneration causes the resorption of a young blastema, or prevents further growth by cellular proliferation but allows redifferentiation in older regenerates. It is also well established that grafts of unirradiated limb tissues into the irradiated amputation site annul the inhibiton and permit the occurrence of regeneration, although such regenerates are often defective. Four completely different explanations have been offered to account for these results and they need to be considered carefully, for the use of irradiation as an experimental technique to analyse regeneration usually involves one or another of these explanations as an implicit premise.

Explanation 1. Irradiation prevents the division or survival of blastemal cells

This explanation is based on current radiobiological theory, and most observations on irradiated amphibian limbs are in complete accordance with it. In particular, the observations mentioned earlier of continued dedifferentiation in irradiated larval limb-stumps and of contined differentiation in limb-buds and advanced blastemata restrict the radiosensitive activity to the formation and growth of the blastema. Several objections to this explanation have been raised. Firstly, occasional mitotic figures can be seen in irradiated and amputated limbs, indicating that cell divisions occur (Trampusch, 1951, 1959; Rose and Rose, 1965) even if some of them are abnormal (Puckett, 1936; Horn, 1942; Stinson, 1964a; Tuchkova, 1967a). Such cell division must be stringently curtailed, balanced by cell death or both, however, as there is no accumulation of blastemal cells in irradiated limb stumps. In fact the explanation has been modified to meet this objection by admitting that irradiated cells may sometimes divide, provided that the division usually results in defective daughter cells with a low mean survival (as implied by Figure 4.6).

A second objection has arisen from the knowledge that epidermis is

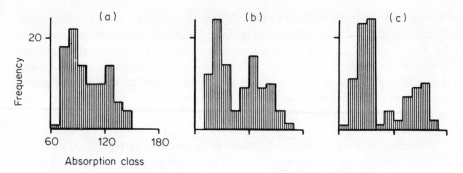

Figure 4.6. Frequency distribution of nuclei from (a) normal cone blastema, (b) 2 days and (c) 4 days after irradiation with 2 krad X-rays (from Maden, 1979a). Note the reduction of intermediate absorption classes (S-phase cells) without any accumulation of postmitotic G1 cells, such as seen in Figure 2.7. (Reproduced by permission of the Company of Biologists Limited.)

continually replaced from its proliferating basal layer and so should be especially susceptible to irradiation, yet it remains intact for months or years afterwards. This has led to the unfounded belief that the epidermis is unaffected by doses which prevent regeneration, particularly as Rose and Rose (1965) recorded normal amounts of DNA synthesis and cell division in the epidermis of *N. viridescens* arms several weeks or months after exposure to 2 krad X-rays. In contrast to these observations, Figure 4.5 shows that 2 krad drastically reduces the epidermal mitotic index in amputated arms of axolotls for the first few weeks after treatment, with subnormal values persisting for about 6 weeks. By that time the irradiated epidermis has thickened considerably by an accumulation of gland cells and it remains visibly thickened for months afterwards. Furthermore, there are a considerable number of incidental comments cited earlier drawing attention to a delayed inflammation and skin ulcers on arms exposed to 7–10 krad X-rays. The long-term recovery from such radiation burns remains something of a mystery even after extensive studies in mammals. The normal process of epidermal replacement is probably adapted to effect a cryptic version of wound healing. When a basal cell divides, one daughter cell typically remains in the basal layer while the other is displaced outward to eventually die and form part of the cornified surface, but this daughter cell can sometimes also remain attached to the basal membrane thus replacing a dead or defective adjacent cell. It is quite likely that defective daughter cells would be preferentially pushed out after a dose of irradiation which allows some basal cells to recover and divide, while occasional viable cells remain to repopulate the basal layer of the surrounding area. Similarly, epidermis from neighbouring shielded regions should gradually encroach into the basal layer of the irradiated zone and eventually replace all of its epidermis. That would provide an explanation of the normal mitotic index and DNA synthesis noticed 3 weeks or more after irradiation (Tuchkova, 1967a, 1969). Epidermal cells certainly are damaged by X-rays, however, and the epidermis seems more vulnerable than internal tissues of the intact limb as predicted from the present explanation. The initial stages of regeneration must be even more susceptible to irradiation, merely because the establishment of a viable blastema entails a prompt and rapid cellular proliferation in contrast to the more leisurely process of epidermal replacement.

The third kind of objection is based on the belief that regeneration is only reversibly inhibited by irradiation. Several investigators have supposed that a graft of unirradiated tissue restores the ability to regenerate in the irradiated host cells by a process like embryonic induction. To my mind, this contradicts the well established long-term local action of irradiation, for it predicts that a wave of reactivation should spread into the irradiated zone from neighbouring shielded regions—and there is ample evidence that this never happens (Brunst, 1950). Yet Polezhaev and his associates (1960–77) have reported that injections of tissue extracts, RNA or protein will each reverse the inhibitory effects of X-rays.

Similarly, Desselle (1968, 1974) stated that an acellular homogenate of cartilage could reactivate irradiated limb cells to divide and form a blastema, but he has not yet offered any evidence either that the homogenate was genuinely free of living cells or that he had obtained reproducible results with it. There are even more dubious aspects concerning the experiments described by Polezhaev and his colleagues. Polezhaev (1960, 1972; Polezhaev *et al.*, 1960–64) persistently reported their use of 210 volt X-rays, a voltage only capable of producing ultra-violet light, none of which would escape from the X-ray tube, until first Teplits (1962) and then Tuchkova (1969) finally realised they were dealing in kilovolts. They exposed both hind limbs of juvenile axolotls to 7–10 krad of fairly hard X-rays and usually amputated them twice, at 10 and 100 days after irradiation. Without any other treatment, 2–13% of the irradiated stumps regenerated after the first amputation and 5–57% did so after reamputation. Under similar conditions and presumably with the same X-ray apparatus, Tuchkova's (1969, 1973a) results demonstrate that 6 krad completely suppresses regeneration. The only possible conclusion is that 7–10 krad consitutes an excessive dose, as witnessed by the inflammation and ulcers which these investigators described on the exposed legs. The manner in which these ulcers are healed is entirely a matter of conjecture, but it might well involve a migration of shielded cells down the leg. Such a migration would occur on a larger scale after additional damage, either inflicted deliberately or attendant on a series of injections, and be more complete at the time of reamputation. That provides a coherent explanation of the results obtained with a variety of injected material, any of which seemed to enhance the frequency of regeneration. These results are considered in detail later (Table 6.3 and 6.4) with a novel interpretation which is quite compatible with the present account of how irradiation inhibits regeneration.

The present explanation does not preclude a minor contribution of irradiated cells to a regenerate, provided they have not been subjected to the strain of proliferation. In the presence of an unirradiated graft, for instance, some neighbouring irradiated cells might dedifferentiate and become incorporated into the base of a blastema only to redifferentiate without having divided. Such cells probably would survive but it might be difficult to detect them. The most that can be claimed for the following investigations is that they have identified a dwindling number of irradiated cells through early phases of regeneration. Desselle (1968) believed he had traced irradiated host muscle cells into regenerates formed by labelled cartilage grafts. The basis of this belief seems to be that the regenerated muscle was unlabelled, but that is an unsatisfactory criterion because the radioactivity of graft cells becomes diluted as they divide until their descendants are also unlabelled. Tuchkova (1973a) reversed this experimental procedure by employing axolotl arms which were both radioactively labelled and irradiated. When grafts of unlabelled and unirradiated skin, skeleton or muscle promoted regeneration in about half of these limbs, or in the few regenerates stimulated by injections (Tuchkova, 1976), a few labelled cells were carried into the blastema. They became less frequent and less intensely labelled during growth to a 15 day cone stage. Desha (1974) performed a virtually indentical experiment using skin autografts to stimulate regeneration of irradiated and labelled *N. viridescens* arms. He also observed a transient population of labelled cells occupying a small zone at the base of the blastema. In neither case could host cells be traced much further, as they lost their label either by being destroyed or by dividing repeatedly. Dunis and Namenwirth (1977) examined regenerates obtained from triploid axolotl skin grafts on irradiated diploid arms. The regenerates were predominantly composed of triploid cells, according to the proportion of cells showing three nucleoli, but this proportion was consistently less than that estimated from control regenerates of entirely triploid unirradiated arms. Table 4.3 shows the data and a reconstruction of their χ^2 test which led Dunis and Namenwirth to think the regenerates must have contained appreciable numbers of diploid cells from the irradiated host arm. A more appropriate statistical test, however, suggests that only three tissues (dermis, connective tissue and cartilage) exhibited a significant deficiency of cells with three nucleoli. That deficiency does not necessarily imply the presence of diploid cells, for it could be explained by enhanced nucleolar fusion or even that the sections of these malformed experimental regenerates were less perfectly oriented than those of the controls. Table 4.3 also includes comparable data from another triploid skin graft regenerate and a triploid control, to illustrate the inconsistency of scoring cells with three nucleoli in this material. Triploidy is undoubtedly the best kind of marker for such experiments, even though the efficiency of scoring triploid cells ranges from 15% to 60% in different tissues. A comparison between the experimental and control data (Table 4.3) indicates that at least 75–90% of the experimental regenerate cells are derived from the triploid graft. A more

Table 4.3 Percentage of cells with three nucleoli in experimental regenerates (E) from 3N skin grafts on X-rayed 2N limbs, and control regenerates (C) of entirely 3N limbs.

Tissue		Cells with three nucleoli (%)*		Analysis of (2) group means		Analysis of limb scores used in (2)		
		(1)	(2)	$\chi^2_{(1)}$	P	t	d.f.	P
Epidermis	E	37	16.8 ± 2.2	1.64	>10%	0.861	8	>40%
	C	21	18.0 ± 2.0					
Dermis	E	21	14.2 ± 1.5	32.35	≪0.1%	4.983	9	<0.1%
	C	18	19.3 ± 1.2					
General connective	E	—	19.1 ± 1.1	7.16	<1%	3.56	9	<1%
	C	—	21.9 ± 1.1					
Cartilage	E	20	21.0 ± 1.8	14.55	<0.1%	3.575	10	<1%
	C	20	25.0 ± 1.7					
Joint connective	E	—	18.7 ± 1.7	15.21	<0.1%	2.119	11	>5%
	C	16	22.7 ± 0.8					
Myoblasts	E	38	44.2 ± 5.2	10.47	<1%	3.334	4	>2%
	C	34	56.2 ± 3.4					
Muscle connective	E	38	30.5 ± 5.1	7.66	<1%	2.099	4	>10%
	C	40	38.4 ± 4.0					

*All the estimates were from identically stained 10 μm sections: (1) from 200–800 cells of single regenerates (Namenwirth, 1974); (2) mean and standard deviation from 200–1000 cells in 3–8 regenerates and 3–4 controls (Dunis and Namenwirth, 1977). The χ^2 test ignores sampling errors, which are illustrated here by comparing columns (1) and (2).

accurate estimate might be obtained by using the more precise control of a regenerate formed from a triploid graft on an irradiated triploid arm, but it would still leave a considerable margin of error. The best way to identify host cells in such regenerates would be to graft diploid skin on to irradiated triploid arms, and that has not yet been attempted.

The most recent evidence for reactivation of irradiated cells has several perplexing features. Desselle and Gontcharoff (1978) grafted phalangeal cartilage from *Desmognathus fuscus* into irradiated legs of *T. vulgaris* to obtain regenerates in which they detected about twice the original number of graft cells among a larger number of host cells, estimated as forming 86–95% of the differentiated tissues. By discrimating between graft and host cells on the basis of their nuclear DNA content, Desselle and Gontcharoff would have classified most post-replicative (S and G2) graft cells as G1 host cells and thus gravely underestimated the proportion of graft cells in the growing blastema. The most curious aspect of the results, however, is that they did not measure any post-replicative cells at all—implying a miniscule level of cellular proliferation. This leads me to suspect an immunological destruction of graft cells matched the mortality of irradiated dividing cells, leaving two 'minor' contributions to regenerates which seem to have been very small indeed.

The most recent evidence against reactivation was obtained by Maden (1979b) who grafted white axolotl blastemata on to irradiated dark arm stumps. This combination provided the supposed conditions for reactivation and ample opportunity to detect it on a gross scale, as the grafts were positioned in a way which provokes normal arm stumps to produce duplicate regenerates. No host contribution was evident, meaning that only a trivial fraction of regenerate tissue could be derived from reactivated irradiated cells.

Explanation 2.   Irradiation destroys the nervous control of regeneration

Different forms of this explanation have been offered by Rose (1962–1974) and Trampusch (1964), on the basis of the strikingly similar effects of denervation and irradiation on amputated limbs. Both these treatments inhibit regeneration and growth, cause the resorption of young blastemata and the regression of larval arm stumps, as described in detail previously. Trampusch (1964) quoted an experiment of Vergroesen (1958) which, he believed, demonstrated that unirradiated axons were sufficient to cause the regeneration of an otherwise irradiated limb stump. Considering this explanation to be unsatisfactory, I repeated Vergroesen's experiment with an increased variety of control operations and could not confirm his results (Table 4.4). Far from supporting the thesis of radiosensitive neural connections, the nerve supply seemed relatively resistant to ionizing radiation (Wallace *et al.*, 1971). The defects of the original experimental design were further exposed by the subsequent discovery that Vergroesen's operation of temporarily deflecting a nerve trunk to shield it from irradiation established shielded cells of the nerve sheath and probably from nearby skin at the site of amputation, and that either of these could cause the limb to regenerate (Conn *et al.*, 1971; Wallace, 1972).

Table 4.4   Results of operations involving limb nerves of juvenile axolotls followed by irradiation of the limb and amputation.

Operation	Dose (krad)	Regenerated	Inhibited	Reference
Sciatic nerve deflected	0.96	16	0	Vergroesen (1958)
under lead shield via a	1.44	11	0	Vergroesen (1958)
skin slit during X-ray	2	6	0	Wallace *et al.* (1971)
Sciatic nerve deflected	0.96	5	4	Vergroesen (1958)
as above but replaced	1.44	0	8	Vergroesen (1958)
prior to irradiation	2	5	0	Wallace *et al.* (1971)
Skin slit as above	2	3	2	Wallace *et al.* (1971)
No operation	2	0	5	Wallace *et al.* (1971)
No operation	0	21	0	Wallace *et al.* (1971)
Unirradiated nerve graft	1.44?	0	12	Vergroesen (1958)
in irradiated limb	2	8	1	Wallace (1972)

Rose's (1962) concept of 'tissue arcs' has evolved over the years from a generalized idea of reflex arcs (relating a wound stimulus to a regenerative response) into a specific theory concerning a control of regeneration by changes of electrical potential and polarity (Rose, 1970a; S. and F. Rose, 1974). The part which concerns us here is the assertion that nerve fibres cannot establish proper epidermal connections in the wound epithelium if both have been irradiated (F. and S. Rose, 1967), and the hypothesis that it is the interruption of those connections which prevents regeneration. It would be quite difficult to demonstrate an interference with epithelial nerve endings by radiation, particularly as the normal density of nerve endings in the epidermis is not well established and epithelial innervation is not required for regeneration (Sidman and Singer, 1960; Thornton, 1960a). The brief note on this subject furnished by Rose and Rose (1967) provides little further information. They appear to have observed a dense innervation of the apical cap in regenerating limbs of *N. viridescens*. The much sparser innervation observed at the end of irradiated limbs might be the cause or effect of, or even coincidental to, the absence of an epidermal cap. Fortunately for this argument, there are plenty of examples where both the epidermis and the nerve supply of the limb have been irradiated sufficiently to prevent regeneration, yet unirradiated tissue grafted into the amputation site allowed regeneration to occur. Experiments of this kind have been performed with two species of adult newts and both adult and juvenile axolotls using a variety of tissue grafts (Umanski, 1937, 1938; Thornton, 1942, Trampusch 1951, 1958a, b; Desselle, 1968; Wallace, 1972; Wallace *et al.*, 1974; Namenwirth, 1974) and even dissociated cells (Skowron and Roguski, 1958). In each case, we must conclude either irradiated neuro-epidermal connections function adequately or their disruption has no material influence on regeneration. F. and S. Rose (1974) attempted to fortify their explanation by offering a demonstration that aneurogenic limbs can regenerate after high doses of radiation. This experiment produced some curious results which cannot yet be interpreted reliably. Local irradiation of a developing arm in young *Ambystoma* larvae apparently did not always inhibit its ability to regenerate. Rose and Rose found this applied up to the time when three digits had just formed, and observed that nerves were still entering the arm at this late period. They also found that aneurogenic arms on older and larger parabiotic twins could often regenerate after doses of X-rays which inhibited regeneration in the innervated arms of their parabiotic partners (Table 4.5). These arms are not strictly comparable in terms of development. Aneurogenic parabionts are smaller than their normal twins, with relatively small and slender arms (Piatt, 1942; Yntema, 1959a). It is not evident from the report of F. and S. Rose (1974) that any of the aneurogenic arms which regenerated had reached a four digit stage prior to amputation, while some irradiated aneurogenic arms showed the typical regression of a larval limb which cannot regenerate. These ambiguous results suggest to me that irradiation inhibits regeneration of aneurogenic arms just as effectively as it inhibits normal limb regeneration, provided they

Table 4.5   Results of exposing normal and aneurogenic arms to X-rays prior to amputation.

Larval length	No. of digits	Innervation of fore-arm*	Regenerated	Inhibited
(A) A. maculatum and A. opacum exposed to 2–7 krad (F. and S. Rose, 1974)				
<25 mm	2–3	±	24	24
>25 mm	3–4	+	0	17
<25 mm host	2–3	±	6	11
Aneurogenic	2–3	Sparse	5	12
>25 mm host	3–4	+	0	30
Aneurogenic	?	Sparse	19	11
(B) A. mexicanum exposed to 2 krad (Wallace and Maden, 1976a)				
<25 mm	2–3	+	0	45
>25 mm	3–4	+	0	20
Aneurogenic	3–4	Sparse	0	20

*No nerves were detected in about 30% of the aneurogenic arms.

have the same developmental status and reasonably compact tissue. In any event, that is a permissible interpretation of the results shown in Table 4.5. Rose and Rose, however, preferred to draw a conclusion which suited their general explanation and claimed aneurogenic arms were not affected by X-rays, even though almost half of their cases did not regenerate.

Despite these reservations, the experiment seemed worth repeating to check if the results could support some nervous intervention in the effects of irradiation. A repetition performed on axolotls yielded completely opposite results (Wallace and Maden, 1976a). Irradiation here consistently prevented the regeneration of larval arms, either before or after gaining a functional innervation, as well as when aneurogenic or sparsely innervated (Table 4.5). If these two sets of contrasting results reflect a difference between the species employed, which seems unlikely although limb development occurs rather later in axolotls, then we could still conclude that the basis of Rose's interpretation had no general validity. It is more probable that the differences were experimental ones, particularly those concerning the precise localization of irradiation over limbs that were only about 4 mm long. F. and S. Rose carefully restricted radiation to the limb and perhaps did not irradiate it completely, while we irradiated the entire arm and part of the shoulder. It would be easier for dedifferentiation to encroach into shielded tissue of the smallest limb stumps in the former experiment, and that may explain why some cases of regeneration occurred only under those conditions. In both experiments, irradiation inhibited the regeneration of at least some arms at all stages of development tested and in all states of innervation, a result which plainly contradicts the interpretation offered by Rose and Rose. The neurotrophic factor is evidently not affected by conventional doses of X-rays, for tissue grafts allow irradiated limbs to regenerate (see Tables 7.1 and 8.1).

Wertz and Donaldson (1979) have also argued that nerves cannot be the primary target of X-rays, for doses exceeding 500 rad inhibited regeneration as stringently in the lower jaw as in the limb of *N. viridescens*. The significance of this comparison rests on a previous demonstration (Finch, 1969) that the lower jaw is poorly innervated and regenerates even after virtually all its nerve supply has been severed. Based on these arguments, we can be quite confident that local irradiation acts directly on the cells concerned in regeneration without any nervous mediation. Much higher doses of X-rays applied to the brachial or lumbar neurons in the central nervous system will eventually cause peripheral axons to degenerate and consequently impede regeneration (Lazard, 1968), but that is a long-term response quite distinct from the production or transport of the trophic factor by intact nerves.

Explanation 3. Irradiation prevents dedifferentiation

This explanation could be seen as subsidiary to the previous one if intact innervation were required for the process of dedifferentiation. Dedifferentiation has been observed so often in denervated larval and adult arms, even though partly denervated stumps tend to regress less than fully innervated ones (Karczmar, 1946; Tweedle, 1971), that we have to consider this as an independent explanation. Rose (1962) considered it as a subsidiary observation when he remarked that X-rayed limb stumps of adult newts lost less structure than did unirradiated ones. If that is true, it can only indicate that dedifferentiation is quickly halted, for some mesenchymal cells undoubtedly accumulate at the tip of an irradiated stump (Desha, 1974). Irradiated larval limb stumps regress by continued apical dedifferentiation for several weeks after amputation. Polezhaev (1972, 1977) also maintained that irradiation prevented dedifferentiation in axolotls. His opinion was based partly on the belief that the inhibitory effects of X-rays were reversible and partly on the descriptive studies of Tuchkova (1964a, b). Tuchkova observed a prolonged destruction of internal tissues at the end of legs which had been amputated 19 days after exposure to 10 krad of X-rays. Cells liberated by this process accumulated under the wound epithelium to form new connective tissue 30 days after amputation, and later thickened into a scar or callus. Tuchkova insisted the liberated cells were not dedifferentiated, because they appeared darkly stained with pycnotic or condensed nuclei, unlike the 'activated' appearance of normal blastemal cells. Their behaviour and the occurrence of division figures, however, leave no doubt that they were dedifferentiated. These results thus conform to the bulk of other observations on adult urodeles (e.g., Wertz and Donaldson, 1980) in detecting a limited amount of dedifferentiation by irradiated cells.

Explanation 4. Irradiation destroys the morphogenetic field

Although morphogenetic fields and regeneration territories will be considered in more detail in later chapters, it is already clear that some regional specificity exists so that limbs regenerate limbs and tails regenerate tails. Trampusch (1951, 1958a, b, 1959) for a time championed the explanation that irradiation destroys a cell's territorial memory: such amnesic cells could not decide into which organ they should regenerate and consequently did not respond to amputation. Since morphogenetic fields and memories are such intangible characteristics, it is not surprising that no mechanism for their destruction has even been contemplated. Trampusch adopted this explanation as a less distasteful alternative, when he found a graft of unirradiated tissue caused an irradiated limb or tail to regenerate into an organ which displayed the characteristics of the donor tissue (see Table 8.1). Experiments of this kind are essentially ambiguous. Consider the case of tail skin transplanted onto an irradiated limb which is then amputated through the graft and regenerates into a tail-like structure, possessing a fin but lacking digits (Trampusch, 1958a). One may conclude either that grafted skin cells can be converted into muscle and skeleton of the regenerate while retaining their own morphogenetic instructions and ignoring any exerted by the inert tissues of the irradiated stump; or one may adopt the converse conclusion—the graft provides a stimulus which revives the regenerative capacity of irradiated host cells, which then retain their tissue specificity but must become subservient to a new morphogenetic field imposed by the graft. Plainly, Trampusch could not stomach the former alternative and so felt obliged to accept the latter. Umanski (1937) and Luther (1948) had previously chosen the more direct former interpretation on the basis of similar results. Umanki (1938a) vindicated his choice by means of genetically marked skin grafts between white and dark axolotls. Trampusch (1972) has since revised his interpretation of these experiments to agree that the morphogenetic field is a permanent characteristic of those cells which form the blastema, while their tissue specificty can be altered during regeneration. At the same time, I obtained an unequivocal demonstration of the latter point (Wallace, 1972). When pieces of brachial nerve were transplanted into irradiated arms, reciprocally between dark and white axolotls, the amputated arms regenerated quite well and acquired the colour of the graft donor (Figure 4.7). The analysis of this situation is complicated by the virtual absence of pigment cells from the grafts and by the unusual fact that the genetic control of this colour difference acts through surrounding tissues on pigment cells of all genotypes. Irrespective of whether the pigment cells were provided by the host or by the graft, the colour of the regenerates implies that their mesodermal tissues were largely or entirely derived from grafted cells (Wallace and Wallace, 1973). The same colour difference shows that grafted blastemata develop identically on irradiated or on normal stumps (Maden, 1979b, 1980a). If the stump has any morphogenetic influence on the blastema, which I doubt, then its influence is not abolished by radiation.

Rose and his associates have not appreciated the force of this argument and have occasionally expressed their support for the idea of a morphogenetic field which can be destroyed by ionizing radiation (e.g. Rose and Rose, 1965,

Figure 4.7. Regenerates obtained by grafting brachial nerves into irradiated arms (from Wallace, 1972). A nerve from a dark specimen was inserted into the irradiated left fore-arm of the white example shortly before amputation near the graft; a nerve from a white specimen was similarly implanted into the dark example. Both examples show the nerve genotype dictates the colour of the regenerate, which is known to be a response by melanophores to the genotype of surrounding tissue.

1974). One of Rose's students undertook the invidious task of investigating the supposed field effect when two arms of a newt are sutured together and amputated to create a double amputation plane (Oberpriller, 1968). She demonstrated that the normal interference between these joint arm stumps was abolished when one of the arms had been irradiated previously (see Table 8.14). On the premise that radiation destroys nervous connections—explanation 2 which I have already criticized—Oberpriller developed a circular argument to assert the results showed this aspect of the morphogenetic field must be susceptible to X-rays. There is a more simple and logical interpretation of her results. Nobody has yet demonstrated that a morphogenetic field extends beyond the living tissue which expresses it, and it is only too obvious that irradiated limbs do not regenerate. Therefore, the irradiated limb stump could not interfere with the single regenerate produced by the unirradiated arm, except by increasing the total wound surface and thus, perhaps, increasing the number of digits formed.

4.6 Gamma-rays, neutrons and ultra-violet light

The effects of X-rays described previously correspond quite closely to the effects of denervation and hypophysectomy, probably implying that all three treatments inhibit a common process such as cellular proliferation. Similar effects can be obtained by other ionizing radiatons, according to the little evidence available. Schaper (1904) reported that γ-rays from radium inhibited regeneration of larval newt limbs. The larvae were totally irradiated and did not survive as long as 3 weeks, but the results seem much the same as obtained later with X-rays. Horn (1942) calculated that neutrons were more efficient than X-rays in preventing limb regeneration in *A. maculatum* larvae. He observed loosely-spaced putative blastemal cells in the tip of the regressing irradiated limb. These cells did not increase in number although some were fixed while dividing. Horn recorded that clearly aberrant division figures were present and concluded that the daughter cells usually died.

Ultra-violet light has a much more limited influence on regeneration because it cannot penetrate far into living tissue and causes practically no ionization. Thornton (1958) applied daily doses of ultra-violet light, calculated to penetrate three cells deep, to the recently amputated arms of *A. talpoideum* and *A. tigrinum* larvae. This repeated irradiation of the wound epithelium prevented both the formation of an apical cap and any accumulation of blastemal cells for the 3–4 week period of treatment, but the limbs subsequently recovered to regenerate normally. When Thornton delayed the onset of treatment until an apical cap six cells thick has already appeared, he found that the same daily doses of ultra-violet light retarded regeneration but did not prevent it. Besides supporting Thornton's contention that an epidermal apical cap might be a prerequisite for the establishment of a blastema (see Chapter 7, section 4), the results imply that ultra-violet light either inactivates or destroys cells at some unknown dose, as is well known for tissue cultures and micro-organisms. Butler and Blum (1963) produced local damage to the elbows of *A. maculatum* and *A. opacum* larvae by exposing them to a single dose (1.9×10^8 erg cm^{-2}) of ultra-violet light while shielding the rest of the body under aluminium foil. Superficial damage and regression of the exposed region was followed by the formation of a lateral blastema just above the elbow, and the eventual growth of a supernumerary fore-arm.

4.7 Conclusions

Several forms of radiation can be used to inhibit regeneration. Hard X-rays, γ-rays and neutrons penetrate living tissues easily and produce a permanent inhibition, although only X-rays have been used regularly for this purpose. Ultra-violet light is completely absorbed by superficial layers and consequently only inhibits regeneration temporarily and under special conditions. X-rays have a direct effect upon the cells which are responsible for regeneration, and have been used to demonstrate that only cells lying within a few millimetres of the wound surface are normally involved in regeneration. Increasing doses of locally applied X-rays cause progressive delays of regeneration and reduced growth of the regenerate, culminating in a dose which completely and permanently inhibits regeneration. This dose is less than 2 krad of high voltage or filtered X-rays; higher doses increase the likelihood of undesired side-effects. The mechanism underlying this inhibition has been generally disputed, but the detailed analysis given here leaves little room for further argument. Irradiation prior to amputation does not prevent wound-healing or a certain amount of dedifferentiation and redifferentiation in internal tissues, but it virtually eliminates cellular proliferation. Independent evidence from such systems as tissue cultures indicates that X-rays delay or prevent cell division, and produce genetic lesions causing the death of many cells which do divide, at doses which leave the function of differentiated cells ostensibly unimpaired (secretion, nervous conduction, muscular contraction). The data on limb regeneration conform to this pattern and are best explained on the basis that local irradiation prevents any cumulative proliferation of mesenchymal cells. The few weaknesses or inconsistencies in this argument are probably a measure of our ignorance. Alternative explanations which are plainly inconsistent with our present understanding have not been entirely abandoned, however, so that investigations based on these unlikely premises may be expected to continue for several years.

Chapter 5

Metabolic Changes

The advent of molecular biology has made redundant much of the older literature on this topic, besides making recent reports seem tediously obvious in that the metabolism of blastemal cells has generally turned out to be quite predictable. Consequently, it will be sufficient to provide a brief survey concentrating on contributions of the last decade, and to cite Schmidt's (1968) exhaustive review of the earlier literature.

5.1 Blastemal metabolism

The bleeding which follows amputation is usually soon checked by a muscular contraction which occludes blood vessels at the end of the limb stump. The subsequent destruction of injured cells, the dedifferentiation of others and the accumulation óf blastemal mesenchyme thus all occur under relatively anaerobic conditions and in isolation from an external energy source. Degenerating muscles liberate glycogen which apparently furnishes a major source of energy, but the breakdown of specialized structural proteins such as myosin and collagen must also supply additional material. All this material is degraded anaerobically to a large extent, resulting in respiratory quotients below unity and a concomitant accumulation of lactic acid. That accounts for the observed drop in pH from a normal 7.2 to about 6.7 in the cone blastema. The blood circulation is restored shortly afterwards (cf. Peadon and Singer, 1966) and respiratory metabolism then returns to normal.

Nucleic acid metabolism has been traced by most available methods but with little precision. DNA synthesis begins very soon after amputation (Lazard, 1970), as cells dedifferentiate, and increases overall with their proliferation. The DNA content of blastemal cells varies within the normal limits imposed by replication and mitosis, when allowance is made for errors of measurement (Desselle, 1974; Mescher and Tassava, 1975). The DNA content of the whole blastema could thus be taken as an estimate of the cells it contained, and probably a more accurate one than the conventional assays

99

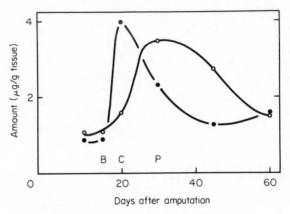

Figure 5.1. RNA (●) and DNA (○) content of
axolotl regenerates, expressed as μg/g fresh tissue
(data of Teplits, 1964b). The values for 60 day
regenerates are slightly less than those found in
mature limbs. The increase in DNA reflects intense
proliferation of cells with little growth in the cone
stage (C); it declines after the palette stage (P) as
differentiating cells enlarge. Expressed on a cellular
or unit DNA basis, the RNA content would reach a
maximum as the blastema (B) grows into a cone,
and diminish during redifferentiation.

of dry weight or total protein nitrogen, to standaridize other measurements
by discounting overall growth (Figure 5.1). RNA synthesis is stimulated even
earlier than DNA synthesis, for the enlarged nucleolus, by which some
investigators have recognized 'activated' cells at the onset of
dedifferentiation, is a token of enhanced ribosomal RNA synthesis. This is
the predominant feature of RNA metabolism (Morzlock and Stocum, 1971,
1972) just as the prominent cytoplasmic RNA denotes a dense population of
ribosomes, as found in proliferating and redifferentiating blastemal cells.
Bodemer (1962b) and more recently Kelly and Tassava (1973) have followed
RNA synthesis in autoradiographs showing the incorporation of radioactive
uridine and found, as expected, that it increases steadily during blastemal
growth. Bantle and Tassava (1974) laboriously identified putative messenger
RNA in advanced regenerates. There is an obvious expectation that each
phase of regeneration should be marked by a changing population of
enzymes and structural proteins whose production depends upon the
activation of different genes which are transcribed into specific messenger
RNA molecules. No means are available to follow the events leading to one
particular state of differentiation apart from detecting a few tissue-specific
proteins, and even then the sensitivity of detection remains a problem.

Protein synthesis occurs throughout regeneration, of course, and total
protein content obviously increases with growth. Some changes in the rate of

incorporation of amino acids have been observed by autoradiography (Bodemer & Everett, 1959; Anton, 1961, 1965). An early rise in cathepsin activity, soluble proteins and free amino acids may be taken to indicate an extensive degradation of structural proteins during dedifferentiation. The diversity of soluble proteins has been demonstrated by electrophoretic fractionation. Schmidt (1966, 1968) displayed a sample of 40 electrophoretic bands, 12 of which persisted throughout regeneration. Some bands were characteristic of differentiated tissues and absent from early blastemata, while others were restricted to particular phases of regeneration. Donaldson (*et al.*, 1974) obtained similar results but involving rather fewer bands from limb and tail regenerates. Dearlove and Stocum (1974) also repeated and extended this type of analysis on proteins extracted from the terminal millimetre of stump tissue, from blastemata and later regenerates of *N. viridescens*. They resolved 24 protein bands from normal limb tissue, 14 of which were present at all stages of regeneration. The remainder disappeared during dedifferentiation and reappeared later, together with eight additional bands which could not be detected in mature tissues. Most of this category of regenerate-specific proteins were nerve-dependent, as they and some other proteins were absent from regenerates which had been denervated 3 days earlier. Denervation also elicited the precocious reappearance of several proteins, presumably in relation to the early redifferentiation observed in nerveless arm stumps, and even produced a couple of novel bands which might be unique proteins or aberrant forms of other nerve-dependent ones. The investigators argued that most of these soluble proteins must be enzymes, and further supposed the nerve-dependent ones were connected with DNA synthesis and mitosis. Although these are two major features which are known to be curtailed by nerve deprivation, it may be recalled that denervation also diminishes RNA and total protein synthesis (Morzlock and Stocum, 1972; Singer and Ilan, 1977), implicating a colossal number of enzymes.

Both nervous and hormonal influences on regeneration have alerted investigators to the potential significance of cyclic AMP in some part of the process, especially following reports that cyclic AMP stimulates DNA and protein synthesis in explanted blastemata (Babich and Foret, 1973; Foret and Babich, 1973b). Two apparently conflicting sets of measurements are availabe for regenerating arms of *N. viridescens* but it may be possible to reconcile them as shown in Figure 5.2. If this composite result is roughly correct, it suggests an accelerating synthesis of cyclic AMP during early blastemal growth, followed by a rapid degradation as cells begin to redifferentiate. The latter phase bears the hallmark of phosphodiesterase which may be synthesized or activated then. Such changes are compatible with the evidence for a slight stimulation of blastemal growth or morphogenesis following treatments calculated to enhance the concentration of cyclic AMP (Foret, 1973; Sicard, 1975b). The concentration of GMP also fluctuates during regeneration but is out of phase with cyclic AMP (Figure

102

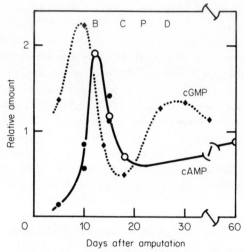

Figure 5.2. Concentration of cyclic nucleotides in *N. viridescens* regenerates, B, C, P, and D indicate blastema, cone, palette and digits stages. Measurements in pmole/mg DNA are expressed as a proportion of values found in mature arm tissues. The solid line is drawn using data for cAMP from Jabailly *et al.* (1975) (O) and adjusted data of Sicard (1975a) (●). The dashed curve shows values for cGMP from adjusted data of Liversage *et al.* (1977).

5.2). The fluctuations may well be more complex than depicted here, for Taban *et al.* (1978) have identified two peaks of cyclic AMP during blastemal growth and an even earlier one associated with wound healing in *T. cristatus*. Since these three peaks correspond to periods when cyclic AMP synthesis is most stimulated by noradrenaline, Taban suggested it might be one neurotrophic factor while cyclic GMP might respond to a neurotrophic protein. It is difficult to reconcile this hypothesis with the accepted view that noradrenaline is delivered by sympathetic nerves which can be cut without prejudice to regeneration (Singer, 1942).

A variety of common enzymes have been demonstrated in blastemata by histochemical tests but the results are not particularly informative. Lactic acid dehydrogenase provides a suitable example, being abundant throughout the blastema but showing alterations in the pattern of isozymes (Schmidt, 1968). The isozyme changes tell us very little, however, for they are probably related both to the alterations of respiratory metabolism and to the specific patterns of differentiated tissues. The abundance or activity of cholinesterase falls abruptly and then increases again in the growing blastema (Figure 5.3). This change closely follows the state of muscle differentiation and may monitor the formation of motor end-plates. The concentration of ATP follows

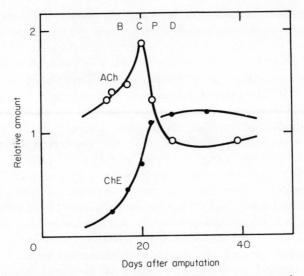

Figure 5.3. Acetylcholine (ACh) content and cholinesterase (ChE) activity in *N. viridescens* regenerates, both expressed relative to the values found in mature arm tissues (data of Singer, 1959; Singer *et al.*, 1960). ACh was estimated by a heart bioassay with control values of 6.5 ± 2 μg/g dry weight; ChE was measured colorimetrically.

a similar temporal sequence (Faustov and Carlson, 1967). Acetylcholine accumulates in the growing blastema, presumable reflecting a rich motor innervation, but then decreases dramatically as cholinesterase activity increases (Figure 5.3).

A few structural proteins and mucopolysaccharides which are characteristic of particular tissues have been traced as a means of testing the determination of blastemal cells. DeHaan (1956) prepared rabbit antiserum against myosin and then tested its ability to react with extracts of axolotl regenerates. Blastemal mesenchyme extracts did not react, although palette stages which already contained some redifferentiated muscle provided a precipitable extract. This test is incapable of demonstrating the absence of myosin and Hay (1970) doubted if the technique's sensitivity surpassed that of a direct search for myofilaments with the electron microscope. Ogawa (1959, 1962) detected actin 1 week earlier than myosin in advanced regenerates of *Cynops pyrrhogaster*, although one would expect redifferentiating myoblasts to exhibit coordinated synthesis of the two molecules. Characteristic myosin structures have been reported in single myoblasts even before they fuse into myotubes (Holtzer, 1961) but myosin has never been detected in earlier blastemal mesenchyme. Linsenmeyer and Smith (1976) traced the synthesis of a collagen specific to cartilage in the growing arms of *Ambystoma* larvae and in redifferentiating cartilage of arm regenerates, but they could not find any synthesis in young blastemata. They suggested the special synthesis ceases during dedifferentiation and early blastemal growth, and interpreted the change as one involving gene repression. The argument applies equally well to actin or myosin, or probably to any other molecule or structure which is characteristic of a particular differentiated cell. Mailman and Dresden (1976) obtained comparable data on total collagen in the regenerates of *N. viridescens*, but preferred to interpret the changes in terms of constant synthesis and variable degradation. Chapron (1974) recorded a preferential incorporation of fucose into the epithelium of the young blastema, perhaps indicating a local secretion of glycoproteins and production of a hyaluronate matrix to replace this degraded collagen. There is considerable evidence for the destruction of collagen and its temporary replacement by a hyaluronate matrix. Grillo (*et al.*, 1968) detected continued degradation of collagen for 20 days or more after amputation. Toole and Gross (1971) recorded the successive increasing syntheses of hyaluronate which peaked at 10 days, of chondroitin sulphate associated with redifferentiating cartilage, and finally of dermatan sulphate of the dermal and tendon collagen. Collagen synthesis in the redifferentiating regenerate is a nerve-dependent process (Mailman and Dresden, 1979).

5.2 Chemical disturbance

Quite a variety of substances have been applied to regenerating limbs or tails, often with the intention of causing a specific inhibition or stimulation which might identify metabolic peculiarities of the blastema or provide analogies to the growth of tumours. Perhaps as a consequence of the latter objective, many of the experiments are only recorded as unconfirmed preliminary notes scattered in paramedical journals.

Wolsky's (1974) gallant attempt to extract some general sense from such experiments, by classifying them according to which step between gene transcription and protein synthesis he considered to be affected by the molecule concerned, has been considerably modified in Table 5.1 which includes a fair sample of the published studies but falls far short of providing a definitive list. Clearly any serious interference with the synthesis of DNA, RNA or protein, or with the energy supply, will inevitably curtail cell division and reduce regenerative growth. Most of the experiments shown in Table 5.1 indicate the external concentration or injected quantity of each substance which caused a consistent disturbance of regenerative growth, almost always a retardation or inhibition, but the mechanism involved could be general debilitation rather than interruption of a specific metabolic pathway. Some of the substances are known to be specific metabolic antagonists under appropriate conditions (cf. Kihlman, 1966) but there is little evidence that those conditions obtained in these particular experiments. The efficacy of an inhibitor here depends to a large extent on its permeability, so that concentration in the external medium has little meaning. Besides being quite well protected against external solutions, amphibians are able to metabolize or excrete a surprising range of organic compounds so that an injected inhibitor may be rapidly eliminated. There are limits to the protective systems, of course, and it is likely that almost any molecule can cause some metabolic upset if it is administered in sufficient quantity.

Colchicine treatment provides a good example for the preceding argument. Dilute solutions penetrate easily if applied immediately after amputation, but considerably higher external concentrations are required to inhibit regeneration if the wound has healed before the treatment begins. In both cases colchicine almost certainly acts by disrupting microtubules of the mitotic spindle so that cells are unable to divide. Wolsky (1974) supposed a blastema could form without cell division and consequently postulated a direct action of the drug on nucleic acid synthesis. If that were so, it would be most unlikely that the observed accumulation of arrested metaphases could occur, especially with the rapidity witnessed by Luscher (1946). Local infusions of colchicine also disrupt the microtubule system associated with axonal fast transport and could cause the resorption of a blastema by depriving it of neurotrophic support. The same external concentration of colchicine which inhibits regeneration in young larvae also arrests the growth of hind limb buds, however, suggesting a direct effect independent of nerves. Vinblastine inhibits cell division in the same way as colchicine and presumably affects a regeneration in the same manner. Mercaptoethanol is also known to cause a rapid and reversible disorganisation of the mitotic spindle (Mazia, 1958) and has been included in the same category of mitostatic drugs for this reason, although it was originally tested as one of a series of sulphydryl compounds (Lehmann, 1961).

Aminopterin interferes with both purine and thymidine syntheses and, like thalidomide, provides a rare example of chemically disturbed morphogenesis. Most treatments retard growth and result in hypomorphic regenerates with a reduced number of digits, but an oral dose administered soon after amputation in the upper limb sometimes yields polydactylous regenerates (Gebhardt and Faber, 1966). The two best known specific inhibitors of DNA synthesis, hydroxyurea and fluorodeoxyuridine, have apparently not been tested on regenerating systems, but fluorouracil is probably converted into the latter which blocks thymidylate synthetase and thus arrests DNA replication and causes chromosome fragmentation (Kihlman, 1966). That provides the most likely explanation of how fluorouracil arrests regeneration, although like thiouracil (Beck and Howlett, 1977) it may also be incorporated into messenger RNA in place of uridine but be read as cytidine during translation of aberrant proteins. Radiomimetic alkylating agents attach to DNA to impede its replication, prolong S phase and produce chromosome aberrations. Subsequent delayed mitoses may involve chromosome bridges and fragments: successful divisions produce defective nuclei and micronuclei, while failure of division results in giant polyploid

Table 5.1 Some chemical inhibitors of regeneration (others are listed in Table 2.4).

Substance	Subject	Application	Effect on regeneration	Reference
(1) Substances that prevent cell division				
Colchicine	*A. mac.* arms	Immersed in 2.5 μM	Inhibits, stump regresses	Thornton (1943)
Colchicine	*N. v.* arms	Local infusion	Inhibits, destroys nerves	Singer *et al.* (1956)
Colchicine	*Xenopus* tail	Immersed 25 μM–1 mM	Inhibits growth	Luscher (1946)
Vinblastine	*R. pip.* tail	Immersed up to 20 μg/ml	Inhibits growth with inconsistent side effects	Francoeur (1968), Francoeur and Wilber (1968)
Mercaptoethanol	*Xenopus* tail	Immersed 50–500 μM	Reduces growth	Hahn and Lehmän (1960) Descotils-Heernu *et al.* (1961)
(2) Substances that interfere with nucleotide synthesis (and thus cell division)				
Aminopterin	*R. pip.* tail	Immersed	Retards growth (delayed effect)	Bieber and Hitchins (1959)
Aminopterin	*A. mex.* limb	Oral 1 mg	Retards growth, affects number of digits	Gebhardt and Faber (1966)
Fluorouracil	*R. pip.* tail	Injections 0.5 mg	Arrested during treatment	Dumont and Sohn (1963)
(3) Alkylating Agents (radiomimetic, interfere with replication)				
Nitrogen mustard	*A. mac.* arm	Immersed	Inhibits, stump regresses	Karczmar (1948)
Hydroquinone	*A. mac.* arm	Immersed	Retards	Karczmar (1948)
Myleran	*A. mex.* tail	Immersed 40–200 μM	Reduced growth and mitosis with abnormal nuclei	Basleer *et al.* (1963) Matagne-Dhoossche (1964)
Uracil mustard	*N. v.* tail	Immersed 333 μM	Retards growth	Yankow (1965)
(4) Base Analogues incorporated into DNA during replication				
Mercaptopurine	*R. pip.* tail	Immersed 1 mg%	Reduces growth	Bieber *et al.* (1954), Bieber and Hitchings (1959)
Thioguanine	*R. pip.* tail	Immersed 10 mg%	Reduces growth	Bieber *et al.* (1954), Bieber and Hitchings (1959), (also Yankow, 1965)
Br-deoxycytidine	*R. pip.* tail	Immersed 333 μM	Reduces growth	Yankow (1965)
I-deoxyuridine	*R. pip.* tail	Immersed 333 μM	Reduces growth	Yankow (1965)

(5) Substances that interfere with RNA synthesis or stability

Actinomycin-D	R. pip. tail	Immersed 1–10 mg/ml (almost toxic)	Retards or inhibits growth and differentiation	Wolsky and Van Doi (1965)
Actinomycin-D	A. mac. tail			
Actinomycin-D	N. v. limb	Toxic injection 2–10 µg/g	Inhibits blastema	Carlson (1967a)
Actinomycin-D	N. v. limb	Skin 10 µg/ml, 3 hours	Delays regeneration	Carlson (1969)
Ribonuclease	N. v. limb	Infusion 0.8 mg	Delays differentiation	Bodemer (1962a)

(6) Substances that inhibit protein synthesis

Chloramphenicol	N. v. limb	Injection 5 mg daily	Retards and reduces with abnormal differentiation	Burnett and Liversage (1964)
Puromycin	N. v. limb	Injection 10 mg daily		Liversage and Colley (1965)
Streptomycin	R. clam. tail	Immersed 0.9–1.3 mg/ml	Disturbs growth	Procaccini and Doyle (1970)
Streptomycin	N. v. limb	Injection 40 µg/g	Reduces growth and no. of digits	Procaccini and Doyle (1972)

(7) Supposed Competitors for endogenous RNA

Tail RNA	N. v. tail	Injection 50 µg	Stimulates growth	Wolsky and Wolsky (1969)
Liver RNA	N. v. tail	Injection 50 µg	Inhibits growth	Wolsky and Wolsky (1969)
Liver RNA	R. pip. tail	Immersion 120 µg/ml	Inhibits growth	Wolsky and Wolsky (1969)
Liver RNP	R. pip. young frog arms	40 daily injections	Stimulates blastema	Smith and Crawford (1969)
Rat liver RNA	A. mex. X-rayed leg	Injections	30% regenerate	Polezhaev (1972a) (see Table 6.3)

(8) Miscellaneous

Cyclic AMP	N. v. arm	Injections	Accelerates morphogenesis	Sicard (1975b)
Theophylline	N. v. arm	Injections	Accelerates morphogenesis	Sicard (1975b)
Chlorpromazine	N. v. arm	Injections	Retards morphogenesis	Sicard (1975b)
Trypan Blue	N. v. arm	Injections	Stops dedifferentiation	Dearlove et al. (1975)
Thiourea	R. pip. leg	Immersed 50 mg%	Delayed arrest of growth	Peadon (1953)
Perchlorate	Bufo leg	Immersed 0.05%	Prolongs regenerative capacity	Michael and Aziz (1976)
Vitamin A	Bufo leg	Immersed 15 I.U./ml	Accelerates dedifferentiation, suppresses growth	Saxena and Niazi (1977)
Semicarbazide	N. v. arm	Infusion 50 mM	Retards blastema formation	Deck and Shapiro (1963)
Beryllium	R. temp. tail	Immersed 66 mM	Completely inhibits	Needham (1941)
Beryllium	A. mac. arm	Immersed 66 mM	Inhibits, stump may regress	Thornton (1949–51)
Beryllium	N. v. arm	Infusion	Inhibits, stump may regress	Schueing and Singer (1957)
Thalidomide	N. v. arm	Oral 3 mg	Like human phocomelia	Bazzoli et al. (1977)

cells. These effects have been observed in myleran treated regenerating tails. Retarded growth is an obvious consequence of the prolonged cell cycle and abnormal divisions. Nitrogen mustard and hydroquinone seem to have similar gross effects but have not been studied in as much detail. Several base analogues can be incorporated into DNA causing some modification of its structure and stability. It is by no means certain, however, that this alteration is responsible for the general reduction of regenerative growth recorded for specimens immersed in high concentrations of the base analogues. Mercaptopurine and thioguanine act as competitive inhibitors, however, being rendered ineffective in the presence of the corresponding natural purines (Bieber *et al.*, 1954).

Actinomycin-D binds to DNA and preferentially inhibits gene transcription, especially affecting the synthesis of ribosomal RNA. When present in the culture medium, it retards the growth and differentiation of larvae immersed in it, but clearly toxic amounts are needed to completely arrest regeneration. Carlson (1967) found that injecting newts with 2 $\mu g/g$ body weight both caused a strong inhibition of regeneration and also reduced the median survival period to 30 days. Merely soaking excised limb skin in actinomycin before replacing it and amputating at the same site caused a temporary delay of regeneration (Carlson, 1969). Ribonuclease presumably degrades nascent RNA as well as destroying functional RNA molecules. The consequent reduction in cytoplasmic RNA and protein synthesis have been observed in limb blastemata (Bodemer, 1962a) together with a rapid recovery leading to normal regenerates delayed by about 2 days. Two inhibitors of protein synthesis, chloramphenicol and puromycin, produce severely retarded and hypomorphic regenerates after a prolonged series of daily injections. Shorter treatments designed to cover either the period of dedifferentiation or redifferentiation allowed Liversage and Colley (1965) to confirm the assumption that protein synthesis is needed during both phases of regeneration. Streptomycin interferes with the function of prokaryote ribosomes but is generally held to have only a negligeable effect on protein synthesis in eukaryotes (Franklin and Snow, 1975). As might be expected, only nearly toxic doses disturb regeneration consistently, and the mode of action is suspect.

The group of studies involving RNA treatment (Table 5.1) can be traced back to a curious report by Wolsky and Fogarty (1962) that transplanting a newt tail blastema to the flank accelerated the second regeneration of the tail stump. This curious discovery has been pursued by others (Skowron *et al.*, 1963; Weber and Maron, 1965) who concluded that injections of blastemal homogenates and extracts had the same effect on axolotls. If there is a genuine stimulation, it should also affect regeneration after a single amputation which could be measured more accurately, but no such test has been reported. Attempting to substitute RNA for blastemal extracts may be thought of as searching for a transmissible messenger, particularly where the type of RNA is held to be organ-specific. Table 5.1 reveals mutual inconsistencies among the claimed results but also grossly exaggerates the trivial changes which were originally recorded. Considering the difficulty in obtaining precise measurements of growth from the baseline of a dedifferentiating tail stump, the small and temporary stimulations noticed are probably insufficient to establish the phenomenon. The minor changes recorded in frog arms after a course of ribonucleoprotein (RNP) injections, such as cartilage nodules after 3 months (Smith and Crawford, 1969), are no greater than found occasionally among controls or after a variety of traumatic treatments. RNP injections which had been pretreated with ribonuclease appeared to be mildly inhibitory. That does not mean RNA was the effective component of the injection, however, for the residual activity of the ribonuclease would have the same effect (Bodemer, 1962a). Polezhaev's protracted series of experiments on irradiated axolotl legs certainly make a clear distinction between normal regeneration and healed limb stumps. Injections of RNA or protein or complete homogenates of rat liver each apparently induced regeneration in about one-third of the irradiated limbs, but a recombined injection mixture of RNA and protein did not cause any regeneration. Similarly, DNA injections resulted in regeneration in some experiments but not in others. These internal inconsistencies and others described by Polezhaev (1972a, b) suggest that his procedure of delayed and repeated amputation after massive doses of X-rays has such erratic consequences as to defy conventional interpretations, although an unconventional one is offered in the following chapter.

Cyclic adenosine monophosphate (cAMP) is now known as the intracellular second messenger which mediates a variety of hormonal stimuli and some nervous functions (Lentz, 1972; Dretchen et al., 1976); theophylline protects it from degradation. Both substances have little or no effect on the formation of blastemata but accelerate later regeneration so that digits appear earlier than normal (Sicard, 1975a, b). Conversely, a neurotransmitter antagonist chlorpromazine retards morphogenesis. Combined injections of cAMP and chlorpromazine can cancel each other's influence, perhaps because the latter alters the levels of hormones whose effects are mediated by cAMP. The teratogenic dye trypan blue has been found to halt regeneration in its initial stages, perhaps even arresting dedifferentiation, as it also inhibits hyaluronidase. The gross observations were made on adult newt limbs, where dedifferentiation is so limited that it has frequently escaped detection, however, so the conclusion remains rather tentative. Thiourea and perchlorate are well known inhibitors of the thyroid gland and eventually retard or arrest both regenerative and normal limb growth in metamorphosing tadpoles, besides retarding or suppressing metamorphosis and thus prolonging regenerative capacity. Vitamin A both enhances dedifferentiation, presumably by a direct action like that found in tissue cultures, and acts as a thyroxine antagonist to retard metamorphosis and the regeneration of tadpole legs, commonly reducing the number of digits or distorting the regenerate. Repeated infusions of semicarbazide retard blastema formation on newt limbs, but perhaps at concentrations which have more general effects than interfering with histamine synthesis. Beryllium nitrate acts as a potent poison when it can penetrate to internal tissues, as happens when it is applied directly to a wound surface (Thorton, 1949–51) or when it is infused into a blastema. Needham (1941, 1952) suggested beryllium might inactivate a wound hormone which constituted the initial stimulus for regeneration. Localized poisoning and cell death provides a simpler explanation which is supported by the regression observed in treated arms both of young larvae and, more remarkably, of adult newts (Schueing and Singer, 1957). Additional inhibitors of growth of regenerating tadpole tails are listed by Bieber and Hitchins (1959) and by Lehman (1961). Most of them were selected in the belief that they principally interfered with purine or protein metabolism. Even if that were true, it is unlikely that they could provide additional useful information as some acted synergistically when applied in combination, others did not, and there was no obvious basis for predicting synergistic combinations (Hahn and Lehmann, 1960).

5.3 Cultured blastemata

More than any other experiments concerned with regeneration, those employing chemical treatment are liable to arouse the righteous anger of antivivisectionist and conservationist lobbies. It is admittedly cruel to subject animals to drugs up to their limits of tolerance and, in the main, it has turned out to be a rather futile exercise. The aims of the investigators can usually be defended on ethical grounds, for many of them were genuinely pursuing the goal of tumour therapy. They can only be criticized for the universal human failings of naivety and incompetence: with the wisdom of hindsight, it is alarming how few appreciated either the action of the drugs they employed or the normal course of regeneration. The charge of depleting stocks of increasingly rare species may prove more serious eventually, and could be avoided by restricting such experiments to species maintained as laboratory colonies. This practice would also enhance experimental precision by reducing the genetic disparity and the hazards of disease and acclimatization found among captured specimens. Although only a few amphibian species are widely kept in laboratories, and they have been used in only about 10% of the experiments considered here, there is no doubt that their advantages have become increasingly appreciated during recent years.

The pharmacological attack on regeneration itself can be diverted into a less distressing and more sophisticated path by the use of isolated blastemata as test material. Young blastemata have been successfully cultured for a week or so, which should be ample time to assess a response to chemical treatment in terms of their metabolic changes. The technique originally invented for limb buds (Wilde, 1950) was first used for naked blastemal mesenchyme by Stocum (1968a) and has lately been refined by the provision of a grid to act as a permeable support for the blastema. Deprived of its normal innervation and hormone supply, the isolated blastema shows little growth or morphogenesis but is capable of serving as a model system for testing the metabolic requirements of regeneration. Increasingly sophisticated culture media have been employed in attempts to discover the full range of requirements. According to Vethamany-Globus and Liversage (1973a), a diluted Leibowitz medium formulated for mammalian cell cultures is improved by the addition of growth hormone, hydrocortisone, thyroxine and insulin. They subsequently found that only insulin and a neurotrophic factor improved synthesis in cultured blastemata (Vethamany-Globus *et al.*, 1978). Foret and Babich (1973) used short-term cultures to demonstrate that acetylcholine reduced protein synthesis, but cyclic AMP stimulated DNA and protein syntheses. Culturing blastemata for several days in a medium enriched with hormones has allowed two groups to gain evidence of a neurotrophic factor which stimulates cell division (Globus and Globus, 1977) and protein synthesis (Choo *et al.*, 1978). Even though these examples mainly confirm conclusions which had been reached previously from the study of whole animals, they show that isolated blastemata can provide more precise information than much of that summarized in Table 5.1. The provision of a substitute innervation allows explanted blastemata to show some further growth and differentiation (Liversage and Globus, 1977). Both a brain homogenate and cyclic AMP can stimulate cell division in cultured blastemata (Carlone and Foret, 1979).

5.4 Electrical phenomena

Ever since morphogenetic fields were conceived by analogy to magnetic fields there have been repeated attempts to relate the polarity of growth and regeneration to potential gradients, with considerable success in coelenterates. More recently, several investigators have invoked electrical changes as an essential early phase of limb regeneration. Monroe (1941) measured minute potential differences over the surface of adult *T. cristatus*, while the newts were fully anaesthetized and kept moist in air. He found a rather irregular distribution of positively and negatively charged regions along the dorsal midline, with a negative potential of about 1 μV at the tail tip relative to a reference electrode on the nose. The tip of each limb was always more positive than its base, by up to 5 μV. Monroe traced an outline on the flank of points which were isopotential to the base of the limb, arguing these might delimit the limb's regenerative territory. He found the general pattern of potential differences was unaffected by destruction of the central nervous system but could be perturbed locally by amputating one arm. Relative to the base of that arm, the cut end had risen to 16 μV next day and took about 10 days to relapse to a normal value (Figure 5.4). Lassalle (1974a, b) also found a fairly regular potential gradient along the arm of anaesthetized *P. waltl*, although measured in mV presumably because

112

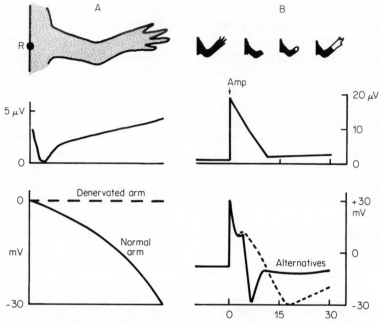

Figure 5.4. Measurements of surface potential along newt arms. A recording electrode is moved along the arm (A) or applied to the tip of a regenerating arm stump every few days (B) to show potential differences from a reference electrode R on the dorsal midline. Upper graphs, fully anaesthetized *T. cristatus* deduced from Monroy (1941). Lower graphs, *N. viridescens* recovering from anaesthesia, adapted from Becker (1961a, b), Schmidt (1968) and Rose (1970). The dashed line under A was recorded from a denervated arm; the dotted curve under B may be a case of delayed healing.

of the different characteristics of the recording system. The peak of the gradient near the wrist registered an average of 15 mV above the base of the arm. This gradient could be reduced or even reversed by any superficial wound but was restored quickly after healing. It recovered within 2 days of denervating the arm and was thus independent of a normal nerve supply.

These surface potentials must reflect a sodium pump operating through the skin, like the better known one in frogs, perhaps modified by the efficiency of clearance to the local blood circulation. The temporarily increased potential at the end of an arm stump would then indicate leakage of current across the wound epithelium. That remained a speculative explanation until Borgens *et al.* (1977b) applied a sensitive 'vibrating probe' device to measure the current entering or leaving the limbs of anaesthetized *N. viridescens* submerged in water of known salinity. They recorded current densities in the range of 10–100 $\mu A/cm^2$ leaving the amputated end of a limb until it had grown an obvious blastema. This current was rapidly reduced by placing the newt in water lacking sodium ions, but seemed to be actually increased by denervation prior to amputation. Similar currents have been recorded from damaged human finger tips (Illingworth and Barker, 1980).

The effect of the skin's sodium pump seems to be obscured by larger potential differences over the surface of active newts. Becker (1960, 1961a) recorded positive regions over the brain and in the dorsal midline at the pectoral and pelvic girdles when specimens of N. viridescens were entering or recovering from anaesthesia. Relative to these high points, the potential fell away gradually along the tail and limbs to reach values as low as −30 mV at their tips (Figure 5.4). The ventral midline showed a more uniform potential, positive to that of the adjacent flank but unfortunately not fully related to the dorsal pattern. Becker stated that the potential gradient along an arm was virtually eliminated within 1 minute after denervation, and concluded the gradient must be an attribute of normal nervous activity. A mock-denervation control involving a similar wound seems needed to validate that conclusion (cf. Lassalle, 1974b). Amputation caused a dramatic reversal of the potential at the wound surface, followed by a recovery and over-shoot, then a final recovery to −10 mV as the blastema formed. The two examples shown Figure 5.4 may refer to extreme discrepancies between the time taken to complete wound healing. The initial reversal of potential is surely a composite effect of the open wound and the elimination of nerve endings. Becker (1961b) claimed the later recovery must indicate a nerve-related direct current which could be a necessary stimulus for regeneration. It is quite likely that the richly innervated wound epithelium could account for the over-shoot preceding a return to normality as the apical cap thickens, but no causal relationship to regeneration has been established. S. and F. Rose (1974) attempted to demonstrate such a relationship by measuring the potentials along regenerating, denervated and irradiated forearms of active N. viridescens. They partly supported Becker's observations in obtaining more negative measurements distally but also found inconstant alterations of positive and negative regions along the length of the arm. Early blastemata present 10–18 days after amputation showed potentials 0–60 mV below those of adjacent stump surfaces. Arms which had been amputated and denervated for the same period showed an equally variable potential (−30 to +30 mV) between the healed tip and a similar reference point. While confirming that local potentials are partly related to nerve function, these results eliminate any notion of a regular proximo-distal potential gradient. Only by averaging repeated measurements on the same arms could the investigators discern patterns of potential which seemed more characteristic of regenerating (and irradiated) arms than of denervated ones. The only possible deduction from this comparison is that a negative tip is not a meaningful parameter of regeneration, but that conclusion did not appeal to the investigators who noted an additional complication. Their irradiation procedure, a conventional 2 krad at 110 kV, caused a local reduction in blood circulation some 2 or 3 weeks later. This was sometimes accompanied by epidermal thickening and ulceration, and by a reversal of the arm's general potential gradient. Amputation prior to recovery from this reaction, presumably the inflammatory reaction noted by Brunst (1960), resulted in wound surface potentials as much as 100 mV above the neighbouring irradiated skin.

The characteristic proximo-distal gradient of surface potential along the arm of an adult newt is certainly a property of the skin, for Lassalle (1977) has recorded it from isolated pieces of skin. The altered gradient noted after amputation or other wounds surely reflects change in the sodium pump or permeability of wound epithelium and their gradual recovery as the skin redifferentiates. Two arguments convince me these changes are merely coincidental to regeneration. Firstly, the gradient is swamped by the interplay of nerve potentials and circulatory changes during normal activity. Secondly, there is accumulating evidence from tadpoles (Taylor and Barker, 1965), urodele larvae (Lassalle, 1979) and axolotls (Thornton, 1968) that their limbs either do not exhibit surface potential gradients or do not possess the means to establish one. There is no cause to postulate a specific regeneration potential, even in adult newts, until it is shown to differ from any other wound potential. Yet this unnecessary postulate and the derivative hypothesis of an electrical control of regeneration have provoked several attempts to induce limb regeneration in postmetamorphic frogs and in mammals.

Bodemer (1964) reported he had induced 'incipient regeneration' in the arms of adult *R. pipiens* and *catesbiana* by subjecting the main brachial nerve to a 20 minute dose of intermittent 300 mV shocks. The response can fairly be described as minor and sporadic. Although a second series of shocks 1 or 2 weeks later did not help, the single treatment seemed to act synergistically with immersions in a strong salt solution. Smith (1967, 1974) generated direct currents down frogs' arm stumps, at first by inserting a silver–platinum couple under the skin and later by placing a mercury battery and megohm resistor in the dorsal lymph sac with an insulated negative lead ending at the amputation surface. The battery delivered a constant current of rather less than 1 μA for a month. Smith (1974) stated that he obtained one case of perfect regeneration after this treatment in addition to several cases with 1–3 digits containing flexible phalanges. The latter seem to have been very imperfect structures, however, and cannot counter the widespread belief that his solitary perfect case was one of mistaken identity. Becker (1972; Becker and Spadaro, 1972) implanted silver–platinum couples with a variety of resistances into amputated rat arms and claimed to have obtained results which greatly exceeded their (unspecified) expectation within 7 days treatment at 3–6 nA. Their arbitrary classification of apical growth up to a supposed blastemal stage did not reveal any clear difference between the functional bimetallic couple and either silver or platinum wires used separately. Such isolated wires might short out a natural potential gradient but could not impose a new one. Similarly, Smith (1967) had found such couples to be equally effective whichever way they were inserted into the frog arm and hence with currents in either direction.

Any growth beyond normal wound healing was identified in the preceding investigations as incipient or partial regeneration. Only a minority of treated limbs produced perceptible blastemata, which sometimes deteriorated later. The most likely explanation of these results is that implanted wires and perhaps currents create enough damage to augment the usual number of dedifferentiated cells left after amputation. Even when such cells accumulate at the tip of the stump they cannot establish a growing blastema but redifferentiate into scar tissue containing occasional islands of cartilage, bone or even muscle. Responses of this type may be uncommon among adult controls of these species but have been observed in related species (Michael and Al Sammak, 1970). One study using batteries implanted into adult *R. pipiens* (Borgens *et al.*, 1977a) has finally achieved consistent results and is reported in enough detail to command respectful attention. An insulated negative lead from the battery, causing a potential drop of 100–200 mV down the arm stump and a steady discharge of some 0.2 μA, provoked excessive nerve growth and indications of partial regeneration (Table 5.2). When this lead was connected to the positive battery terminal, however, it produced extensive destruction which even spread to the humerus well above the level of amputation. The quite limited growth and differentiation achieved by cathodal stimulation probably followed from the new nerve growth which, at twenty times the amount found in control arms, must greatly exceed the threshold levels previously postulated for regeneration. Borgens and his colleagues have confirmed this discovery by finding that a similar cathodal stimulation of adult *Xenopus* arm stumps enhances the growth of hyperinnervated spikes, and argued that the lead establishes a circuit through arm muscle which is normally shunted by conduction through the subdermal lymph spaces of frogs but not of urodeles (Borgens *et al.*, 1979a, b).

As an alternative to studying battery-driven frogs, Borgens *et al.* (1979c) attempted to impede regeneration in adult urodeles by reducing the current which emerges from their arm stumps. The simplest test is how well an arm can regenerate in water containing little or no sodium ion, which should starve the skin sodium pumps that generate the local currents. *N. viridescens* maintained a reduced stump current in such conditions, presumably by pumping other ions, and their limb regenerates were retarded for several weeks but eventually caught up with controls kept in pond water. Adult *A. tigrinum* gave similar results but regenerated normally when kept in humid air. It is surprising that the latter salamanders did not also show a delayed regeneration (at least no delay was reported), for they too were deprived of external sodium although their moist skin might support local currents. Borgens used these as a control series and applied amiloride to the arm stumps of similar salamanders on alternate days, in order to block the local sodium pumps. About half of the treated arms regenerated normally. Such unaffected specimens need not invalidate the other results, for they might have moulted or rubbed off the amiloride during an early critical stage. The remaining treated arms were usually quite unable to regenerate, although a

Table 5.2 Evidence of partial regeneraion after 3–4 weeks cathodal stimulation of frog fore-arm stumps (data of Borgens *et al.*, 1977a).

Stimulation at wound surface	Observations			Number of cases showing neoformations					
	Months	Cases	Growth (mm)	Protuber-ances	Excess nerve	Muscle	Skeleton	Callus	Degener-ated tissue
Cathodal	5–11	9	+0.5 to 4.7	9	9	9	9	0	0
Anodal	3–6	4	−2.0 to 1.0	1*	1	2	0	4	4
None (sham implant)	6–10	5	+0.7 to 3.5	0	0	1	0	5	0

*This anomalous case contained small amounts of new cartilage and muscle with considerable nerve growth, despite a 2 mm shortening of the arm.

Table 5.3 Capacitance of skin of an adult axolotl measured in pF (data of Umanski
et al., 1951).

| Body region | Measured relative to long (PD) axis | | |
	(1) Parallel	(2) Perpendicular	Difference, (1) − (2)
Left arm	4.930	4.870	0.060
Right arm	5.248	5.208	0.040
Left leg	5.413	5.228	0.185
Right leg	5.140	5.040	0.100
	Measured relative to cephalocaudal axis		
Belly	5.190	5.190	0.000
Head	5.450	5.450	0.000

few severely hypomorphic regenerates occurred. The complete failure of
regeneration after amiloride treatment is so different from the temporary
retardation caused by sodium deprivation, however, that one may suspect
amiloride has other effects in addition to reducing stump currents. These last
studies have at least established a plausible argument for electrical events
having some indirect bearing on regeneration, but there are still several
inconsistencies to be resolved. In the current state of the art, it might prove
more profitable to depolarize cell membranes (cf. Moolenaar *et al.*, 1979)
than to impose potential differences across whole organs.

The limb skin of axolotls has been reported to possess slight dielectric
properties, with a capacitance along its main axis some 2% greater than that
measured in a perpendicular direction (Table 5.3). Although these
measurements still await an independent confirmation, they provide a timely
reminder of Monroy's original concept that electrical properties and
morphogenetic fields should reflect an underlying molecular organization
such as a polarized arrangement of epidermal cells or of the basement
membrane beneath them. Umanski *et al.* (1951) related the dielectic
property to the action of skin transplanted on to an irradiated limb.
Amputation through such a graft elicited regeneration when it was oriented
normally or turned 180°, but not when it had been rotated 90°. Such results
have been confirmed on several occasions (see Chapter 8) but the underlying
mechanism still remains a mystery.

5.5 Conclusions

Several fundamental metabolic changes occur during regeneration. These
may be summarized as involving the conversion of cells bound in tissues and
exercising specialized functions into undifferentiated proliferating
mesenchyme. Considerable structural breakdown occurs anaerobically,
nucleic acid synthesis accelerates markedly and differential gene transcription

can be deduced from the changing population of proteins as regeneration proceeds. According to the essentially negative evidence of tests for constituents of specialized cells, blastemal mesenchyme consists of genuinely undifferentiated cells.

Being a region of rapid metabolism, the blastema is relatively susceptible to any growth inhibitor. Mitotic poisons, drugs which interfere with nucleic acid or protein synthesis, and probably many others can retard regenerative growth or even arrest it at doses which approach toxic levels. The morphogenetic pattern of regeneration is less easily disturbed, although inadequate growth typically produces a hypomorphic structure with few digits. The use of chemical antagonists has revealed virtually nothing about blastemal metabolism which has not been deduced more reliably from descriptive studies. Neither method has indicated any metabolic peculiarity in the blastema which might serve as a guide for analysing morphogenesis.

Local electric currents and potentials probably stem from differences in the skin's ionic pumps and permeability, together with residual effects of nerve impulses. There is no evidence that these have more than a tenuous influence on regeneration. Prolonged direct current stimulation seems to increase nerve growth and consequently to enhance the growth of frogs' arm stumps, but the net result falls far short of replacing the missing structures.

Chapter 6

Genetic Aspects

The metabolic changes outlined previously, especially the changing spectrum of proteins in growing regenerates, provide indirect evidence of a changing pattern of gene transcription. Nobody would dispute the general principle that regeneration can be envisaged as a controlled sequence of selective repression and activation of genes, although many would doubt the utility of the principle. Furthermore, the occurrence of histological transformation considered in chapter 7 implies that dedifferentiated cells can be reprogrammed to transcribe previously unused sets of genes. The nature of the programme and the mechanism by which it can be imposed on undetermined or partially determined cells are the common basic problems which link development and regeneration.

6.1 Morphogenetic mutants and genetic markers

One obvious approach to this problem consists of exploiting genetic defects which cause developmental abnormalities, preferably those showing a simple mendelian inheritance and a constant phenotypic expression. There is no doubt that tissue transplantation between mutant and normal chick embryos has proved invaluable in analysing limb development there. Most inherited limb abnormalities probably arise through metabolic defects in the mesoderm, e.g. the polydactylous mutant *duplicate* (Zwilling and Hansborough, 1956). The *wingless* mutant also involves an ectodermal defect which enabled Zwilling (1956, 1974) to demonstrate the mutual interdependence of the two tissue layers and to deduce the existence of an 'apical ridge maintenance factor'. No analogous experiments to these have been attempted on regenerating limbs, although they could be equally rewarding. Amphibian genetics is admittedly in a primitive state and the long generation time is quite an effective deterrent, but even the known mutants have been strangely neglected. Several of the forty or so genes known in axolotls affect limb development to some degree (Malacinski and Brothers, 1974; Humphreys, 1975). Most of them are recessive lethals which

119

Figure 6.1. A homozygous lethal 'peg-leg' mutant axolotl. This large example shows the characteristic small limbs and terminal oedema.

cause death before regeneration could be tested easily, but one notable exception is the recessive *ova-deficient* (*o*). The viable *o/o* homozygotes are identified by their slow and defective limb regeneration (Humphrey, 1966). Other potentially useful mutations occur in Humphrey's (1975) category of late lethals—homozygous mutants which feed and grow for a considerable period but whose limbs grow poorly and often show skeletal deformities. *Phocomelia* (*ph*) and *short toes* (*s*) (Humphrey, 1967) belong to this category, as does a novel recessive mutation provisionally called *peg-leg* (*pl*). Homozygous *pl/pl* larvae can be recognized by the retarded development of their arms, which become distorted and rarely form more than three fingers (Figure 6.1). Their legs grow equally slowly into flippers or stumps which occasionally bear five toes. Although extravagant oedema may occur at any time after the larvae begin to feed, even bloated larvae usually survive for several weeks and their limbs regenerate just as slowly and defectively as they developed. The peg-leg mutation was discovered quite recently and will need to be characterized in developmental terms before it can be applied to regenerative problems. The only genetic difference which has been used extensively is that between dark and white axolotls, exclusively as a means of marking grafts without impairing their capacity for regeneration. It is not a particularly precise marker, leaving the boundary between host and graft in some doubt and being useless for identifying individual cells. The colour distinction could be improved, either by using *melanoid* (*m*) homozygotes or a pituitary implant to intensify the dark phenotype or, in some circumstances perhaps, by substituting a genuine *albino* (*a*) for the *white* (*d*) marker.

A considerable number of mutants have been discovered in the aquatic toad *Xenopus laevis* (Gurdon and Woodland, 1975) including two suitable recessives, one causing polydactyly and the other interfering with the thyroid gland and thus preventing metamorphosis. It would be interesting to learn if the latter results in giant tadpoles which are still capable of perfect limb regeneration, even though that might only confirm the conclusions drawn from thyroidectomy and from chemical suppression of thyroid activity. An albino gene may provide a marker for tissue grafts when its effects are better known (Hoperskaya, 1975; Tompkins, 1977). The recessive lethal *nucleolar marker* (variously abbreviated as n, N, or nu) has been widely used to mark transplanted nuclei and occasionally employed for tissue grafts. The two experiments which used this marker in connection with regeneration were not particularly informative. Barr (1964) grafted skinned $1n$ legs on to wild-type $(2n)$ tadpoles so that the grafts became covered with binucleolate host epidermis. He examined regenerates of these legs for the presence of cells with two nucleoli in internal tissues, and found none. That result is compatible with the belief that epidermal cells do not enter the blastemal mesenchyme, but would only prove it if equivalent $2n$ control regenerates regularly contain cells displaying both nucleoli. Unfortunately, the high incidence of nucleolar fusion typical of proliferating and differentiating cells (Wallace, 1963) caused enough difficulties to deter Barr from pursuing the subject to its logical conclusion. Burgess (1967) used leg blastemata of $1n$ tadpoles to provide marked donor cells for nuclear transplantation. None of the recipient eggs developed far, perhaps because the nuclei came from static blastemata arrested at metamorphosis of the donor larvae. Dasgupta (1970) obtained a young larva when performing the same experiment on axolotls.

6.2 Transplantation immunity

Many of the experiments devised to analyse regeneration have involved transplanting pieces of tissue, blastemata or even whole limbs. When they are transplanted to different regions of the same individual, such grafts usually heal in their new position and persist indefinitely. Sometimes, however, it is more convenient or even expedient to transplant tissue from one specimen to another. This type of operation is equally successful when performed on embryos and a conventional terminology has grown up to describe the various types of grafts. Orthotopic grafts are those placed in their normal position on the same or another specimen, heterotopic grafts are those placed in an abnormal position. Autoplastic grafts (abbreviated to autografts) are those returned to the same specimen from which they were obtained. All transplantations between individuals can be termed allografts: homografts between two members of the same species; heterografts between closely related species; or xenografts between different genera or families. Although all these kinds of embryonic grafts usually survive permanently, there is an obstacle to postembryonic transplantations between different individuals at least for vertebrates. An individual can detect and destroy foreign tissue, even that of its close relatives, as demonstrated most clearly by skin grafts between mammals. This homograft or allograft reaction involves the production of antibodies to foreign substances or antigens, such as proteins which are produced by the graft but not by the host, and culminates in an invasion of the graft by sensitized host lymphocytes. The graft is said to be rejected, which is a polite way of saying it is destroyed. The genetic basis of graft rejection and transplantation immunity is clearly demonstrated by the virtual restriction of successful reciprocal transplantation to identical twins and to members of extremely inbred populations—isogenic individuals providing 'isografts'.

Tolerance of allografts can be induced either by operating on embryonic or perhaps early larval stages before the immune system has fully developed, or by eliminating the cells which produce antibodies and the stem cells of lymphocytes. Such procedures have rarely been attempted in connection with regeneration studies, yet both urodeles and anurans reject allografts readily when adult and to some extent even as larvae. Cohen (1966) has described in considerable detail the fate of skin grafts in adult *N. viridescens*. First there is a latent phase lasting about 2 weeks at 23°C while both autografts and allografts heal and become vascularized. Allografts then degenerate during a period of 4 to 6 weeks, with a sequence of vasodilation, haemostasis, haemorrhage and depigmentation. These symptoms accompany a massive invasion of the graft by small lymphocytes which engulf dead cells and debris, besides probably attacking live graft cells. A second allograft from the same donor tends to be rejected more rapidly than the first as the latent period is reduced. Rejection occurs much more slowly or incompletely at lower temperatures, however, or if the host immunity system has been weakened by prior exposure to 500 rad X-rays. Skin grafts between randomly collected newts are sometimes retained indefinitely, suggesting the existence of a reasonably small number of antigenic differences (Cohen and Hildemann, 1968). Perhaps the skin is particularly sensitive to the immune reaction, for homografts or even xenografts of pituitary glands can survive and function for many months (Dent, 1966, 1967; Cohen, 1971).

Skin homografts are rejected with identical symptoms but at different speeds in other urodeles and anurans (Cohen, 1968, 1969; Simnett, 1964, 1965; Horton, 1969). Both invading lymphocytes and circulating antibodies have been recorded in axolotls (Glade, 1963; Ching and Wedgewood, 1967). The homograft is usually destroyed slowly enough for host skin to replace it without leaving any open wound during rejection. The speed, nature and probably the genetic basis of rejection in amphibians thus resemble the chronic response of mammals rather than the acute response caused by differences at the latter's major histocompatibility locus.

According to a series of abstracts (Delanney 1961, 1978; Delanney *et al.*, 1967) laboratory axolotls possess a series of at least five alleles controlling transplant immunity and quite young larvae can reject homografts. The frequency of rejection increases with age, however, up to metamorphosis in *A. tigrinum* (Cohen, 1967, 1969). De Both (1970a) exchanged blastemata between juvenile axolotls obtained from several European colonies, populations that had been isolated for 20–30 years or perhaps only 5–10 generations. Such grafts were usually retained, as were homografts between members of the same colony. Unrelated specimens from different importations rejected blastemal grafts more frequently while xenografts of newt blastemata were rejected regularly and more rapidly by the same axolotls. De Both considered each axolotl colony as a random breeding population, which is an unlikely proposition considering their history in captivity. Many colonies have been deliberately inbred to maintain the white mutant discovered among the progeny of specimens imported to Paris in 1864. The common occurrence of white specimens indicates that the vast majority of colonies are descended from this stock, many founded from the gift of a breeding pair or a single clutch of embryos and maintained incestuously to preserve the white mutant. This stock spread across Europe at least as far as Moscow. Some were taken from Poland to America by R. G. Harrison in the 1930s. A white stock at the Wistar Institute in Philadelphia, and thus R. R. Humphrey's original stock now at the University of Indiana, are descended from this second Atlantic crossing. I obtained some remnants of a spawning (which Humphrey designated mating 3228), brought them to England and mated them to members of a local colony which also carried the white marker. These two stocks had probably been isolated from each other for a century, or 30–40 generations. The inter-colony hybrids seemed completely tolerant of each other's grafts and so, in retrospect, their parents must have each been homozygous for all histocompatibility factors. Subsequent back-cross generations, however, frequently rejected grafts from their sibs. In one case, only 15% of the recipients tolerated cartilage grafts, a proportion expected if the two stocks differed at six or seven co-dominant loci (Maden and Wallace, 1975).

This deduction contains several hidden assumptions which cannot be completely justified at present. The number of incompatibility classes among a single spawning certainly demands a radically different explanation than that offered by the major histocompatibility locus of mammals (reviewed by Munro and Bright, 1976). The simplest assumption is that axolotls carry several unlinked allogenic loci, which are each expressed independently without dominance or epistasis. The consequences of this assumption are easily predicted for the simple cases of mating two homozygotes which differ at n allogenic loci (and are thus mutually incompatible), for breeding their F_1 hybrid progeny to give a second generation (F_2), and for mating the F_1 to either parent to produce a back-cross generation. The predictions include how many incompatibility classes (or groups of isogenic individuals) should be found in each generation, the relative frequency of individuals among these classes, and thus the expected success rate for graft retention after random transplantations between sibs or the mutual retention of reciprocal grafts. The last test is the most useful one, for any pair of sibs which are mutually compatible can be taken as identical for the allogenic loci. The predictions given in Tables 6.1 and 6.2 are readily tested by grafting skin or

Table 6.1 Predicted crosses for n loci each with two alleles (a, b) starting with homozygous parents P_1 (aa etc.) and P_2 (bb etc.).

Parents	Offspring Alleles at each locus	No. of classes	Proportion identical to P_1	P_2	F_1	Proportion of grafts retained by sibs Single	Reciprocal
$P_1 \times P_2$	F_1 all ab	1	0	0	1	1	1
$P_1 \times F_1$	B_2 aa or ab	2^n	$\frac{1}{2}^n$	0	$\frac{1}{2}^n$	$\frac{3}{4}^n$	$\frac{1}{2}^n$
$F_1 \times F_1$	F_2 aa, ab or bb	3^n	$\frac{1}{4}^n$	$\frac{1}{4}^n$	$\frac{1}{2}^n$	$\frac{5}{8}^n$	$\frac{3}{8}^n$

digits (Figure 6.2) and find support in the results of the few tests that have been carried out to date. One example of a completely independent test has been provided by Charlemagne and Tournefier (1974). They demonstrated a single gene difference affecting histocompatibility in *Pleurodeles waltl*, but their finding that 73% of random grafts were rejected in the original stock also suggests the presence of other alleles or at least three loci which govern antigenic differences in that species.

The full implications of graft rejection have emerged too recently to be considered in all but a handful of regeneration studies. In the circumstances, it seems most useful to consider the likely implications in terms of the following four generalizations which can be elaborated by means of examples.

(1) Homografts conducted on axolotls have used partially inbred stocks, yet graft rejection interfered with several of these studies.

Table 6.2 Number of incompatibility classes and frequency of graft tolerance among sibs, calculated from Table 6.1.

Loci n	B_2 (back-cross progeny) No. of classes	Percentage of grafts retained Single	Reciprocal	F_2 (2nd hybrid generation) No. of classes	Percentage of grafts retained Single	Reciprocal
0^*	1	100.0	100.0	1	100.0	100.0
1	2	75.0	50.0	3	62.5	37.5
2	4	56.3	25.0	9	39.1	14.1
3	8	42.2	12.5	27	24.4	5.3
4	16	31.6	6.3	81	15.3	2.0
5	32	23.7	3.1	243	9.5	0.7
6	64	17.8	1.6	729	6.0	0.3

*The expectation for isogenic sibs, the progeny of matings between any homozygotes.

(2) Embryonic homografts are expected to persist as a result of induced tolerance. Their persistence is an established fact.

(3) Similarly, homografts performed on larval stages may have induced tolerance and there might be little genetic disparity between larvae obtained from a single spawning.

(4) Homografts conducted on adult newts collected from the wild, however, must be expected to suffer frequent rejection.

The early homografts performed on adult axolotls were rarely considered in terms of the frequency or regularity of regeneration. They allowed simple conclusions based on successful cases, ignoring rejection or any other causes of failure in other specimens. Liosner and Vorontsova (1937), however, abandoned their attempt to graft limb muscle into the tails of other axolotls after obtaining only one *temporary* three-digit regenerate out of 35 cases. It is likely that the few anomalous cases of unpigmented regenerates resulting from implantation of genetically dark tissue into irradiated limb stumps (Skowron and Roguski, 1958; Trampusch, 1959; Wallace *et al.*, 1974) could be attributed to an incipient rejection. Lazard (1959, 1965, 1967) obtained infrequent and defective regenerates after grafting diverse tissues into irradiated legs of juvenile axolotls, and noted that about 10% of the regenerates regressed abruptly with symptoms of rejection. Glade (1963) identified invading leucocytes in regenerates containing homografted skin, and considered that only timely fixation had prevented their rejection. Judging from these reported incidents and De Both's (1970a) survey, most major axolotl colonies are still sufficiently heterogeneous for graft rejection to present a problem of the kind experienced most recently by Tank (1978a).

Figure 6.2. Finger grafts on axolotls. Grafts can be identified by their position and rejected ones leave gaps in the series.

Tissue grafts between embryos and young larvae do not permit such a consistent interpretation. Anton (1956, 1965) transplanted limb primordia between embryos of *T. cristatus, vulgaris* and *alpestris* in order to study their regeneration both in larvae and after metamorphosis. The survival of these heterografts strongly suggests an induced host tolerance. Heath (1953) exchanged the arms of *A. tigrinum* larvae with those of the slower growing *A. maculatum* and *Taricha torosa*. He recorded autonomous growth rates without commenting on rejection even of the xenografts. Liversage (1959) performed heterotopic limb homografts between larvae of several *Ambystoma* species. The limbs persisted up to 6 months, well beyond metamorphosis, and were able to regenerate. The regeneration of transplanted larval limbs or blastemata is well enough advanced in 3 or 4 weeks for an experiment to be terminated before any rejection has occurred. That may explain why Stocum (1975a) found no rejection in 25 days after performing blastemal homografts in young *A. maculatum*, and how Thornton and Tassava (1969) obtained regeneration of larval limbs transplanted from *A. maculatum* on to axolotls.

There have been some experiments on adult newts where graft rejection probably affected the results, but rarely enough to invalidate their interpretation. One example is Bischler's (1926) conclusion that skeletal homografts in *T. cristatus* only provided mechanical support without contributing cells to a blastema. Dedifferentiated cells liberated from the graft would have been most susceptible to destruction by an immune reaction. That surely applies to the cartilage xenografts employed by Desselle and Gontcharoff (1978) and casts doubt on their conclusions (see section 4.5). Thornton (1942) obtained remarkably few good regenerates, 10% plus an additional 10% defective ones, after transplanting muscle into irradiated limb stumps of adult *N. viridescens*. Although rejection very probably contributed to the majority of arrested stumps, this only strengthens Thornton's conclusion that graft-specific regeneration could occur. Homograft rejection may even have been an essential part of the mechanism whereby Bodemer produced supernumerary arms in *N. viridescens*. He found that a brachial nerve deflected to the pectoral muscles could stimulate this atypical form of regeneration either if the muscle was damaged (Bodemer, 1958), or if fragments of cartilage, muscle, nerve, liver or lung had been implanted there (Bodemer, 1959). Some of these implants must have been obtained from other specimens, and they were probably all homografts. Bodemer suggested the inflammation and later destruction of the implants could substitute for the mechanical trauma of his earlier experiments, without expressing any interest in the cause of inflammation. Ruben (1960) offered an immunological explanation of such results and showed that a genetic disparity between graft and host had a considerable influence on the formation of accessory limbs (Ruben and Stevens, 1963; cf. Carlson, 1967b; Carlson and Morgan, 1967).

Only two investigations have actually used the immunity response to distinguish between host and graft tissues during regeneration. Stinson

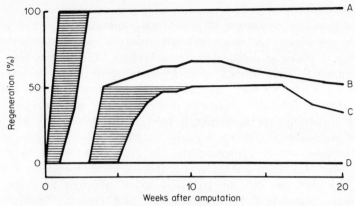

Figure 6.3. Cumulative incidence of regeneration in 5 cm axolotls (from Maden and Wallace, 1975). A, shielded control arms, with shading between cone and three-digit stages. B, C, cone and three-digit stages of irradiated arms containing normal cartilage grafts. Note the arrest and resorption of some of the experimental regenerates. D, irradiated control arms showing no regeneration.

(1964c) compared the regeneration of fore-arms which had been transplanted as autografts to the irradiated upper arms of *N. viridescens* to similar homografts. About half of the autografted arms produced stable regenerates, but the homografted ones only started to form blastemata which were quickly arrested and resorbed. Stinson argued that such a recognition of

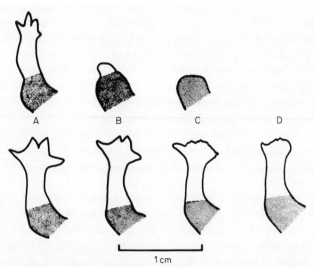

Figure 6.4. Rejection of regenerates formed by cartilage grafts in irradiated axolotl arms (from Maden and Wallace, 1975). A–D, camera lucida drawings at 2 week intervals, 18–24 weeks after amputation. The host arm is stippled.

alien tissue implied that any regenerate obtained under these experimental conditions must be composed of graft tissue: his autografts had actually formed the regenerates. After implanting cartilage homografts into irradiated arm stumps of young axolotls, Maden and Wallace (1975) found that half of the experimental arms grew regenerates but most of them were later destroyed by a characteristic immune reaction (Figure 6.3 and 6.4). We happily adopted Stinson's argument concerning the graft origin of these regenerates, although the epidermis is surely derived from the host and it would be impossible to exclude a minor contribution of host cells to other tissues. It might even be contended that the graft only formed the skeleton, and when that was destroyed the rest of the regenerate inevitably collapsed. In fact, the skeleton was the last part to be affected, persisting while surrounding musculature became inflamed and eroded. This operation certainly left irradiated host tissue occupying much of the amputation surface, yet its contribution to the regenerate was at most trivial. Both of these experiments thus showed the inability of irradiated cells to participate in regeneration, while the latter also demonstrated that a single pure tissue can produce a complete regenerated forearm.

6.3   Immune reactions in regeneration

Apart from the arguments raised about immunity during the growth of supernumerary arms (Ruben, 1960), there is little evidence of any direct connection between the immune response and normal regeneration. Brunst (1960) described an inflammatory reaction and tissue destruction by macrophages as symptoms of the regression which occurs in irradiated regenerates, and a similar but milder reaction intervenes between the phases of wound healing and dedifferentiation of all amputated limbs. This similarity first attracted my attention to the possible immunological basis of the following anomalous case of regeneration by irradiated limbs, and then led me to wonder if an immune response may not be a fundamental aspect of regeneration which has been almost entirely ignored.

Polezhaev (1972a, b) has written long accounts defending his view that regenerative capacity is a fundamental property of living things which therefore (if I understand his philosophical arguments) may be evoked experimentally even in postmetamorphic anurans and in all higher vertebrates. His views have not aroused much interest outside Russia and, in the absence of acceptable experimental support (see Polezhaev, 1946), I have little sympathy for them. As part of his general thesis, however, Polezhaev claimed he had restored a normal regenerative capacity to irradiated axolotl limbs by a series of injections. All the injections consisted of homogenates or extracts of alien tissues which must be expected to elicit an immune reaction in the recipients. Despite Polezhaev's denial, this aspect of the experiments provides a means of interpreting otherwise perplexing results. Both the experiments considered here (Tables 6.3 and 6.4) were performed on large juvenile axolotls, certainly old enough to be immunologically competent. Their hind limbs were exposed to 7 krad X-rays while the rest of the body was protected by a 4 mm lead shield, so their immune systems would not have been greatly damaged. The legs were amputated twice, first at 10 days and then 100 days after irradiation. Seven daily injections in 1 ml of physiological saline were applied to the right leg of each specimen starting on the day of the initial amputation. The limbs were allowed 3 months to regenerate, an adequate time to reach the criterion of 2–5 digits or to mount an immune response.

Two general features of the results deserve particular emphasis. Firstly, an appreciable proportion of uninjected control specimens managed to regenerate one irradiated leg after the initial amputation, and the later amputation provoked an equal or greater incidence of regeneration (Tables 6.3 and 6.4). Irradiation clearly did not inhibit regeneration completely, therefore, and there seemed to be a spontaneous recovery from its inhibition. These are surprising results, for such a drastic dose of X-rays should prevent regeneration permanently. The claimed stimulatory effects of the injections must consequently be measured against a particularly uncertain control baseline. Secondly, the uninjected legs regenerated just as frequently as the injected ones in experimental specimens: either leg might regenerate after the initial amputation and usually both did so at the second attempt. Whatever effect the injections had must therefore have been a systemic one, as an immune response would be. Since no difference could be detected between left and right legs, they are both included in the results to effectively double the number of systemic responses which could be recorded for each specimen. Subject to these qualifications, which might persuade the reader to ignore the results entirely, Table 6.3 reveals that homogenates of rat muscle and liver may have slightly stimulated regeneration after the first amputation

Table 6.3 Results of injecting homogenates of rat tissues into irradiated right legs of juvenile axolotls (data of Polezhaev and Ermakova, 1960; further details in text).

Material injected as homogenate	First amputation (10 days)		Second amputation (100 days)	
	Limbs*	Regenerates†	Limbs*	Regenerates†
None (control)	44	1 (2%)	40	2 (5%)
Fresh muscle	34	5 (15%)	34	30 (88%)
Boiled muscle	12	2 (17%)	12	12 (100%)
Frozen muscle	20	1 (5%)	20	18 (90%)
Fresh skin	20	1 (5%)	20	20 (100%)
Fresh liver	16	5 (31%)	14	14 (100%)
Fresh testis	12	2 (17%)	12	12 (100%)

*Both left and right limbs were irradiated and scored, as regeneration occurred equally often on either.
†Regenerates scored on a criterion of 2–5 digits; 15% or more was taken to show a restoration of regenerative capacity.

Table 6.4   Results of injecting homogenates or extracts of rat tissues into irradiated legs of juvenile axolotls (data of Polezhaev *et al.*, 1961; further details in text).

Material injected	First amputation (10 days)		Second amputation (100 days)	
	Limbs*	Regenerates†	Limbs*	Regenerates†
None (control)	30	4 (13%)	30	17 (57%)
RNA	20	6 (30%)	17	13 (77%)
Acid protein	18	4 (22%)	18	16 (89%)
Liver protein	20	4 (20%)	20	18 (90%)
Freeze-dried muscle	20	4 (20%)	20	15 (75%)
Freeze-dried spleen	20	3 (15%)	20	15 (75%)
Freeze-dried bone marrow	20	3 (15%)	20	12 (60%)
Liver nuclei	20	3 (15%)	20	12 (60%)
Liver nuclei again	20	2 (10%)	18	10 (55%)
DNA	20	3 (15%)	20	12 (60%)

*Injections into the right leg but both legs were irradiated and scored, as regeneration occurred equally often on either.
†2–5 digit regenerates; regenerative capacity was claimed to be restored most effectively by RNA after the first amputation and by proteins after the second—but neither treatment was consistently much more effective than no injection.

and all injections did so after re-amputation. The changed control values found in Table 6.4 might be ascribed to the use of a different X-ray machine, but the injections then produced no reliable differences from these new control values, apart from possible marginal stimulation by RNA and protein injections. Polezhaev *et al.* (1961) provided two intriguing details about this second experiment. They attempted to repeat their previous successful test with a rat liver homogenate, this time using a concentrated freeze-dried preparation, but nine of the ten recipients soon became hyperaemic, oedematous and died. Such mortality surely could be described as anaphylactic shock following an overdose of foreign protein. The investigators also noted that all injections protected the axolotls from ulcers which affected the irradiated legs of control specimens. The ulcers imply excessive superficial radiation damage, so the irradiated skin of the injected axolotls must have been either repaired or more probably replaced.

The poor reproducibility of the results among control or injected series precludes any firm interpretation embracing these and subsequent experiments (Polezhaev *et al.*, 1960–64). There are some indications that extracts of foreign tissue may enhance regeneration after a partial inhibition: perhaps any foreign protein (certainly not a tissue-specific one) could have a mild influence on some part of the process. By comparison, there is relatively strong evidence of an immune response which would have the observed delayed systemic effect on regeneration. An immunological activation of leucocytes might hasten the destruction of damaged and irradiated cells, thus encouraging their replacement by healthy ones. The control series suggest the presence of some viable cells in these irradiated legs, but severe inflammation might stimulate cellular migration from the shielded base of the limb in the same way that mechanical damage seems to do (Tuchkova, 1966, 1976).

The sensitive nature of the immune response supplies an argument in favour of the speculation that it might be implicated in normal regeneration. The ability to detect a diverse array of novel molecules as antigens and to respond to them has obviously not evolved by selection to prevent allografts, and seems surprisingly refined to combat infectious diseases. The concept of immunological surveillance involves the ability to recognize and destroy any cells whose secretions or surfaces betray them as abnormal for that individual, whether they are genetically different or accidentally distorted. Presumably cancer cells survive because they have not changed externally beyond limits tolerated by the immune system. Severely damaged and defective cells resulting from either amputation or irradiation, however, often are destroyed by leucocytes. If these leucocytes are sensitized by an immune reaction, it could account for several poorly authenticated observations (some cited by Needham, 1942) of relatively rapid regeneration occurring after a second amputation or the regeneration of one limb being accelerated by amputating another. Perhaps the possibility of stimulating regenerative growth in its early phases by blastemal autografts, homogenates, extracts, or even RNA was too casually dismissed in the previous chapter, for they might affect regeneration by provoking an enhanced immune response. Conversely, one may ponder whether the immune system achieved by anurans at metamorphosis and even earlier in amniotes may not be too sophisticated to allow the perfect limb regeneration displayed by urodeles.

6.4 Conclusions

Genetic aspects of regeneration have been tragically neglected. This chapter advertises the potential usefulness of several amphibian mutants, in an attempt to rectify such short-sightedness and in the hope than other mutants will be speedily discovered and exploited. The genetic basis of amphibian immunity systems is briefly explored. This system develops during larval life and enables amphibians to destroy allografts. Graft rejection occurs after a considerable delay, however, which is often sufficient for the grafted tissue to participate in regeneration. The delay and the tolerance induced by grafting on to embryos or young larvae perhaps account for the belated recognition that tissue incompatibility can interfere with regeneration studies. Fortunately, graft rejection is unlikely to have invalidated the conclusions derived from most of these studies, and it has been used to advantage in two of them. An immune response may even be involved regularly in normal regeneration, during the destructive phase between wound healing and dedifferentiation. If so, it is likely to be accentuated by virtually any operations, injections of foreign material, or irradiation.

Chapter 7

The Mechanism of Regeneration

The mechanism of regeneration can be analysed most simply by considering in turn successive stages of the process: the origin of the blastema, its growth and redifferentiation, and the kinds of controls which may be exerted to regulate its morphogenesis even where it is uncertain at what stage these controls are applied. One aspect or another of this analysis has usually supplied the motivation for the descriptions and experiments set out in the preceding chapters and several of these studies will be mentioned again where they are crucial to this analysis.

7.1 Origin of blastemal cells

Both descriptive studies and experiments using localized irradiation concur in showing that internal tissues within 1–2 mm of the amputation surface liberate dedifferentiated cells which accumulate to form a blastema. There is general agreement nowadays that all mesodermal tissues can contribute cells to the blastema in this way, and some evidence that ectomesenchymal Schwann cells also enter the blastema (detailed by Maden, 1977a). It is a majority opinion that epidermal cells either do not normally enter the blastema or do not transform into mesodermal cells there, and perhaps they are incapable of such a transformation under any circumstances. This is a particular case of the general argument about the degree of tissue-specific determination which cells may retain when they have lost their differentiated characteristics, an argument epitomized in the phrase 'dedifferentiation or modulation'. Although blastemal mesenchyme cells show no trace of differentiation even when subjected to discriminating tests by electron microscopy (Hay, 1959, 1962) or for tissue-specific molecules (DeHaan, 1956; Laufer, 1959; Linsenmayer and Smith, 1976), dedifferentiation here means the reversion to an undetermined state which allows a variety of subsequent pathways for redifferentiation. Modulation, in contrast, implies that the apparently undifferentiated cells remain completely determined and can only revert to the same type of tissue from which they originated when conditions permit, as seems to occur commonly in short-term tissue cultures. As all mesodermal tissues of the limb stump contribute cells to the blastema, it is obviously quite possible that each cell and its descendants could retain an original tissue-specific determination and eventually revert to the same type of tissue. There is also a further possibility: the observed dedifferentiation of specialized cells might merely indicate their degeneration, while the blastema really originated from undifferentiated reserve cells hidden in the connective tissues. These three possibilities cannot be distinguished by direct observation of normal regeneration and, in fact, the distinction has stubbornly resisted experimental investigations.

The experiments devised to examine tissue specificity during regeneration have all involved grafting some tissue, whose cells could be identified subsequently, into a limb either before or soon after the latter was amputated. Several means of distinguishing between graft and host cells are available but all have limitations. Grafts of haploid or polyploid tissue allow some of the descendant cells to be recognized by nuclear size or by the number of nucleoli (Hertwig, 1927; Hay, 1952). Although this appears to be the most discriminating technique used, the majority of cells in a regenerate cannot be distinguished and occasional patches of polyploid cells can arise spontaneously in host tissues to give false positive results. The polyploid tissues are necessarily homografts and so are liable to be rejected. Radioactive labelling of either the graft or the host DNA with tritiated thymidine has been used repeatedly and with some success, but requires considerable caution in execution and interpretation. The labelling always becomes diluted during successive divisions so that cells are only marked temporarily and for an uncertain number of generations, depending on both background label and the original non-uniform incorporation of the tracer. Excess labelling causes radiation damage which reduces viability and cell division of the marked cells, and that may apply even to normal labelling. Grafts of different genotype, recognised in terms of the colour differences between dark and white axolotls or by histocompatibility differences, only identify the bulk of the regenerate without identifying individual cells. The limitations of all these techniques usually restrict the number of cells whose origin can be identified in the regenerate, to a degree which depends on the relative contributions of host and graft to the blastema. Small grafts intended to cause little departure from the normal conditions of regeneration cannot be expected to produce more than a trivial proportion of the regenerate and their detection is correspondingly difficult. If the host arm has been irradiated, however, the graft must provide all the mesenchymal cells of the blastema and whatever tissues they form by redifferentiation. The former controversy over this point has already been examined in great detail in order to arrive at the firm conclusion that irradiation acts on a cellular basis: adequately irradiated cells cannot proliferate to the extent demanded during regeneration and cannot be rescued by adjacent cells of a healthy graft (Maden and Wallace, 1976). It should be perfectly feasible, therefore, to graft some pure tissue into an irradiated limb stump and to examine the regenerate for the presence of other tissues, thus testing for the occurrence of histological transformation or metaplasia. In formal terms, that would be equivalent to proving the observed dedifferentiation entailed a loss of determination (or increase of potency) required by the theoretical concept of dedifferentiation but not of modulation.

The vast majority of experiments devoted to this topic have failed to achieve conclusive results. Those performed without irradiation yielded regenerates which consisted principally of host tissues. Those employing impure grafts could not exclude the possibility that cryptic reserve cells were responsible for any regeneration which occurred. Skeletal cartilage can be prepared free of perichondrium and its adhering connective tissue, however, to provide a graft of pure differentiated chondrocytes. When such a graft has been obtained from the limb skeleton it should be capable of participating in regeneration. Only four studies have employed pure cartilage grafts in irradiated limbs and these are described below in some detail, as the ones most likely to reveal the occurrence of metaplasia.

Eggert (1966) irradiated one arm of adult *N. viridescens* with 1.5–2 krad X-rays and amputated it above the elbow. He then implanted a pure fragment of scapular cartilage from the same specimen into the arm stump. Breakdown of the graft led to the formation of a blastema which only redifferentiated into irregular masses of cartilage. That result can be interpreted quite reasonably as the modulation of permanently determined chondrocytes, but it could also be explained by the inability of those particular grafts or conditions to support regeneration. Steen (1968) has since stated that scapular grafts show only a limited degree of dedifferentiation in axolotls. Furthermore, the grafts were obtained from a shielded region but were mildly irradiated and hence would have released a subnormal amount of viable cells, perhaps not enough to establish an adequate blastema.

Namenwirth (1974) exposed the entire pelvic region including both hind limbs of juvenile 10–18 cm axolotls to 3 krad X-rays, and then replaced the distal end of the femur with a homograft obtained from the humeral or femoral epiphysis of an unirradiated triploid specimen. The surface of the grafted epiphysis had been trimmed off until only a core of cartilage remained. She amputated the limbs 1–14 days later and allowed them to regenerate for 2–6 months. Out of 45 such operations, 25% regenerated two or more digits and another 50% produced spikey outgrowths. Eleven cases, presumably the best 25%, were sectioned and examined for the presence of triploid cells by nucleolar counts. Triploid cells were detected in the cartilage, perichondrium and connective tissue of the joints and dermis (Table 7.1). Only one of the regenerates contained any muscle, so Namenwirth cautiously refused to place much reliance on the few triploid

Table 7.1 Percentage of cells showing three nucleoli in 10 μm sections of different tissues from limb regenerates of axolotls (data of Namenwirth, 1974).

Tissue examined in regenerate	Origin of regenerate			
	Triploid (3N) control limb	Irradiated diploid (2N) limb with graft of		
		3N cartilage	3N muscle	3N cartilage + 2N muscle
Cartilage	20	16	13	15
Perichondrium	16	8	10	3
Connective tissue:				
(1) joints	16	10	13	8
(2) fibroblasts	18	10	13	3
Muscle:				
(1) fibroblasts	40	—*	28	9
(2) myoblasts	34	—*	28	0.8
Epidermis	21	0	0	0

*Little or no muscle was detected in these regenerates.

muscle cells it displayed. In an attempt to stimulate the formation of muscle, Namenwirth grafted a mixture of triploid cartilage and diploid muscle into the irradiated host legs. She identified a few triploid muscle cells in the regenerates from these operations, but again discounted them as amounting to less than 1% of the scored myoblasts and possibly representing a misclassified minority of the triploid fibroblasts present in the same tissue. The danger of ignoring positive results here seems as great as that of misclassification, although only the investigator can really assess the difficulties encountered. Leaving aside muscle as an undemonstrated possibility, these results eliminate any doubt about the ability of chondrocytes to dedifferentiate and transform into perichondrium and connective tissue. Histological transformation certainly can occur during regeneration, therefore, but perhaps only to a limited extent. Namenwirth explored these limits by grafting other triploid tissues into irradiated axolotl legs. Muscle grafts contributed to all internal tissues of the regenerates and probably served as the exclusive source of these tissues. Muscle grafts inevitably contain connective tissue fibroblasts, however, so the occurrence of graft-derived triploid cartilage could still be attributed to the activation of reserve cells. Irradiated legs covered by unirradiated epidermis failed to regenerate, as did irradiated control legs lacking any graft, confirming that grafts were responsible for the regeneration which occurred in the other series. On that basis, epidermal cells either cannot transform into mesodermal tissue, or superficial cells do not penetrate into the blastema. Also on the same basis, any regenerated muscle must be suspected of having a graft origin. Some of these results are shown in Table 7.1, which reveals the low efficiency of detecting triploid cells. Only 16–40% of the cells in

136

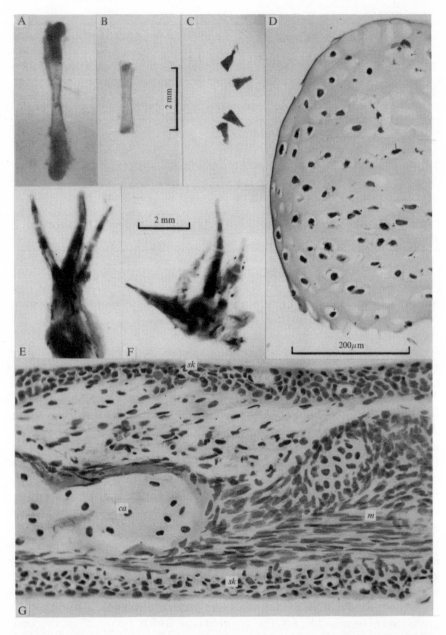

Figure 7.1. Procedure and results of grafting cartilage (from Wallace *et al.*, 1974). Extirpation of a humerus (A) and snapping off both epiphyses leaves a relatively clean diaphysis (B). The perichondrium is peeled off the latter, leaving cones of pure cartilage (C, D). The regenerates obtained by inserting such cones into irradiated arm stumps are defective (E, F) but contain both cartilage (ca) and muscle (m) in their digits (G).

different tissues of completely triploid control regenerates showed all three nucleoli in one section. Much the same percentage of cells in the experimental regenerates could be identified as triploid in the same way in the series where exclusively triploid grafts were employed, surely demonstrating that all the regenerate apart from its epidermis had been derived from the graft.

My colleagues and I concluded that chondrocytes can transform into all other internal tissues, including muscle, on the basis of similar experiments on younger axolotls (Wallace *et al.*, 1974). We peeled the perichondrium away from humeral diaphyses and grafted the pure cartilage remnants into previously irradiated (2 krad) forearms which we amputated 2 days later just distal to the graft. Figure 7.1 illustrates some aspects of this technique and some of the better results obtained from it. One series of operations on 7–8 cm specimens yielded little over 50% of spikey outgrowths, but a repetition on smaller specimens yielded small regenerates with genuine digits in the majority of cases (Table 7.2). Several of these regenerates contained strips of muscle which must have been derived from the grafted chondrocytes, as the irradiated host limbs were incapable of providing blastemal cells. Admittedly, that conclusion could be reached more elegantly by using some positive identification of the grafted cells: the colour marker used in these experiments failed to operate, probably because the grafts were partly incompatible. We only became aware of that explanation following yet another series of operations, in which pieces of pure cartilage were inserted into the tips of amputated irradiated arm stumps (Maden and Wallace, 1975). As described in the last chapter (Figures 6.3–4), 50% of these specimens yielded early regenerates which were destroyed by the host immune system. Such rejection could only occur if the bulk of the regenerate were recognized as foreign tissue by the host, which confirms our previous conclusion concerning its origin from the graft.

The frequency and normality of the regenerates obtained in these four experiments declines noticeably with the age of the specimens (Table 7.2). This provides a ready explanation of the frequent failures of regeneration

Table 7.2 Degree of regeneration achieved by irradiated arm stumps containing pure cartilage grafts.

Specimens			Percentage in each category			
Age/Size	Species	No.	Stump	Spike	Hand	Reference
Adult	*N. viridescens*	47	96	2	2	Eggert (1966)
10–18 cm	*A. mexicanum*	45	25	50	25	Namenwirth (1974)
7–8 cm	*A. mexicanum*	20	45	55	0	Wallace *et al.* (1974)
5–6 cm	*A. mexicanum*	20	20	10	70	Wallace *et al.* (1974)
5–6 cm	*A. mexicanum*	30	35	15	50	Maden and Wallace (1975)

and a means of avoiding the conclusion which many have drawn from Eggert's (1966) results that chondrocytes can only modulate. It is even possible that the age of the host has more influence than that of the graft, but it seems more likely that the ability to dedifferentiate, proliferate and transform into other cell-types is best expressed by the chondrocytes of growing limbs and declines with age as the skeleton is converted into bone. These experiments are the culmination of a large number of studies performed during the last half-century which, when considered together, suggest that well-differentiated cells retain at least a considerable metaplastic flexibility. There are still indications that their potency may not be complete, readily expressed, or apply to all tissues. Several other studies, for instance, support the conclusion reached by Namenwirth (1974) in finding no conversion of epidermis into internal tissues or vice versa. Desselle and Gontcharoff (1978) confirmed this point and also detected some transformation of xenograft chondrocytes into muscle, although they considered irradiated host cells were also reactivated.

Since the blastema consists of dedifferentiated cells with at least some metaplastic flexibility, it is reasonable to enquire if metaplasia occurs regularly during the normal process of regeneration. That would provide one mechanism whereby the blastema could regulate its form to correspond to that of the limb stump. Alternatively, mesenchymal cells are probably able to move past each other and might aggregate into groups with a mutual affinity in order to produce a normal arrangement of tissues. This question can be approached by tracing cells of a small graft which should make only a minor contribution to the blastema produced by an unirradiated limb. Pure cartilage is also the most suitable graft material here, as it cannot contain any reserve cells. The following two studies come closest to meeting the criteria for obtaining a clear conclusion. Patrick and Briggs (1964) replaced the distal end of the humerus in diploid axolotls with cartilage obtained from a triploid humerus. The arms were amputated at the elbow 5–7 days later and allowed to regenerate until digits had formed after 35–40 days. The majority of the regenerates (28 out of 39) revealed no triploid cells according to counts of nucleoli in sections. Four additional cases retained some of the grafted cartilage in the upper arm stump, but again no triploid cells could be detected in regenerated tissue. The remaining seven regenerates contained abundant triploid cartilage cells, but so few in the muscle that they could have arisen as spontaneously polyploid host cells like some that were noticed in the epidermis. Thus grafts which were estimated to provide up to 20% of the regenerated cartilage did not yield any clear evidence of transformation into other tissues.

Steen (1968) grafted pure cartilage marked by triploidy or radioactivity, or sometimes by both markers, into amputated axolotl arms, and analysed the graft's contribution to the regenerate by counting nucleoli and autoradiography. He could only confidently identify the appreciable numbers of marked cartilage cells by these criteria, although he encountered a few labelled muscle cells as well. Steen (1970b) has since emphasized that neither his own study nor any previous ones had demonstrated a metaplastic change of cartilage cells. That is a perfectly accurate admission: the experiments were intended to reveal metaplasia but failed because they were not adequately designed for the purpose. The labelled grafts occupied the normal position of cartilage in the limb stump. They would be expected to liberate cells in the centre of the blastema, and central cells are likely to redifferentiate into cartilage. It is rather improbable, therefore, that genuinely undetermined graft cells would form any other tissue when surrounded by an excess of host mesenchyme. When considered in terms of normal regeneration, however, this argument tends to reinforce Steen's other conclusions. Each tissue of the stump probably provides the bulk of the corresponding tissue in the regenerate, if only because that entails a minimum amount of cellular displacement.

While convincing evidence now exists of histological transformation in at least one limb tissue, and there is plausible evidence that modulation provides an apt description of the events in normal regeneration, neither of these conclusions can eliminate the possible existence of undetermined reserve cells. There is no evidence that they do exist either: they remain as a rather annoying hypothesis, a hangover from the early inconclusive arguments about this subject. When a limb is amputated in a region from which the skeleton has been extirpated, a normal skeleton still forms in the ensuing regenerate and sometimes extends back into the old limb stump for a short distance. Bischler (1926) provided a particularly thorough analysis of this situation in *T. cristatus* in addition to reviewing similar studies conducted over the preceding 60 years which we can conveniently ignore. Thornton (1938b) and Goss (1956a) have performed the same operation on *N. viridescens*, while more recent experiments of Steen (1968) and Foret (1970) achieved much the same end by inserting labelled muscle into the cavity formed by extirpation of skeleton from the limbs of *Ambystoma* larvae. All these investigations demonstrate that non-skeletal tissues can provide blastemal cells which redifferentiate into cartilage and perichondrium. The labelled implants reveal that such pluripotent cells are present in muscle. Such results cannot be interpreted strictly as modulation, but are often regarded as a 'minor transformation' of fibroblasts into chondrocytes with both included as variants of connective tissue. Adherents of this curious view admit that myoblasts are so different from chondrocytes that their interconversion would be a major transformation. Even a minor transformation presented a considerable dilemma to embryologists who had long accepted the convention that cellular differentiation entailed a progressive loss of potency and was thus irreversible (discussed by Grobstein, 1959). This theoretical obstacle was neatly hurdled by Weiss (1939), who proposed that connective tissue might contain reserve cells which remained undifferentiated and thus retained the capacity to differentiate into any mesodermal tissue of the limb when required to do so. The postulated reserve cells or neoblasts were almost universally accepted as the basis of regeneration in invertebrates, until a direct transformation of differentiated cells was demonstrated in *Hydra* (Normandin, 1960; Burnett, 1968) and shown to be quite likely in planarians (Hay, 1968). Students of amphibian limb regeneration have never displayed much enthusiasm for reserve cells, understandably enough, as the observed dedifferentiation of tissues suggests a continuity between differentiated cells of the stump and of the regenerate. Furthermore, the classical case of Wolffian regeneration, where the dorsal margin of the iris in some newts can regenerate a replacement for the extirpated lens, has always seemed to be a clear case of histological transformation, even though considerable finesse was required to finally establish the point (Yamada and McDevitt, 1974). Similarly, the few experiments which employed pure cartilage in irradiated axolotl limbs have only demonstrated that the reserve cell theory is inapplicable in this particular case.

There are several other pertinent studies involving radioactively labelled tissues grafted into regenerating limbs. Although these grafts contained several cell-types and thus are unable to fully test the concept of dedifferentiation or to exclude the reserve cell theory, they do provide a measure of the lability of grafted cells besides indicating some differences of interpretation. Oberpriller (1967) labelled 15–30% of the blastemal cells growing on amputated tails and intestines of *N. viridescens*. He inserted pieces of the labelled mesenchyme into 14 day limb blastemata, a stage prior to redifferentiation, and followed the graft cells through 3–4 divisions into condensing skeletal cartilage. As Oberpriller pointed out, intestinal blastemata have not originated from cartilage and would not form it normally. The ability of such grafted cells to be assimilated into limb cartilage can only be described as a capacity for metaplasia. Steen (1970b) criticized this experiment on the grounds that host cells could have become labelled by incorporating radioactive breakdown products from dying graft cells. That is highly improbable, for radioactive nucleotides would be quickly diluted to undetectable levels. Radioactive muscle (Foret, 1970) and radioactive triploid grafts of muscle or tail-fin connective tissue (Steen, 1968, 1970a) can also contribute cells to all internal tissues of a regenerating limb. Tuchkova (1973b) has repeated Oberpriller's experiment and confirmed his results.

A considerable number of investigations have involved grafting composite tissues into irradiated host limbs close to the site of amputation. With our present knowledge that such grafts form the overwhelmingly major or exclusive source of the regenerate's internal tissues, identification of the grafted cells becomes unnecessary. Several of the studies have used marked grafts, however, and provide further support for that conclusion. Triploid grafts of skin or muscle (Namenwirth, 1974; Dunis & Namenwirth, 1977, Table 4.3) besides grafts of skin or nerves which produced donor-coloured regenerates (Umanski, 1938a; Wallace and Wallace, 1973) demonstrate the point in axolotls. Unmarked grafts of limb skin, skeleton or muscle have each produced fairly complete regenerates on irradiated limbs of axolotls (Umanski, 1937; Trampusch, 1951), as have muscle grafts in *N. viridescens* (Thornton, 1942) and skeletal implants in *T. cristatus* (Desselle, 1968). A variety of internal organs have also been recorded as promoting some degree of regeneration when implanted into irradiated axolotl limb stumps, including lung, liver and heart (Siderova, 1951), testis, spleen and embryonic tissue (Lazard, 1967). As might be expected, the variety of tissues employed in these studies hardly exceeds the variety of conclusions drawn from them. The experiments provide an accumulation of cases where regeneration of a complete set of limb tissues cannot be explained merely in terms of a modulation of pre-existing cell-types. That leaves available the alternative explanations of metaplasia or reserve cells. Since a blastema arises by the dedifferentiation of tissues adjacent to the site of amputation, and grafted cartilage cells are now known to enter a blastema and later emerge in all other internal regenerated tissues, their histological transformation evinces all the specific properties which have been postulated for reserve cells. Consequently, there is no way of establishing the existence of reserve cells in any tissue, nor any prospect of disproving their existence in most tissues: the reserve cell theory must remain an idle speculation. By this process of elimination, it seems most likely that dedifferentiation results in a homogeneous population of undetermined blastemal cells capable of transforming into other types of tissue even if they retain some tendency to revert to the tissue from which they arose.

7.2 Clonal modulation

The preceding section assumed that blastemal mesenchyme redifferentiates into all tissues of the regenerate except for its epidermis. The majority of descriptions and experiments support that view but there have been dissenting opinions, according to which the blastema contains several discrete populations of undifferentiated but determined cells or that some stump elements enter the blastema without even losing their differentiated characteristics. Schmidt (1968, p. 37) quotes with evident approval his own earlier statement that striated muscle is irreversibly determined and at no time loses its genetic identity. Holtzer seems to have reached the more extreme position, based on his studies of tail regeneration (Holtzer, 1959; discussion following Hay, 1970), in believing that myoblasts do not even contribute cells to the blastema but grow into the regenerate independently. Others have described the formation of the blastemal circulatory system as a growth of stump vessels rather than by an assimilation of blastemal cells into new vessels. Schwann cells have been observed moving along regrowing axons and thus might act as the exclusive source of new myelin sheaths.

The argument for an independent regeneration of muscle has been completely refuted. Striated muscle sometimes differentiates from the mesenchyme of a blastema cultured in isolation from the cone stage, or transplanted to the fin some distance from the myotomes (Stocum, 1968a, b), results which require a modulation of myoblasts as a minimum explanation. Chondrocytes can transform into myoblasts when all neighbouring tissues have been thoroughly irradiated (Wallace *et al.*, 1974). The independence of blood vessels and nerve sheaths is less easily tested, especially as the blastema rarely grows normally unless it gains a blood supply and innervation. In one case at least blood vessels seem to differentiate from blastemal cells. Regenerates obtained after grafting brachial nerves from dark axolotls into irradiated arms of white axolotls show a typical dark coloration. A major part of that donor coloration can be traced to the affinity of melanocytes for blood vessels which carry the wild-type gene D (Figure 7.2). These blood vessels must have arisen from graft cells and not from the host's circulatory system. Both descriptions and indirect experimental evidence indicate that Schwann cells dedifferentiate and enter the blastemal mesenchyme (Chalkley, 1954; Wallace, 1972; Maden, 1977a). It is likely that they retain the ability to redifferentiate as a neurilemma around newly grown axons, and quite possible that any other blastemal cell could also do so. Yet a migration of Schwann cells from higher up the limb would serve as an equally plausible source of the new neurilemma, so perhaps remyelination stems from a mixed population of cells. There is a means of distinguishing between these possibilities, as the experiments cited above can be adapted to achieve a selective destruction of Schwann cells in the entire limb, a technique which has not yet been exploited.

As explained previously, classical analyses of development portrayed a totipotent fertilized egg dividing into blastomeres which segregate cytoplasmic factors and influence each other, so that a gradual process of determination or chemodifferentiation culminates in the visibly different cells of each tissue. When tested in terms of isolated embryonic regions or novel combinations of such regions, the process can often be separated into discrete steps such as inductions. At each step, a new determination to form one tissue inevitably includes the loss of potency to form another. Hence progressive determination necessarily involves at least a temporary limitation of potency. Although regeneration provides an independent test of this concept, it seems only reasonable to adopt the provisional hypothesis that tissues obey the rules according to which they developed (and that we understand those rules to some extent). There was considerable justification, therefore, in accepting the view that blastemal cells might modulate for the purpose of proliferating, or even that reserve cells might be set aside as a precaution against the need to regenerate. We can also see a logical progression as increasing knowledge forced us to abandon these preconceptions. The initial concept of tissue-specific modulation has been greatly weakened by the experiments described here. Its essence is retained by some who claim there are only three types of tissue in the limb (epidermis, contractile muscle and connective tissue) each of which contains

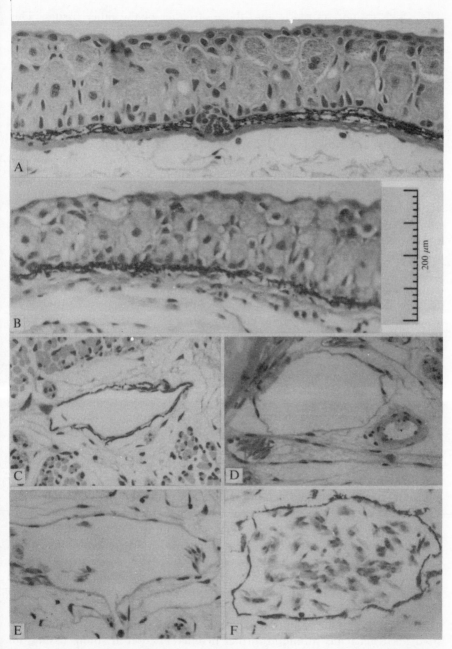

Figure 7.2. Pigmentation of regenerates obtained from nerve grafts in irradiated arm stumps (from Wallace and Wallace, 1973). The dermal pigment layer in a regenerate from a dark donor nerve grafted into a white host (A) is identical to that in a normal dark regenerate (B). Pigment also surrounds blood vessels in corresponding cases (C, F), but not in white control regenerates (D) nor in regenerates formed from a white donor grafted into a dark host (E).

cells which cannot transform into those of either of the other tissues. This view accepts the evidence that fibroblasts can be converted into chondrocytes during regeneration, and perhaps allows the reverse process, but regards such a conversion as being only a minor transformation. It strikes me as an illogical view, for connective tissue includes several well-characterized types of cell which are apparently stable end-products of differentiation and no less unique than muscle fibres, neurons and erythrocytes. In any event, this view cannot be maintained now that chondrocytes have been demonstrated to transform into myoblasts. These cells must have undergone histological transformation or metaplasia, by means of dedifferentiation accompanied by loss of determination, as they passed through the blastema. An analogous conclusion has been reached several years ago from nuclear transplantation studies: the specialized synthetic activity of a nucleus is effaced as it begins development again in an egg (Gurdon, 1974). Even though transplanted nuclei show a wider range of differentiated products, it is even more striking to find the same kind of flexibility displayed by intact cells. One consequence of this view is to concede that epidermis might be capable of transformation into mesodermal tissues, if these cells could be forced to enter the blastema under suitable conditions. The actual occurrence of such a transformation has only been claimed on contestable evidence (Rose, 1948; Rose and Rose, 1965, 1967) but has mainly been disputed because of prejudice.

7.3 Blastemal growth and regulation

Mitosis first occurs as cells dedifferentiate and escape from stump tissues (David, 1934) and it continues among the mesenchyme cells which accumulate under the wound epidermis. According to the estimates of cell numbers and mitotic indices shown in Figures 1.3–4, continued dedifferentiation releases additional cells to augment those already present in the early blastema and most dividing cells occur in the stump until the blastema has grown considerably. This assessment of initial growth by cellular proliferation contradicts earlier opinions: Polezhaev (1936, 1972) and Wolsky (1974) supposed a blastema could form without cell division, perhaps based on their observations of regenerating tadpole limb buds which may be capable of morphallactic rearrangements, while others have claimed the existence of amitotic cell divisions. These early opinions can probably be disregarded as they were largely formulated on negative evidence, the inability to detect a reasonable number of dividing cells which can be demonstrated by more refined methods. It is quite likely that a moderately sized blastema tends to inhibit further dedifferentiation in stump tissues, for a blastema grafted onto the amputated end of a denervated larval arm prevents it from regressing (Figure 7.3). The presence of a blastema is also believed to impede the regression of irradiated limbs (Butler and Puckett, 1940). At about the time when a moderately sized blastema has formed, redifferentiation begins in the stump and progresses distally into the base of

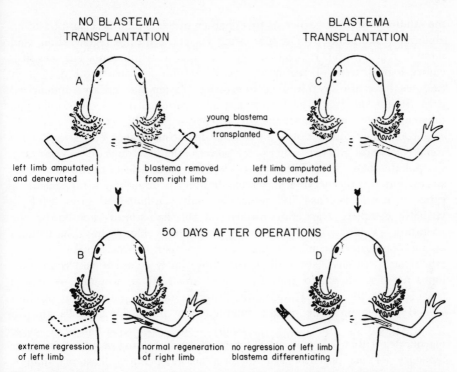

NO BLASTEMA
TRANSPLANTATION

BLASTEMA
TRANSPLANTATION

A

C

young blastema
transplanted

left limb amputated
and denervated

blastema removed
from right limb

left limb amputated
and denervated

50 DAYS AFTER OPERATIONS

B

D

extreme regression
of left limb

normal regeneration
of right limb

no regression of left limb
blastema differentiating

Figure 7.3. Cartoon version of their experiment from Schotté, Butler and Hood (1941). A larval left arm which is amputated and denervated is completely resorbed (A, B). Transplanting a blastema on to such a denervated limb stump prevents its resorption (C, D), although the blastema only forms a defective regenerate. (Reproduced by permission of The Society for Experimental Biology & Medicine.)

the blastema which must then support its own further growth by cellular proliferation. These features led Faber (1960) to postulate the existence of an apical proliferation centre, to which he attached considerable theoretical importance. Perhaps it is only a matter of emphasis or terminology, but no investigation has yet revealed a distinct proliferation centre. Cell division occurs throughout the blastema including the redifferentiating tissues of the regenerate, although the frequency of division may be reduced in the latter region (Chalkley, 1954; Maden, 1976; Smith and Crawley, 1977).

The two widely used inhibitors of regeneration, denervation and irradiation, both seem to act by interfering with cell division. Both allow wound healing and dedifferentiation to occur normally, and the latter may continue for a prolonged period without any significant accumulation of mesenchymal cells. It seems likely that hypophysectomy has a similar but far less effective action on blastemal cell division and growth (Sato and Inoue, 1973).

There is no doubt that the cells which compose a blastema must interact with each other in order to form a structurally normal regenerate. Several of the studies described earlier in this chapter provide suitable examples of an abnormal juxtaposition of tissues being corrected during regeneration, and normal regeneration is itself a form of defect regulation. The classical embryological tests of regulation, compensation for structural defects or excesses, have also been applied to growing blastemata and are summarized here. Bryant and Iten (1974) systematically examined the effects of removing different halves of a blastema or regenerate on the arms of adult *N. viridescens*. They found dorsal or ventral half cone stages could each form complete hands, whereas anterior or posterior halves could do so even after the palette stage. The ability to regulate for such defects declined at later stages and the regenerated skeletons showed an abnormal arrangement or extra elements, because the remaining half continued to grow while a virtually complete regenerate formed on the half-stump exposed by the operation (Table 7.3). The growth of complete hands from half-stump amputation surfaces confirmed an earlier demonstration performed by Weiss (1926) on *T. cristatus* (see section 8.6). Since fairly complete supernumerary limbs are also known to grow from quite small lateral wounds, we should consider that the normal amputation surface has the potential to produce several blastemata. The mutual interaction of mesenchymal cells and an apical epidermal cap, however, usually results in regulation to a single regenerate. Table 7.3 may be interpreted as indicating a rather stronger tendency for the posterior side to form an independent blastema than is evinced by the anterior side of the arm. Curiously enough, a similar posterior dominance has appeared in results of skin grafts (Lheureux, 1975a) and transplanted limb buds (Harrison, 1918; Swett, 1932). Since this feature could be related to an axial determination or the posterior polarizing zone of developing limb buds, it deserves further investigation in the context of experiments described in the following chapter.

Regulation of excessive blastemal material is less completely documented. Several blastemata can coalesce into a single one when implanted into a confined wound (Polezhaev, 1936; de Both, 1970b; Michael and Faber,

Table 7.3 Skeletal regulation in regenerates formed from half blastemata (data of Bryant and Iten. 1974: reproduced by permission of Springer-Verlag).

Stage when operated	Anterior half removed				Posterior half removed			
	Total	Normal	Extra elements	Deformed	Total	Normal	Extra elements	Deformed
Blastema	17	17	0	0	16	16	0	0
Cone	5	5	0	0	6	6	0	0
Palette	3	3	0	0	2	2	0	0
3-digit	6	3	0	3	6	2	2	2
4-digit	6	2	2	2	7	0	7	0
Mature limb	6	0	6	0	6	0	4	2

1971), but the initial excess there barely compensates for the drastic destruction of cells which occurs in most implants. A small lateral blastema may sometimes fuse with an apical one to form a unified regenerate (Guyénot *et al.*, 1948; Thornton and Thornton, 1965; Wallace *et al.*, 1974). The insertion of an extra skeletal element prior to amputation often results in the appearance of a corresponding extra cartilage in the proximal portion of the regenerate, as shown by Goss (1956a) for *N. viridescens* arms containing an extra or transposed ulna, but distal regions of the regenerate show the normal skeletal pattern and are hence considered to have regulated. Skeletal defects in the limb stump are usually corrected throughout the regenerate, as demonstrated repeatedly for both larval and adult urodeles (Bischler, 1926; Thornton, 1938b; Goss, 1956a; Foret, 1970). Defective limbs and tails show a mild improvement on average after repeated cycles of amputation and regeneration (Swett, 1924; Newth, 1958), although initially normal limbs tend to reveal increasingly frequent trivial defects after repeated amputation (Dearlove and Dresden, 1976).

The occurrence of defective regenerates emphasizes the limitation in their ability to regulate and points to the mechanism involved in regulation. Only young blastemata seem capable of regulating directly; palette stages and older regenerates can only do so when partially resorbed or showing a secondary dedifferentiation as described in the next chapter. Regulation thus seems to be a feature of undifferentiated proliferating cells. Similarly, little regulation occurs in blastemata grafted to a site which curtails further growth (Stocum, 1968b; Stocum and Dearlove, 1972; Michael and Faber, 1971). Hypomorphic regenerates, for instance, can often be attributed to conditions which restrict growth without preventing redifferentiation, as when an established blastema is irradiated or denervated (Butler and Schotté, 1949; Schotté and Liversage, 1959; Singer and Craven, 1948). Although a blastema can assimilate small amounts of added tissue into its normal structure, larger heterotopic grafts tend to produce a distorted chimaeric regenerate, especially in cases where the graft was inserted prior to amputation. Such chimaeras contradict the formerly accepted impression that an early blastema can regulate completely to conform to its new surroundings after transplantation. Blastemal transplantations performed on *T. cristatus* (Milojevic, 1924; Weiss, 1927) and larval *S. salamandra* (Giorgi, 1924) have been the subject of a splendid review by Guyénot (1927b), who concluded that the early blastema must be induced by basal tissues, which thus dictate the eventual form of the regenerate. It now seems likely that these early blastemata were completely eroded and replaced by a new blastema derived from the underlying host tissue. Consequently, these studies provide no grounds for contesting the chimaeric regenerates generally obtained after grafting muscle from tails into amputated limbs (see section 5). The examples suggest that regulation is limited to interactions within the blastema itself. Dedifferentiation is therefore interpreted here as entailing a complete loss of tissue determination without affecting each cell's memory of its former position within the body or even within the limb, an interpretation which obviously depends on the known local origin of blastemal cells. The blastema usually requires some innervation and a blood supply, aptly described as 'conditions banales' by Guyénot (1927b), for they support but do not direct its growth. The blastema seems quite independent of any other stump influence and thus corresponds precisely to the embryonic limb bud which Harrison (1918) characterized as a harmonious equipotential system in Driesch's terminology. This cannot be claimed as the prevalent view of blastemal morphogenesis, yet it seems to be gaining acceptance gradually. For instance, Stocum (1968b) grafted partial blastemata into the dorsal fin of *Ambystoma* larvae and deduced from their growth there, when isolated from stump influences, that cone stage blastemata contain a virtually mosaic pattern of autonomous limb regions. Faber (1960) concluded from rather similar experiments that a transplanted blastema tends to form distal limb structures, and hence must normally be induced by the limb stump to conform by shifting to more proximal regions. In subsequent reviews, however, both Faber (1971) and Stocum (1975b) attached more importance to the regulative autonomy of the blastema revealed in their original results. When assessing experiments which are concerned with the establishment of characteristic axial patterns during regeneration (chapter 8), it is worth considering whether or not the results can be explained entirely in terms of regulation within a blastema. I believe that explanation will prove adequate, although few of the investigators concerned appear to share my belief.

7.4 Wound epithelium and apical cap

The mesenchymal blastema was treated in isolation in the preceding sections because there is no known exchange of cells between it and its epidermal covering. There is considerable evidence of some interaction between them, however, which will be considered below. Amputation stimulates a typical wound-healing response in the adjacent epidermis, which migrates over the cut surface to form a thin but complete epithelium, often within a few hours in larvae but taking longer to cover jagged surfaces in adults (Repesh and Oberpriller, 1978). Continued epidermal migration causes the wound epithelium to thicken into an apical cap, detectable 4–5 days later. Local cell division is not resumed until much later but may eventually contribute to a further thickening of the apical cap and to its persistence during the cone stage of blastemal growth (Singer, 1949; Taban, 1949). In practice, the apical cap is identified only by its relative thickness, a feature which is easily confused in poorly oriented sections. That hampers any precise definition of its extent or duration. The negative features of the wound epithelium and apical cap are probably more significant: they lack any underlying dermis or distinct basement membrane and are thus in intimate contact with the dedifferentiating mesodermal tissues and later with the blastemal mesenchyme until the blastema has grown considerably and is starting to redifferentiate (Singer, 1949; Singer and Salpeter, 1961). There seem to have been no investigations of whether or not apical caps persist at the tip of each growing digit, perhaps because they could not be distinguished from the interdigital webbing.

There is some reason for thinking the wound epithelium may be needed for regeneration to occur at all. If a freshly amputated arm stump is inserted into the peritoneal cavity, like putting a hand in one's pocket, the wound surface is not covered by epithelium and no blastema forms (Polezhaev and Favorina, 1935; Goss, 1956c, d). Similarly, a flap of skin sutured over the wound surface immediately after amputation prevents regeneration. Tornier (1906), who discovered this means of inhibiting regeneration, explained it as a mechanical impediment to growth, but Godlewski (1928) and Efimov (1933) found that a similar skin flap sutured 1 or 2 days after amputation, when a wound epithelium had already formed, no longer prevented regeneration. This result permitted several interpretations, most of which have since been eliminated. Godlewski's original suggestion of a vital contribution of epidermal cells to the blastema has been considered earlier and dismissed. Namenwirth's (1974) finding that pure epidermis is the only limb tissue which does not promote regeneration when grafted on to an irradiated limb is only the latest of several counter-arguments. Another possibility, that a temporarily open wound might furnish a necessary stimulus for regeneration, perhaps by providing a wound hormone (cf. Needham, 1941), has been refuted by the following demonstration that regeneration can occur without amputation or any external wound. Thornton and Kraemer (1951) found that internal injury to the muscle or skeleton of a denervated larval arm led to considerable regression. Following sufficient regression to involve the resorption of the digits and apical dermis, re-innervation was followed by regeneration. The apical epidermis had attained the condition of a typical wound epithelium in these cases: its innervation and thickening as an apical cap constituted the first signs of regeneration (Thornton, 1954). Thornton thus recognized the potential importance of the apical cap, especially by analogy to the apical epidermal ridge which had been discovered in chick limb buds a few years earlier (Saunders, 1948). Unfortunately, no ridge can be found on urodele limb buds or blastemata (Sturdee and Connock, 1975; Tank et al., 1977).

The chronological sequence of forming a wound epithelium, its innervation and thickening as an apical cap, and the accumulation of mesenchymal cells underneath it can be envisaged as a causative sequence of events, at least as a working hypothesis. Thornton (1956a, b) obtained results which supported this hypothesis in general terms when he examined the arrested regeneration in metamorphosing tadpoles and the nerve-dependency of transplanted urodele limbs. In both cases, a deficient epithelial innervation seemed to preclude the formation of an apical cap and later phases of regeneration. There are a variety of experiments, however, which dissociate epithelial innervation from the rest of the sequence. Aneurogenic arm stumps provide one example by forming typical apical caps early in their regeneration (Thornton and Steen, 1962), as do arm stumps which lack sensory innervation (Thornton, 1960a; Sidman and Singer, 1960). Singer and Inoue (1964) described the growth of apical caps both on totally denervated arms of adult *N. viridescens* and on arms retaining only the small fifth spinal nerve which provided a normal or excessive innervation of the wound epithelium. No regeneration occurred in either case, and the apical caps subsided after 2 weeks when mesenchymal cells would normally be expected to gather underneath them. Thornton (1962) found that head skin of *A. talpoideum* larvae grafted on to the arm could form a well-innervated wound epithelium but did not grow an apical cap or support regeneration. Normal innervation cannot induce unsuitable epidermis to form an apical cap, therefore, while normal limb epidermis does not seem to need innervation for that or any other aspect of regeneration. The most reasonable scheme to be deduced from these experiments involves a spontaneous origin of the apical cap by a continued distal migration of epidermal cells after the completion of wound healing, and a stimulus from the mesenchymal blastema which maintains the apical cap. Since a nerve supply is usually needed to stimulate mesenchymal proliferation, it would have an indirect effect on the apical cap. This scheme conflicts with the report that aneurogenic arms cannot regenerate when their epidermis has been replaced by skin-grafts from innervated arms (Steen and Thornton, 1963). It is difficult to rationalize this last result with the preceding evidence, however, and perhaps its interpretation should be held in abeyance until it has been corroborated. It is curious that the converse experiment of placing aneurogenic skin on a normal or denervated arm stump has been attempted but without published results (Thornton, 1965). Recent experiments with skin flaps (Tassava and Lloyd, 1977; Tassava and Garling, 1979) suggest the wound epithelium is not needed for dedifferentiation but may stimulate division of the underlying mesenchymal cells.

The apical cap seems to have the properties of stimulating and directing blastemal growth. The first property may be deduced from Thornton's (1958) study on the effects of local irradiation with ultra-violet light, which he found penetrated to a depth of about three cell layers. Exposure of the wound epithelium prevented the formation of an apical cap and also prevented regeneration for the 3–4 week period of daily treatments in *A. talpoideum* and *A. tigrinum* larvae. Identical treatment of the apical cap, beginning at least 5 days after amputation, was unable to reach the inner layers of epidermal cells and only retarded the normal course of regeneration. Thornton (1957) had previously managed to excise the apical cap with little damage to underlying tissue. He found that repeated daily excisions could prevent regeneration provided that no thickened cap was reformed in between the operations, as in 50–60 mm larvae of *A. tigrinum*. Smaller *A. maculatum* larvae formed a new wound epithelium in 4 hours after the excision, erected an apical cap in 12 hours and continued to regenerate slowly despite having to repair the epidermal wound each day. Irrespective of whether the difference in response should be related to the size of the specimens or to the species employed, Thornton observed that dedifferentiation occurred in all cases, and could correlate the presence of an apical cap with the accumulation and growth of a blastema. The notion of growth stimulation also gains support from naked limb blastemata transplanted to either internal or superficial sites in the dorsal fin of *Ambystoma* larvae, for only the projecting grafts acquired a wound epithelium and elongated appreciably (Stocum and Dearlove, 1972). The second property of directing distal growth results from the initial accumulation of blastemal cells under the apical cap. When a small wound is made in the skin at one side of the stump apex 5 days after amputation, closure of the wound displaces the apical cap and mesenchymal cells accumulate in a correspondingly eccentric position so that the regenerate deviates from the major limb axis. This has been demonstrated in both normal and aneurogenic arms (Thornton, 1960b; Thornton and Steen, 1962) and is therefore independent of any nervous influence. Similar wounds made at cone or palette stages heal without causing any distortion of the regenerate. Thornton and Thornton (1965) managed to transplant the apical cap to the base of the same blastema in relatively large larvae of *A. talpoideum* and *A. mexicanum*, and noted that a new apical cap soon formed in the original position. In almost half the cases the two caps remained sufficiently distinct to stimulate partially separate blastemata which grew into a compound regenerate with supernumerary digits.

According to these studies, the apical cap at least provides a focus for the accumulation and proliferation of blastemal cells. It could do so by providing a permissive site for blastemal growth, or by actively attracting mesenchymal cells. The latter mechanism would best account for Carlson's (1969) observation of retarded regeneration in axolotl limbs after their skin had been exposed to actinomycin-D. There are also indications that a growing blastema prolongs the existence of the apical cap. This mutual support seems very similar to the interaction between the mesenchyme and apical epidermal ridge of chick limb buds (reviewed by Saunders, 1977).

7.5 Redifferentiation

The problem of how blastemal cells redifferentiate into particular tissues has received little attention and there is no prospect of solving it at present. According to a popular view promulgated by Guyénot (1927b), tissues of the limb stump induce adjacent blastemal cells to redifferentiate into corresponding tissues, which in turn induce more distal blastemal cells to respond in the same way. Guyénot realized such homoiogenetic inductions could not completely satisfy the experimental results known at the time. He postulated a dominant inductive influence of the stump connective tissue because arms from which the skeleton had been extirpated could still

152

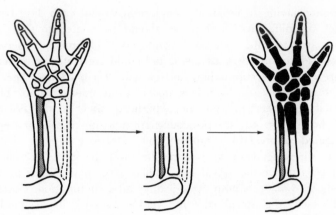

Figure 7.4. The effect of a transposed ulna in *N. viridescens* (after Goss, 1956a). When amputated a week after the transposition, the regenerate contains distal parts of two ulnae but the hand is quite normal. The grafted ulna is cross-hatched, its former site shown by a dotted line. Regenerated skeleton is shown in black.

produce regenerates which contained complete skeletons. Goss (1956a) devised an elegant demonstration of the kind of results which are held to support this view. He transposed an ulna to the anterior side of the radius in *N. viridescens* before amputating through the fore-arm. The regenerate included two ulnae: one as a continuation of the transposed ulna and another corresponding to where the ulna used to be in the stump (Figure 7.4). The immediate conclusion from experiments of this kind is that a blastema can regulate to compensate for defects in the stump but faithfully perpetuates any excess stump structure. That is not completely true, for the more distal parts of the regenerate revert towards structural normality. The influence of the transposed ulna is usually restricted to the base of the regenerate. Guyénot's general thesis of an inductive influence spreading some distance from the stump, in addition to an intrinsic blastemal capacity for regulation, reasonably accommodated such results and further supposed an early blastema to be 'nullipotent' or incapable of redifferentiation when isolated from its base, which Giorgi (1924) had found for young limb and tail blastemata grafted to the backs of *S. salamandra* larvae. Polezhaev (1936) contested this last result, arguing that graft destruction was probably the main reason why these blastemata failed to differentiate. Polezhaev contrived a novel if rather cumbersome means of preserving grafts by extracting the femur and muscles from the thighs of axolotls to leave a collar of skin with intact major blood vessels and nerves. He sutured the collar over the graft and amputated the package a month later in order to stimulate regeneration. What happened to the graft during that initial month is still unknown but probably includes both dedifferentiation and destruction. A single young tail blastema seemed unable to grow or differentiate when grafted in this way, but four or more tail blastemata packed into the same cavity could combine to produce a tail-like regenerate with a fin although lacking any skeleton. Polezhaev also found that finely chopped fragments of muscle implanted in the same manner dictated the nature of the regenerate. Limb muscle produced limbs; tail muscles produced caudiform regenerates identified by abundant fin mesenchyme. Considering that these regenerates had arisen from completely disorganized basal tissue, Polezhaev reasonably concluded that a blastema does not require any structural instructions from stump tissue. This conclusion concerning at least a qualified independence of blastemal redifferentiation has been verified in all the commonly employed urodeles in varying anatomical and histological detail. Conceding that grafts of tail muscle to a limb are likely to result in chimaeric regenerates, Liosner and Vorontsova (1937), Monroy and Oddo (1943), and Roguski (1961) have all obtained confirmatory results. The ability of macerated implanted tissues to dedifferentiate and redifferentiate while retaining their regional properties is thus established beyond question. Where the implantation site had a detectable influence on the nature of the regenerate there is no means of deciding from these experiments whether it actually induced a change or perturbed regeneration solely by contributing cells to the blastema which then behaved as an autonomous regulative unit.

Autonomous redifferentiation would be readily explicable as modulation, which includes the belief that permanently determined blastemal cells must revert to their former state whenever they are allowed to differentiate. The demonstrated inadequacy of modulation, however, compels us to view the blastema as a mass of undetermined cells which is capable of organizing itself in a predictable way. Some blastemal cells redifferentiate into cartilage, others into muscle and so on (but none of them have been noticed to form kidney tubules), and they usually do so according to a regular spatial arrangement which changes in different segments of the regenerate. It is thus possible to resolve the original problem into two separate aspects, tissue-type and tissue arrangement, even though they are almost certainly interrelated. Both aspects of the problem are identical to those posed by limb development but they have not been solved there either.

The first aspect, that of histological redifferentiation in the most restricted sense, remains such an enigma that it still seems profitable to pick out a speculative determinative sequence from the normal course of limb development and regeneration. Both limb bud and blastema start as undifferentiated mesenchyme. A central focus of cells, or perhaps two foci in the fore-arm blastema, first become compacted and secrete cartilage matrix. Each cell is surrounded by a capsule of cartilage but can divide a few more times at least. Continued production of matrix and continued cell division between the diaphysis and epiphyses are responsible for the initial growth of the skeleton. Cells in several peripheral regions of the blastema can be identified as myoblasts when they fuse into long fibres and elaborate myofibrils within their cytoplasm. Muscular growth occurs by the addition of extra myoblasts which apparently cannot divide after fusing into a syncytium. It would over-simplify the problem to dismiss the rest of the mesenchyme as merely forming connective tissue. Some inner cells apply themselves to the diaphysis and secrete a distinctive perichondrium, the outermost cells form the dermis; others invest the muscle and connect it to cartilage in the form of tendons, still others are converted into the walls of blood vessels; both blood vessels and nerve fibres become surrounded by fibroblasts of more general connective tissue. All these types of redifferentiation progress as waves from the base of the limb bud or blastema towards its distal tip. When the procartilage condensations are first detected at the palette stage, they show a fairly complete shortened limb pattern. Muscle blocks form at the same time (Grim and Carlson, 1974, 1979) even in denervated palettes, so there is no reason to suppose their arrangement is dictated by the skeletal primordia or by nerves. The disposition of major nerves and blood vessels, however, might depend on these large blocks of tissue. For instance, Piatt (1942) found that nerves penetrate a transplanted aneurogenic arm to attain a surprisingly normal pattern. The remaining mesenchyme can be envisaged as differentiating according to whether it is most influenced by adjacent cartilage, muscle, epidermis, nerve fibres or blood vessels. There is no evidence for that except the observation that perichondrial sheaths do not form round blood vessels or nerves. The spike regenerates of adult *Xenopus*, however, show nerve bundles surrounded by cartilage.

The epidermis also undergoes a form of dedifferentiation and redifferentiation. The relatively simple epithelium of larval urodeles contains unicellular glands, the Leydig cells which are dependent on a neurotrophic factor (Hui and Smith, 1976), as well as superficial microvilli and a basement membrane, all of which are absent from the wound epithelium. The stratified epithelium of adult urodeles with its keratinized outer layer is even more obviously different from the wound epithelium and thickened apical cap. Late in regeneration, a new basement membrane appears as an extension of the one in the old stump. Epidermal cells are believed to secrete at least part of this new membrane as they redifferentiate into a typical epithelium. The undifferentiated wound epithelium and apical cap probably play only a subservient role in regeneration, even though they may evoke or mediate blastemal growth. The mesenchyme ultimately dictates the nature of the limb, as has been demonstrated by exchanging ectoderm between fore- and hind-limb buds of chick embryos or between mutant and normal limb buds. The investigations of Glade (1963) and Namenwirth (1974) may be cited as justifying much the same conclusion for the regenerating blastema in axolotls, where an amputated leg invested in tail epidermis still forms a typical limb with its appropriate tissues, while an irradiated limb stump covered by normal epidermis cannot regenerate at all.

The second aspect of the problem of redifferentiation, whether or not the eventual arrangement of tissues in a regenerate can be traced back to a pattern in the early blastema and the extent of determination shown by such a pattern, has been approached in a relatively abstract manner in both the limb bud and blastema. Redifferentiated tissues are usually treated as mere markers of the regenerate's orientation and thus of an overall pattern of morphogenesis. A long-standing interest and the recent upsurge of both experimental and theoretical analysis demand that the following two chapters should be devoted to charting morphogenesis in these abstract terms. The underlying mechanism of limb development and regeneration probably cannot be pursued much further until morphogenesis has been clarified in some consistent manner, which helps to justify even the more speculative theories and artificial models which have been proposed to embrace it.

7.6 Conclusions

Amputation elicits a genuine dedifferentiation of cells. The early blastema thus consists of an accumulation of mesenchyme cells, derived from all internal tissues of the adjacent limb stump, which have lost their former histological specificity but retain an organ-specific character. Although many of the mesenchyme cells normally revert to their former tissue-specific differentiated state (modulation), there are now several experimental demonstrations of histological transformation or metaplasia during limb regeneration. Cartilage cells can transform into muscle cells, probably all mesodermal tissues are interconvertible, and a minority of these cells may routinely show such metaplastic alterations even during normal regeneration. The blastema grows predominantly by cell division and, once it is well-established, prevents further dedifferentiation of stump tissues. The growing blastema evinces considerable powers of regulation but does not seem to be subject to any external influences such as stump inductions. Dedifferentiated epidermal cells rapidly migrate over the wound surface created by amputation and their continued migration culminates in a thickened apical cap, which seems to attract the initial accumulation of mesenchymal cells and may stimulate early blastemal growth. The apical cap is subsequently maintained by the blastema and is probably subservient to it, as the types of redifferentiated tissues and their arrangement in the regenerate are decided by the mesenchyme. According to the bulk of evidence, the blastema can be regarded as a self-regulating entity which is autonomously able to differentiate its own characteristic tissues, presumably according to a pattern which is attributed to the regional origin of the mesenchyme cells, by an unknown mechanism.

Chapter 8

Regional and Axial Determination

The mosaic theory of development which underlay the concept of an absolute determination of tissue-specificity was also expressed in the concept of rigid organ-specificity during regeneration. Philipeaux (1866, 1876), for instance, considered that an organ could regenerate from a small remnant but not if it had been entirely extirpated. He claimed to have prevented arm regeneration in *T. cristatus* by removing all the arm and shoulder girdle, presumably with the associated musculature. Fritsch (1911) and Guyénot (1927b) had less success with this operation and the latter sourly suggested that Philipeaux' results should be attributed to extensive necrosis of the exposed flank tissues. Bischler (1926) found that extirpation of the shoulder girdle in *T. cristatus* did not prevent limb regeneration as other tissues could form the blastema and thus eventually provide a new skeleton. She did not doubt, however, that there must be a restricted region from which the arm can regenerate, especially as Schotté (1926c) had prevented tail regeneration in the same species by amputation in the rump. At about the same time, it became widely appreciated that the amphibian embryo contains a larger territory or morphogentic field capable of forming a limb than is normally devoted to limb development. Strictly mosaic theories of both development and regeneration thus became replaced to some extent by regulative concepts.

8.1 Regenerative territories

In her pioneering studies into the nervous control of limb regeneration in *T. cristatus*, Locatelli (1925) deflected a sciatic nerve to the dorsal lumbar region where it induced the development of a supernumerary limb. She rashly concluded that the nerve dictated the type of organ formed from regionally undetermined tissue. This conclusion was immediately contested by Guyénot and Schotté (1926b) who refuted it by the following remarkable experiments.

They deflected the sciatic nerves of *T. cristatus* to various parts of the flank and tail. Those nerves which ended close to the pelvic girdle resulted in the growth of limbs, those which ended further away in the flank did not, while those ending behind the cloaca induced the growth of a tail-like fin lacking any skeleton. The deviated nerve and local wounds were evidently non-specific growth stimulants, as the nature of the growth depended upon the region where it arose. A systematic series of nerve deviations subsequently revealed several distinct regenerative territories (Guyénot, 1927b; Bovet, 1930; Guyénot et al., 1948) which map as a mosaic pattern on the surface of the newt, but which can be regarded as equivalent to the morphogenetic fields of embryos (Kiortsis, 1953). Besides the limb and tail territories and an inert mid-flank region, nerves deflected close to the dorsal midline induced the growth of an accessory dorsal crest and thus demonstrated yet another territory. A possible third dimension to such territories has not been explored, although it might be revealed by grafting internal tissues into irradiated limb-stumps as practised by Lazard (1967).

The regulative view of regeneration had already obtained a stimulus, although probably a misconceived one, when Milojevic (1924) transplanted early blastemata from arms to leg stumps of *T. cristatus* and identified the regenerates as resembling feet rather than hands, implying that a grafted blastema conforms to its new surroundings. Such results can be criticized on several grounds. The delicate blastema could have been damaged and resorbed without trace, whereupon the host leg would regenerate a normal foot. Also the usual distinctions between hands and feet, based on the number and position of skeletal elements, are often indecisive because of the variable structure of experimental regenerates. The conclusions drawn from these early experiments were unreliable in the absence of any independent means of marking the transplanted cells, and subsequent experiments have mostly contradicted them (see section 7.5). The exchange of tissues or blastemata between limbs and tails offers a better opportunity to recognize distinctive regenerate features. Weiss (1927) concluded tail blastemata could be converted by transplantation on to a limb stump, but the blastema was probably eroded and replaced by local cells. Polezhaev (1936), Liosner and Vorontsova (1937) and Monroy and Oddo (1943) have all grafted tail tissue to limb sites and found the ensuing structure to be composite with elements characteristic of both the graft and the host. Glade (1957, 1963) provided some of the most recent examples, using adult *N. viridescens* and *A. mexicanum*. A limb stump covered with tail skin regenerated defective limb structures with additional cartilages resembling vertebrae, segmental muscle and excessive amounts of fin mesenchyme. Similar grafts of tail epidermis had only a minor influence on limb regeneration, perhaps attributable to connective tissue contaminants in the grafts, while grafts of tail dermis caused much the same abnormalities as whole skin or muscle grafts from the tail. Glade concluded that epidermis seemed to act as a neutral covering but dermal connective tissue and muscle are regionally distinct in terms of morphogenetic pattern. He was unable to decide whether such grafts influenced regeneration inductively or by contributing cells to the blastema. As graft cells have been traced into the blastema on several other occasions, they almost certainly entered it here and apparently retained their regional characteristics while dedifferentiating.

Limb tissue grafted into a neutral territory, of course, may be considered to impose its own morphogenetic field locally and thus dictate the growth of a supernumerary limb, as demonstrated by Kiortsis (1953) on *T. cristatus* and by Trampusch and Harrebomée (1961, 1969) on axolotls. The mid-flank of larval or adult urodeles appears to be a neutral or inert territory, for limb nerves deflected to a skin wound there do not provoke any growth. If a piece of limb skin has been previously transplanted to the flank, however, innervating and wounding it establishes the conditions required for growth of a limb. The frequency and perfection of limbs grown in this way increases in the presence of additional implants of limb cartilage or muscle, when local innervation can substitute for the deviated limb nerve, but limb skin seems always to be needed for the formation of a typical blastema and its growth.

The experiments described above are merely alternatives to transplanting tissue onto an irradiated host organ which then cannot contribute to the regenerate. Tail tissues grafted onto irradiated limb stumps produce unmistakeable tail regenerates. The converse operation results in regenerated limbs projecting from irradiated tail-stumps. Such results have been obtained consistently with several tissues of axolotls and have been confirmed to some extent in other urodeles (Table 8.1). Some of the experiments listed in this table were intended to test the constancy of tissue-specificity during regeneration. They were indecisive in this respect, as all the grafts contained connective tissue fibroblasts, but they demonstrate an undeniable regional specificity by determining the morphogenesis of the regenerate. Such regional specificity is not restricted to large grafts of organized tissue, for Skowron and Roguski (1958) produced tail-like regenerates on irradiated limb-stumps of adult axolotls by injecting into them suspensions of dissociated cells obtained from the tail fin. There can be little doubt, therefore, that individual blastemal cells remember the region of the body from which they came and attempt to reproduce the appropriate structure. When a blastema originates from both host and graft tissues it will regenerate into a structural chimaera, probably resembling one or the other organ according to the predominant cell-type. A minor population of blastemal cells might even be assimilated into structures which are absent from their origin. In these examples and others mentioned later, tissues adjacent to the blastema support its growth by providing a blood supply and innervation but they do not appear to influence the form of growth unless they actually contribute cells to the blastema. This idea of a benevolent neutrality on the part of adjacent tissues opposes the well-entrenched concept of the morphogentic field, whereby a limb-stump should induce grafted tissue to conform to the host regional pattern (see reviews of Guyénot, 1927b; Goss, 1961). Since it

Table 8.1 Characteristic form of regenerate obtained by grafting tissues from diverse regions on to irradiated limbs or tails.

Graft donor		Form of regenerate (reference nos.)[*]	
Region	Tissue	On X-rayed limb stump	On X-rayed tail stump
Limb	Skin	Limb (1, 3, 4)	Limb (5, 8)
Limb	Skeleton	Limb (1, 3, 4)	Limb (6)
Limb	Muscle	Limb (1, 3, 9)	—
Tail	Skin	Tail (1, 4)	Tail (12)
Tail	Skeleton	Tail (4, 11)	—
Tail	Muscle	Tail (1, 9, 10)	—
Trunk	Skin	No regeneration (4) Occasional limbs (7)	No regeneration (5)
Trunk	Muscle	No regeneration (2)	—
Head	Skin	No regeneration (1, 7)	—
Head	Skeleton (jaw)	No regeneration (4)	—

[*]1–2, Umanski (1937, 1938); 3–6, Trampusch (1951, 1958a, b, 1972); 7, Lazard (1967); all on axolotls; 8, Luther (1948) on *S. salamandra* larvae; 9, Thornton (1942) on adult *N. viridescens*; 10, Liosner (1941); 11, Liosner (1947) and Siderova (1949) according to Polezhaev (1972b); 12, Umanski (1946) according to Goss (1961).

has been suggested without evidence that irradiation might inactivate a morphogenetic field (Trampusch, 1958), there is some need to demonstrate that blastemal morphogenesis is independent of unirradiated surroundings. The best example is provided by limb blastemata transplanted into the dorsal fin of *A. maculatum* larvae (Pietsch, 1961, 1962; Stocum, 1968b), where the blastema grows autonomously into a complete limb. At the very least, then, we can conclude that the postulated influence of a limb stump's morphogenetic field is unnecessary for regeneration, as a blastema can grow quite normally when isolated from it. The one qualification to this conclusion is that very young blastemata do not develop any further when transplanted to the flank or to an irradiated stump (Giorgi, 1924; Goss, 1961).

The limb territories revealed by deviating nerves or transplanting tissues correspond in many respects to embryonic limb fields. Such fields must have enlarged in absolute terms but become relatively less extensive during embryonic and larval growth. The anterior and posterior limb fields of early embryos are confluent, so that grafted auditory vesicles (Balinsky, 1925; Filatow, 1927) or nasal capsules (Glick, 1931; Balinsky, 1957) sometimes stimulate the growth of extra limbs anywhere in the flank. The mid-flank of adult newts, however, constitutes an unresponsive territory. All parts of the limb territory then seem equally capable of producing an extra limb when incited by a deflected nerve or when transplanted to an irradiated site (Hofmann *et al.*, 1978), but the best growth occurs from the base of the limb

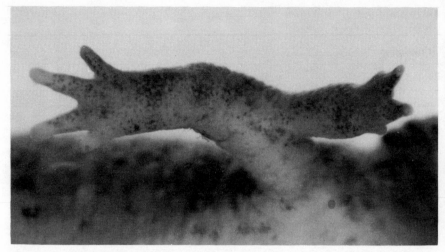

Figure 8.1. Supernumerary arm produced by deviating the inferior brachial nerve to emerge from the surface of the upper arm, in larval *Pleurodeles waltl*.

itself (Figure 8.1) and the boundary between adjacent territories is liable to form a chimaera. The growth of the blastema and the regenerate's orientation are only influenced by the small area of tissue from which it originates, as demonstrated when the latter does not conform to the orientation of surrounding tissue (Kiortsis, 1953). De Both (1971) has criticized the concept of territories, arguing that the sheaths of deflected nerves could supply all the cells which form the blastema. He could only sustain this argument by ignoring much of the evidence cited here whose validity is generally accepted. Nerve sheath cells must routinely enter the blastema but they are outnumbered by other cells and they evidently do not control the pattern of morphogenesis.

8.2 Axial determination

The way in which a blastema grows into a new limb with an obvious asymmetry and orientation is best approached by considering first the better known case of a developing limb bud. The determinative events of amphibian limb development were given a lucid and perhaps definitive interpretation by Harrison and his students, after transplanting the fore-limb primordium of *A. maculatum* embryos so as to rotate each of its three major axes through 180°. The arbitrarily defined axes—anteroposterior (AP), dorsoventral (DV), and mediolateral (ML)—of the whole embryo are shown in Figure 8.2. Although his earlier reports contain misconceptions which are liable to confuse the reader, Harrison (1925) finally established that only the AP axis of the future limb is fully determined in the early tail bud embryo. At that stage, any transplanted bud whose AP axis conforms to that of the

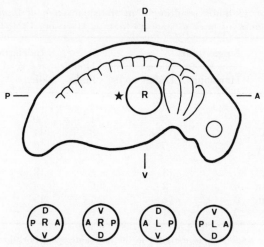

Figure 8.2. Diagram showing the replacement of right limb bud of a stage 29 *A. maculatum* embryo (adapted from Harrison, 1925). The circles below marked R or L represent surface views of right or left limb buds whose orientation is marked by the AP and DV axes. These can replace the host limb bud to effect a reversal of either or both axes. The star indicates a centre of polarizing activity (probably included in Harrison's grafts but not identified at the time, see text). This reverses the flank AP polarity, so that limb buds grafted behind it give different results from the orthotopic grafts shown here (see Table 8.3).

host embryo develops into a 'harmonic' limb showing the normal orientation. When the AP axes of the graft and host are the reverse of each other, the resultant 'disharmonic' limb is recognizably abnormal, as though belonging to the opposite side of the embryo. Despite testing all orientations of the grafts, Harrison found the DV and ML axes of the resultant limbs always conformed to those of the host and concluded that these axes were not determined in the bud at the time of transplantation. The results which best demonstrate these conclusions come from heterotopic grafts where a limb bud was transplanted into the mid-flank of each host embryo. Even then only a small proportion of operations yielded limbs whose orientation could be discerned accurately (Table 8.2). Harrison (1921) had previously grafted limb buds orthotopically into the cavity created by extirpating another bud. It proved impossible to evacuate the site completely, so the host either added mesenchyme to the graft or formed an additional limb independently of it. The vast majority of orthotopic transplantations consequently yielded either single harmonic limbs or mirror-image duplicated limbs (Table 8.3). Yet these results also supported Harrison's interpretation, for the only cases of

Table 8.2 Orientation of limbs resulting from transplantation of limb primordia to the flank in *A. maculatum* embryos prior to stage 33 (data of Harrison, 1921, 1925).

Graft orientation and position*				Single limbs		Double limbs	Indeterminate, resorbed or died	Total operations
				Harmonic	Disharmonic			
AA	DD	MM	hom	3	0	4	12	19
AA	DD	ML	het	3	0	1	9	13
AA	DV	MM	het	7	1†	8	44	60
AA	DV	ML	hom	2	0	0	10	12
AP	DD	MM	het	0	8	2	18	28
AP	DD	ML	hom	0	2	0	4	6
AP	DV	MM	hom	0	11	1	19	31
AP	DV	ML	het	0	1	0	15	16

*AA, DD, MM indicate conformity of the graft to the host anteroposterior, dorsoventral and medio-lateral axes respectively; AP, DV, ML indicate reversal of these axes; hom (homopleural) or het (heteropleural) indicate whether the graft is implanted on the same or the opposite side as that from which it was obtained. Harmonic limbs are oriented appropriately to the side they occupy, disharmonic limbs are not (e.g., a right hand on the left side).
†Explained as an experimental error.

single disharmonic limbs occurred when the AP axes of graft and host were opposed. Duplicated limbs also conformed to this pattern when the primary member, which appeared earlier and often remained larger, was classified as harmonic or disharmonic in the same way as single limbs. Harrison did not provide a complete explanation for the secondary member of a duplicated limb and, indeed that has been a subject of contention ever since. Contrary to the supposition that they must be formed of host tissue, Swett (1932) found that many exhibit growth characteristics typical of the grafted bud. Grafted *A. tigrinum* buds, for instance, grew into duplicated arms which were enormous in comparison to their *A. maculatum* hosts. Despite that, the polarized AP axis of the host must have a major influence on the incidence of duplications, as suggested by Detwiler (1930). Duplications occurred overwhelmingly among orthotopic grafts with a reversed AP axis, and also commonly among heterotopic grafts whose AP axis conformed to that of the host (Table 8.3). Embryonic flank tissues can also be provoked into forming an extra limb by grafting otic or nasal capsules into them and such limbs are usually found to be disharmonic, with a reversed AP orientation. If we explain this by supposing that the flank of the embryo consists of a limb-forming field with a determined AP axis, polarized normally in the immediate orthotopic limb area but with the reverse polarity in the mid-flank, then 70 of the 74 duplications recorded in Table 8.3 can be accounted for as a conflict of AP axes between graft and host. Any grafted limb bud may activate local limb-forming tendencies but these are either assimilated or suppressed when there is no conflict of axes. Discrete conflicting fields, however, may influence labile cells at the wound edges of

Table 8.3   Relative incidence of single and duplicated limbs produced by transplanting limb primordia of *A. maculatum* embryos prior to stage 33 (data of Harrison, 1921 and 1925).

Graft site	Graft and host axes	Polarities*	Single limbs	Double limbs	Total
Orthotopic	AA – same	Coincident	24	1	25
(shoulder)	AP – opposed	Conflicting	2	57	59
Heterotopic	AA – same	Conflicting	15	13	28
(flank)	AP – opposed	Coincident	22	3	25

*Defined on the assumption that the limb-field's polarity is reversed outside the normal orthotopic position.

both graft and host, and consequently tend to result in duplications whose members must be mirror images.

Harrison's account of the determination of the AP axis remained uncontested for many years, although later studies of chick wing buds revealed additional determinative events (see Saunders, 1977). Amphibian limb development has lately been brought into line with the wing bud by the discovery of a polarizing zone which lies somewhat to the rear of the limb primordium and controls AP polarity of the developing arm of *A. mexicanum* and *P. waltl* (Slack, 1976). This zone must have the effect of reversing AP polarity in the mid-flank, as suggested above (and by Rose, 1970b). Slack's results can be easily reconciled with those of Harrison when it is realized that limbs develop much earlier in *A. maculatum* and the timing of axial determination may differ in the same way.

Other experiments mentioned by Harrison (1921, 1925) showed the early limb bud of *A. maculatum* has considerable powers of regulation. The two halves of a bud could be interchanged, or one bud transplanted over another, and a single normal limb would still develop provided the AP axes of the components parts coincided. The AP polarity resided in the bud mesoderm, with the ectoderm being labile or neutral in this respect. According to Swett (1927, 1928), the AP axis is already determined when the limb primordium can first be recognized at neural fold stages in *A. maculatum*; the DV axis becomes fully determined at Harrison stage 35 and the ML axis is established rather later at stage 37. The gradual determination of the DV axis is well documented (Table 8.4) but Swett's evidence concerning the ML axis is less certain. Limb buds invariably grow out from the body, of course, and Harrison had already found that bud mesoderm placed inside out on the flank up to stage 34 still grew out as a typical limb. The difficulty of the operation at later stages forced Swett to include ectoderm with the graft and it may have healed over the original graft apex, causing the bud to rotate slightly and resume its former direction of growth. Swett rarely obtained any growth from the original medial or

Table 8.4 Gradual establishment of dorsoventral and mediolateral polarities detected by heteropleural transplantation of limb buds to the flank of *A. maculatum* embryos (data of Swett, 1927, 1928).

Stage of both donor and host	Orientation of primary limb[*]			Indeterminate, resorbed or died	Total operations
	Harmonic	Disharmonic	Duplications		
(a) *Only DV axis reversed (graft orientation, AADVMM)*					
30–32	12	0	9	13	25
33	12	4	12	24	40
Early 34	2	3	4	5	10
Later 34	1	15	13	10	26
35	0	19	17	7	26
36–40	0	18	16	6	24
(b) *Only ML axis reversed (graft orientation, AADDML)*					
34	23	0	14	9	32
35	11	7	13	29	47
36	12	3	10	27	42
37	0	16	7	12	28
38–42	0	19	8	38	57

[*]When the polarity of the reversed axis was determined prior to transplantation it produces a disharmonic limb (see Table 8.2 for abbreviations).

basal side of the graft (unless such growth contributed to the incidence of duplications in table 8.4) yet such growth can occur from the medial surface of older limbs, as described in section 5. Less extensive studies indicate the early establishment of AP and DV axes in the arm buds of *T. vulgaris* (Brandt, 1924) and leg buds of frog and toad tadpoles (Gräper, 1922).

The importance of these developmental studies in the present context lies in providing two expectations for limb regeneration. The blastema is analogous to a limb bud in several respects and so might be expected to employ the same mechanisms during its morphogenesis. Even if that were not true, the axes of the limb have become determined as it developed and these axes must characterize the tissues exposed by amputation from which the blastema arises. Either the blastema must inherit its axial pattern directly on a cellular basis, or the axes must be imposed on it by adjacent stump tissues, for old and new parts to form a single harmonic structure. The axes are conventionally redefined in regeneration studies to take into account the length and flexibility of a fully developed limb, but they are still arbitrary definitions which correspond closely to those employed for embryos. The original ML axis of the limb bud is now represented by a proximodistal (PD) axis running the length of the limb; the AP axis can be identified by the radius and first digit on the anterior side of the arm; the flexor muscles and palm of the hand then mark the ventral side and thus the DV axis. Identical axes have been defined in different terms for limbs whose palmar surfaces are pressed against the sides of the body in a typical swimming posture (Guyénot *et al.*, 1948).

8.3 Transverse axes of the limb

There have been several investigations concerned with transplanting blastemata to new positions or replacing them after rotation in the same manner used to analyse the axes of limb buds. In all cases, the blastema is placed on a wound surface of a limb stump so as to maintain its normal PD orientation. Rotation about this axis by 180° reverses both AP and DV axes for homopleural (ipsilateral) grafts, and one or the other of them is reversed in the corresponding heteropleural (contralateral) grafts. From some of the earliest experiments of this kind, Milojevic (1924) concluded the early bulb stage blastema had no fixed axial determination, for a rotation was not reflected by any distortion of the regenerate. This result was ambiguous because he had no means of ascertaining that the graft actually formed the regenerate. Later blastemal and cone stages rotated in the same way, however, caused corresponding alterations of the regenerate structure which must be attributed to AP and DV axial determination. Analagous experiments reported by Lodyzenskaja (1928, 1930) demonstrated a determination of the transverse axes from virtually the onset of regeneration, 2 days after amputation. Little graft tissue seems to have survived from such young blastemata but their axial determination could be detected by the formation of duplications when one or both of the graft's transverse axes had been reversed relative to those of the stump. Duplication is best left vague to cover the presence of any extra limb structures, ranging from accessory digits to complete supernumerary arms. Abeloos and Lecamp (1931) also attested to the high frequency of duplications following transplantation of blastemata to effect a reversal between graft and host axes. These experiments are described in more detail in chapter 9, but it is expedient to consider immediately the following investigation which purported to show the AP axis remains undetermined or labile until tissues begin to redifferentiate in the blastema—much later than even Milojevic contemplated.

Iten and Bryant (1975) used four stages of arm regenerates of adult *N. viridescens* for transplantation with reversed AP orientation to the contralateral arm. They distinguished between supernumerary regenerates which arose at the graft–host junction, presumably reflecting an interaction between differently oriented fixed axes, and all structures distal to that junction which they supposed to be predominantly or entirely derived from the graft. Based on this assumption, they considered symmetrical 'double' hands and 'expanded harmonic' hands as intermediate stages in the respecification of the graft AP axis, culminating in completely harmonic hands. They consequently believed only clear disharmonic hands could indicate an irrevocable determination of the AP axis at the time of transplantation. Considering for the moment only those operations which did not alter the PD position of the graft (top and bottom four lines of Table 8.5), the implications of this interpretation become obvious. The increasing frequency of disharmonic hands produced by successively more advanced

Table 8.5 Analysis of regenerates formed after transplanting blastemata on to contralateral arm stumps in adult *N. viridescens* (condensed from Iten and Bryant, 1975: reproduced by permission of Academic Press Inc.).

Axial shift (blastema origin to stump site)*	Stage	No. of cases	Hands classified by structure and orientation (%)					Percentage of arms showing	
			Simple harmonic	Expanded harmonic	Disharmonic	Double	Uncertain	Supernumerary arms	Serial duplications
PROXIMAL TO PROXIMAL	Bulb	14	79	0	0	7	14	0	0
	Blastema	13	8	30	30	8	23	23	0
	Cone	13	0	0	85	15	0	77	0
	3-digit	10	0	0	100	0	0	90	0
PROXIMAL TO DISTAL	Bulb	20	40	25	0	15	20	10	30
	Blastema	13	30	8	23	31	8	23	77
	Cone	13	0	0	100	0	0	100	100
	3-digit	12	0	0	100	0	0	58	100
DISTAL TO PROXIMAL	Bulb	12	100	0	0	0	0	0	0
	Blastema	12	42	33	0	8	17	0	0
	Cone	12	17	50	0	17	17	0	0
	3-digit	13	8	62	15	8	8	15	0
DISTAL TO DISTAL	Bulb	13	38	8	0	31	23	8	0
	Blastema	10	0	30	0	30	40	40	0
	Cone	11	9	36	18	9	27	27	0
	3-digit	9	0	0	78	0	22	100	0

*Blastemata were obtained from fore-arms and upper arms and grafted at the same or other position on the proximodistal axis of the contralateral arm. The proximodistal and dorsoventral axes of graft and host coincided, so only the determination of the graft's anteroposterior axis was tested: it was certainly determined when expanded harmonic and disharmonic regenerates were formed, and possibly also when double or uncertainly oriented regenerates occurred. Supernumerary arms also indicate a conflict between axes and thus determination of the graft anteroposterior axis.

blastemal grafts suggests that the AP axis becomes determined very gradually indeed. The underlying assumption was completely mistaken, however, for double and expanded hands are identical to the duplications distinguished previously by Lodyzenskaja. Double hands are cases with accessory digits (probably derived from the host) either in front of or behind the graft, whose structure can still be discerned as disharmonic, resulting in two forms of composite symmetry. Expanded hands are those where accessory digits formed on both sides of the graft, which may be so reduced that its orientation cannot be discerned and the disposition of the extra host digits necessarily makes the chimaera seem harmonically orientated, but the presence of extra digits indicates an axial conflict and hence determined axes in the graft. This diagnosis is illustrated in Figure 8.3. When the categories of double and expanded hands are allocated to the disharmonic class, Table 8.5 demonstrates the AP axis is already determined in some 'bulb' stages and in the majority of more advanced regenerates. The incidence of harmonic hands is inflated in Table 8.5 by the inclusion of an unspecified number of cases with one or two extra digits which should really be classed as duplications and thus as containing disharmonic elements. Furthermore, the residual harmonic hands may represent instances where the graft was largely or entirely replaced by blastemal cells derived from the host limb. Even though Iten and Bryant did not detect much destruction, it might effectively eliminate the mesenchyme of bulb stage grafts. Since the data shown in Table 8.5 can be interpreted as merely showing variable graft

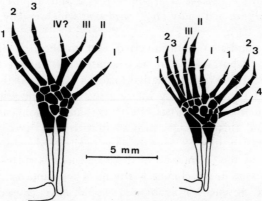

Figure 8.3. Complex axolotl regenerates produced by grafting a left fore-arm blastema on to a right arm stump, with AP axial reversal. Graft digits are identified by roman numerals. Left a 'double anterior hand' with accessory (probably host) digits in front of the graft. Right, an 'expanded right hand' with groups of accessory digits on both sides of the graft.

resorption, there is no compelling reason to doubt that the blastemal AP axis was fully determined at the earliest stage tested. Despite this criticism, it seems likely that dedifferentiated host cells could augment those at the base of the graft and could perhaps reverse the polarity of an early blastema by sheer force of numbers, essentially as an internal democratic decision within a chimaeric blastema without further reference to the stump axes. This possibility might apply especially when the blastema is shifted to a more proximal site, resulting in a massive dedifferentiation and reorganization of the stump and thus tending to swamp the AP polarity of bulb stages without generating many gross duplications (Table 8.5).

The argument advanced here that blastemal axes are precociously determined and perhaps only susceptible to regulation by internal forces is rather heretical. Most previous investigators and commentators have believed the blastema is somehow induced to conform to the stump axes, but only on the basis of evidence whose ambiguity is readily exposed.

There are two further reasons for believing the blastemal axes are always polarized. Firstly, when early bulb stages are transplanted to a neutral site in the flank of axolotls, their orientation dictates that of the regenerate (Faber, 1960). Secondly, the following experiments involving regeneration from chimaeric limb tissues show that the source of blastemal cells dictates the orientation of a regenerate, and thus suggest a determined polarity must be maintained throughout the process of regeneration. These experiments have mainly consisted of transplanting limb skin to alter its position relative to the internal tissues, providing mutually consistent results in three species: *T. cristatus* (Droin, 1959; Rahmani, 1960), *Pleurodeles waltl* (Lheureux, 1972, 1975a, b), and *A. mexicanum* (Carlson, 1974a), and apparently compatible results in a widely quoted study on *N. viridescens* which is only available as an abstract (Settles, 1970).

Droin (1959) obtained both inverted and duplicated regenerates after amputating arms whose skin had been rotated 180° about the PD axis to reverse both AP and DV axes relative to those of internal tissues. Rahmani (1960) confirmed these results by rotating the skin or muscle of legs in the same way. Rahmani and Kiortsis (1961) exchanged skin cuffs between the irradiated left and shielded right arm so that the DV axis of the skin was inverted. Amputation after 18 days to allow healing led to single regenerates on both arms, and the orientation of these regenerates corresponded to that of the shielded tissue in the chimaeric stump. They supposed that conflicting DV axes in the absence of irradiation would have caused many of the regenerates to be duplicated. Subsequent vindication of this belief in another species of newt (see Table 8.7) supports their contention that duplicated regenerates usually arise from a dual source of blastemal cells whose polarities conflict on at least one axis. The use of irradiation here led to results which were more easily classified, but introduces an additional complexity in the interpretation. The principal difficulty alluded to earlier is the possibility that irradiation not only prevents a tissue from providing

Table 8.6 Types of regenerate obtained after transplanting skin cuffs on to the upper arms of adult axolotls (data of Carlson, 1974a).

Difference in polarity of skin cuff from internal tissues	Total cases	Inhibited (No digits)	Hypomorphic (1–3 digits)	Single* (4 digits)	Multiple (>4 digits)
Reversed PD axis	18	0	0	18	0
Reversed DV axis	20	0	1	19	0
Reversed AP axis	20	0	0	8 (5)	12
Reversed AP, X-rayed stump	16	1	5	9 (2)	1
Reversed AP, X-rayed skin	16	0	4	12 (1)	0
PD axis parallel to amputation plane (90° twist), X-rayed stump	5	4	1	0	0

*Hands with abnormal orientation (disharmonic to the stump) are noted in parentheses.

blastemal cells but also abolishes its inductive influence by destroying its morphogenetic field, in the phrase of Trampusch (1958). A second difficulty is that the unirradiated graft tissue present in the stump could produce the results observed by Rahmani and Kiortsis (1961) by imposing its axes on an undetermined early blastema. The latter objection could be avoided by transplanting blastemata onto irradiated limb stumps. It is surprising to find no reports of such experiments prior to Maden's (1979b) demonstration that the blastema develops identically on a irradiated limb stump as on a normal one (cf. Holder et al., 1979).

Despite these difficulties, irradiation has certainly helped clarify the results obtained after skin grafts. Carlson (1974a), for instance, reversed each of the skin axes in turn on axolotl arms and obtained further insight into what was by that time a far from novel situation by irradiating either the skin or the host limb prior to grafting. The results shown in Table 8.6 probably give a spurious impression of normality, owing to their classification mainly by the number of regenerated digits, at the expense of ignoring the orientation of the regenerates. A conflict between skin and internal tissues on the AP axis is easily revealed by the frequent appearance of extra digits. Irradiation of either component removes that conflict, presumably by preventing its normal contribution to the regenerate. Reversal of the DV or PD axes of the skin did not cause equivalent duplications which might be less readily detected by counting digits. A similar series of skin grafts convinced Lheureux (1972) that AP and DV gradients must be present in both the skin and the internal arm tissues of Pleurodeles larvae. Either component could be the dominant influence in determining the orientation of a regenerate or, if neither of them were completely dominant, a conflict between them on one or the other axis resulted in the frequent appearance of duplicated regenerates or supernumerary digits (Table 8.7). Lheureux (1975b) extended this

Table 8.7 Types of regenerate obtained after transplanting skin cuffs on to the upper arms of larval *Pleurodeles waltl* (data of Lheureux, 1972, 1975b, arranged for comparison with Table 8.6).

Difference in polarity of skin cuff from internal tissues	Total cases	Inhibited (No digits)	Hypomorphic (1–2 digits)	Single* (3–4 digits)	Multiple (>4 digits)
Reversed PD axis	20	0	1	19	0
Reversed DV axis	43	0	0	23	20
Reversed AP axis	40	0	0	24	16
AP, DV, PD all reversed	20	0	0	4 (2)	16
Reversed PD, on X-rayed stump	28	1	3	23 (23)	1
Reversed DV, on X-rayed stump	24	1	1	22 (22)	0
Reversed AP, on X-rayed stump	23	0	1	22 (20)	0
AP, DV, PD all reversed on X-rayed stump	46	0	0	44 (43)	2
AP, DV, PD all reversed using irradiated skin	46	0	0	46 (0)	0

*Hands oriented in conformity to the AP and DV axes of the skin cuff in parentheses.

experiment by grafting skin from the shielded left arm on to the irradiated right arm of the same larva, inverting each axis of the graft in turn, together with similar grafts of irradiated skin on to shielded arms. His results confirmed that irradiation of either component of the limb stump virtually eliminated the category of duplications to reveal a uniquely oriented regenerate, whose AP and DV axes almost always conformed to those of the shielded stump tissue (Table 8.7).

Lheureux (1975a, b) argued that, just as axial conflicts between tissues promoted duplications, axial differences within a tissue were required to initiate regeneration. He provided some evidence for this hypothesis by rotating strips of limb skin through 90° so that their original PD axes now lay parallel to the plane of amputation. When such a graft encircled the arm as a complete cuff, amputation through it exposed an edge of the graft skin which corresponded to just one pole of a transverse axis, e.g. the anterior quality as he called it. Whenever the morphogenetic influence of the skin graft predominated over that of internal tissues, as it would when the latter had been irradiated, Lheureux' hypothesis predicts that no regeneration should occur. Prevention of regeneration in this way had already been established for axolotl limbs (Umanski et al., 1951; see Table 8.6), but ascribed simply to perturbation of the skin's PD axis. Although Lheureux could not completely suppress regeneration by means of this operation, he produced a remarkably high frequency of hypomorphic regenerates (Table 8.8), in contrast to the majority of normal ones formed when two or more rotated skin sectors had been grafted on to irradiated arms. In the absence of irradiation, however, the frequency of duplications differed considerably according to which skin sector had been exposed by amputation (Table 8.8), which suggests that additional or alternative explanations may be required for the consequences of 90° skin rotation. Lheureux (1975a, b) performed enough variants of these skin grafts to exclude several alternative interpretations, such as the number of healed edges of the skin grafts having much influence on the form of the regenerate. There is little other evidence against which his hypothesis can be weighed, but he took one previous experiment to contradict it. This involved making a skin incision from the irradiated fore-arm to the shielded upper arm and amputating through the wrist a few days later. Such an operation usually led to perfect regeneration in A. mexicanum and A. opacum larvae (Wallace et al., 1971; Conn et al., 1971). Regeneration here was attributed to the migration of a small number of shielded skin cells down the incision and the cells would be expected to come from a restricted sector of the dermis. That interpretation is debatable, of course, and Lheureux (1975b) found the same operation and more extensive incisions rarely permitted any regeneration in Pleurodeles.

With the data at hand (Tables 8.6–9) it seems possible to reconcile much of the discrepancy between Carlson's impression that only the AP axis can be demonstrated clearly and Lheureux' evidence for separate AP and DV axes. Recalling that these axes are defined in arbitrary terms, a single

Table 8.8 Types of regenerate obtained after transplanting sectors of limb skin, with 90° rotation of the PD axis, on to either normal or irradiated upper arm stumps of *Pleurodeles* larvae (data of Lheureux, 1975a, b).

Skin sector at plane of amputation*	Stump	Total cases	Percentage in each category			
			Inhibited (No digits)	Hypomorphic (1–2 digits)	Single (3–4 digits)	Multiple (>4 digits)
P sector as complete cuff	Normal	70	14	1	36	49
D sector as complete cuff	Normal	86	22	6	44	28
A sector as complete cuff	Normal	82	40	2	40	18
V sector as complete cuff	Normal	59	5	3	85	7
D sector as complete cuff	X-rayed	43	42	58	0	0
V sector as complete cuff	X-rayed	24	29	71	0	0
D and V sectors inverted	Normal	27	0	0	52	48
D and V sectors inverted	X-rayed	27	11	32	54	3
A, P, D, and V on normal quadrants	X-rayed	27	0	7	93	0

*Note that the four skin sectors seem to have different influences on regeneration: posterior skin causes the greatest stimulus. When grafted on to irradiated arms, four sectors virtually ensure normal regeneration but two sectors are much less effective and a single sector never stimulates complete regeneration.

diagonal axis of determination (with one pole centred between the P and D cardinal points for instance) could have been resolved into two components in *Pleurodeles* but fortuitously approximated to the AP axis of axolotl arms. It is noticeable that both species provide higher frequencies of duplications, often exceeding 80%, when the skin has been reversed in terms of both AP and DV axes than when either axis alone has been reversed. Only the former case ensures that an axis of determination occupying any diameter is fully reversed to ensure maximum conflict with the corresponding internal axis. Nevertheless, the evidence from limb buds and other grafts (see chapter 9) indicates that two axes of determination exist in any cross-section of the limb.

Taking the common conclusion from these experiments that duplications result from a conflict between determined axes of the host limb and a rotated or displaced graft, it is reasonable to enquire which tissues show sufficiently stable axes to promote such a conflict. Both skin and muscles of *T. cristatus* limbs must possess determined axes (Droin, 1959; Rahmani, 1960). Carlson (1975b) examined each of the major tissues present in the upper arm of axolotls for the presence of this property and found it only to be expressed by the dermis and muscles. His experiments concerning the dermis and epidermis also suggested that the rotated tissue could only promote duplications if it had free access to the site of amputation and the future blastema, and so warrant a more detailed exposition. Dermis obtained by removing the epidermis from isolated arm skin could be grafted back on to the arm and then became covered by migrating host epidermis. When grafted back with a rotation of 180° to reverse both AP and DV axes, this dermis proved just as effective in causing duplications as a whole skin cuff (Table 8.9). Epidermis could not be grafted in this way apparently, although Glade (1963) had some success with epidermal grafts, so Carlson used its known migratory ability to achieve an equivalent epidermal rotation. He rotated upper arm skin through 180°, allowed it to heal in its new position and then excised the middle 5 mm of the rotated cuff. After a further delay for epidermal migration (and the possible formation of a new dermis), he amputated the arm through the middle of the cuff to obtain only one case of duplication among a decisive majority of normal regenerates. Apart from this single case which will be explained later, the ingrown epidermis had not created an axial conflict of the order shown by similarly rotated dermis or complete skin. The other experiments shown in Table 8.9 involved amputation slightly distal to a skin cuff on the upper arm. A comparison between them indicates that the cuff must be rotated and have direct access to the plane of amputation, without any intervening normally oriented skin, in order to generate major duplications. Minor duplications of a single digit occurred less frequently in the series with delayed amputation and in one control example. The former cases, like the anomalous duplication following epidermal healing within a rotated skin cuff mentioned above, could possibly be attributed to a mild expression of axial determination within the migrated

Table 8.9 Types of regenerate obtained after rotating skin (S), dermis (D) or epidermis (E) around the upper arms of 7–25 cm axolotls (data of Carlson, 1975b).

Operation with indicated rotation and delays of amputation for healing and migration of E	Rotation (°arc)	Delay (days)	Total cases	Single (4 digits)	Multiple (>5 digits)
(1) D replaced after removing E	None	9	16	16	0
(2) D rotated after removing E	180	7	11	2	9 (82%)
(3) Whole skin cuff (S) rotated	180	7	10	2	8 (80%)
(4) As (3) but centre excised, amputated there after E has healed wound	180	17 + 39	16	15	1 (6%)
(5) As (3) but amputated 1–2 mmm distal (host skin)	180	6	10	10	0
(6) As (3), amputated 3 mm distal on skinned arm	180	14	15	9	6 (40%)
(7) As (6) but amputation delayed (E migrates)	180	14 + 30	14	11	3 Minor
(8) As (6) but no rotation	None	14	10	9	1 Minor

epidermis. It is far more likely, however, that enough dermal cells also migrated into the blastema to upset its axial pattern in these cases. The conclusion that epidermis lacks axial determination is insecure for a completely different reason. The determination was only tested in epidermis which had migrated from a rotated skin cuff on the unsupported assumption that it migrated directly along a PD axis. If epidermal cells sensed their positional discrepancy and migrated along a spiral pathway, thus adjusting to their normal APDV sector, then no blastemal conflict need arise. Despite that reservation, these experiments display an appealing ingenuity and provide cogent arguments for an unpolarized epidermis which nicely match the evidence for its lack of regional specificity.

Carlson (1975a, b) also demonstrated an AP axial determination in the musculature of the upper arm. Duplicated regenerates resulted from the exchange of extensor and flexor muscles, much more commonly than when the same muscles were merely rotated and replaced, and even occurred after the muscles had been minced into fragments before being exchanged in position (Table 8.10). This provides confirmation of the conclusion first reached by Polezhaev (1936) that axial determination, like regional specificity, does not depend upon the integrity of the tissue but is surely a property of the individual cells. Rotation or displacement of the humerus, as well as displacement of the brachial nerve, was almost invariably followed by normal regeneration in Carlson's (1975b) experiments. The duplications considered in the previous paragraphs are listed as multiple regenerates in Tables 8.6–10 following Carlson's terminology. The majority of them, however, contained 5–10 digits which can be accounted for by two major axes which subtend an accessory one where they meet. These axes were actually detected in the shape of the apical ridge of those palette stages which later became grossly multiple (Carlson, 1975b).

Tank (1977) extended Carlson's investigations by rotating skin through $180°$ on the upper arm of juvenile axolotls, operating either before

Table 8.10 Types of regenerate obtained after operations involving the muscles of axolotl upper arms (data of Carlson, 1975a, b).

Operation on both extensor and flexor muscle	Amputation delayed (days)	Total cases	Defective (0–3 digits)	Single (4 digits)	Multiple (>5 digits)
Removed and replaced	13	10	0	10	0
Rotated and replaced	14	20	0	14	6 (30%)
Rotated and exchanged	19	20	0	4	16 (80%)
Rotated and exchanged	120	10	0	2	8 (86%)
Exchanged without rotation	14	20	0	2	18 (90%)
Minced and exchanged	5	14	0	1	13 (93%)
Minced and exchanged	30	11	1	1	9 (82%)
Minced and replaced	11	8	1	7	0

amputation or after various periods of regeneration. Skin rotation at blastemal stages produced duplications but later rotations at palette stages rarely had that effect. Rotating the arm skin before amputation resulted in duplications even when it was restored to its normal orientation at either of these later stages. Tank's results are compatible with the knowledge that dermis provides a source of dedifferentiated mesenchymal cells throughout the period of early blastemal growth, and with the present view that such cells retain an axial determination. When Tank performed similar rotations of the regenerate's epidermis, he tended to obtain more duplications with later operations whether they imposed an axial conflict or corrected a previous one. The epidermis was always contaminated by mesenchymal cells, however, and consequently even the supposed correctional rotations must have created an additional conflict among blastemal cells. Tank's own interpretation, that morphogenetic instructions are conveyed to a labile blastema over a protracted period, may be a justifiable one but strikes me as hideously misleading. It is possible to equate his morphogenetic instructions with axially determined mesenchymal cells. The only lability shown by the blastema would then be whether or not it could regulate two determined subpopulations into a single normal hand. Furthermore, it is impossible to demonstrate the occurrence of regulation from this experiment because we only have evidence of the presence of blastemal subpopulations in the duplications which show regulation has failed.

Having already obtained considerable evidence that determined AP and DV axes were present in both skin and muscle of larval *Pleurodeles* arms Lheureux (1977) provided a particularly clear example in the course of extending his theory of axial conflicts to the production of supernumerary arms after nerve deviation. He severed the main brachial nerve at the elbow and diverted it to emerge through a skin wound half-way down the upper arm, usually at a mid-dorsal or mid-ventral position, where novel conflicts were created by grafting reoriented skin cuffs or implanting muscle close to the nerve exit. Table 8.11 compresses the results of these operations into a form which emphasizes action of conflicting axes, for the original 14 series are too many to compare easily, yet a detailed analysis would demand an even greater additional number of control series. Simply diverting the brachial nerve produced a low incidence of extremely hypomorphic growths, irrespective of the wound area but perhaps with a slightly better yield from the ventral surface than from the dorsal side. The hypomorphic growth may be partly related to the fact that the *Pleurodeles* larvae always metamorphosed during the experiment: any delay of growth might contribute to the defective form. The same operation sometimes elicits perfect supernumerary arms in younger larvae and adult newts. In comparison to this particularly stringent baseline or to the similar results of muscle grafted from an identical position on the contralateral arm (Table 8.11, bottom line), an axial conflict imposed by a graft adjacent to the nerve exit often increased the incidence and vastly improved the morphogenesis of these extra arms, and also dictated their

Table 8.11 Supernumerary arms resulting from nerve deviation and adjacent tissue grafts on the upper arms of larval *Pleurodeles waltl* (data of Lheureux, 1977: reproduced by permission of the Company of Biologists Limited).

Original series	Grafted tissue and axial conflicts at nerve exit	No. of cases	Percentage of cases forming supernumerary arms			
			Total	Hypomorphic (0–2 digits)	Normal (3–4 digits)	Duplicated (5–11 digits)
1–3	None, various wound areas	192	26	26	0	0
6–7	Skin cuff, DV conflict	161	28	4	24	0
9–10	Skin cuff, AP and DV conflict	128	53	10	21	22
12–13	Muscle, DV conflict	136	43	21	21	1
14	Muscle, no conflict	40	30	28	2	0

orientation. Conflicts involving two axes increased the number of arms and sometimes caused them to become duplicated. Lheureux considered that a pair of diametrically opposite poles on any transverse axis would be sufficient to promote normal regeneration when brought together at the end of a limb stump or after grafting. Reversing the polarity of a determined axis in the skin relative to the muscle axis resulted in a conflict and hence duplication.

8.4 Proximodistal axis of the limb

Although it could not be ignored completely in the preceding section, the proximodistal axis demands separate treatment for two reasons. Firstly, the blastema always grows outward from its base on stump tissue. No way has yet been devised of altering this directional growth, partly because of the difficulty of securing the distal tip of a blastema to a wound surface and partly owing to the minimal growth shown by explanted blastemata. Secondly, the elongated PD axis of the limb lends itself to transposing a blastema to new sites along that axis to a degree which would be obviously impracticable along the AP or DV axis. In a sense, therefore, the experiments which have been performed involving the limb's PD axis expose different features from those considered previously.

The exchange of blastemata between proximal and distal limb stumps may once again be traced to Milojevic's (1924) pioneering studies on *T. cristatus*. As usual, however, we must doubt whether the youngest 12 day blastemata really conformed to their new position as he claimed or were replaced by stump cells; the regeneration of older blastemata was hardly affected by such displacement. Abeloos and LeCamp (1931) repeated this experiment on older blastemata whose AP and DV axes were reversed at the same time, with results which compelled them to modify their rules concerning the production of supernumerary arms. Duplicate and triplicate regenerates still occurred only when the blastemal and stump AP or DV axes conflicted, but were further restricted to cases where the blastema remained at its original PD position or was shifted to a more distal site. A proximally shifted blastema usually dictated the orientation of the regenerate but rarely provoked any duplication. This exception to the concept of conflicting polarities has been confirmed by similar operations on *N. viridescens* by Iten and Bryant (1975) and can probably be explained by their observations. The combination of a distal blastema with a proximal arm stump results in an exceptional amount of dedifferentiation in both, producing a composite blastema which can regulate to adopt a single axial orientation. With extremely small grafts, the blastemal determination is presumably swamped by dedifferentiated stump cells so that the regenerate's orientation conforms to that of the host with insufficient conflict to generate duplications (Table 8.5). Stocum (1975a) has recorded the same excessive dedifferentiation and delay of regeneration after grafting distal blastemata onto proximal arm stumps of *A. maculatum* larvae. Some of these regenerates possessed extra digits despite the conformity of graft and host AP and DV axes. That might arouse doubts about the conflict of axes being the sole cause of such abnormalities or, equally well, proclaim that minor conflicts occur whenever the grafts are not aligned with absolute precision. The converse experiment of transposing a proximal blastema on to a more distal amputation surface (necessarily of another limb) tends to yield extremely long regenerated arms containing serial duplication of the skeleton, such as a second elbow. Histological analysis suggests that the stump is completely passive when the blastema has not been rotated: its tissues do not dedifferentiate and do not influence blastemal growth to any detectable extent (Stocum, 1975a, Stocum and Melton, 1977). Conflicting AP axes in addition to the distal transposition, however, tend to result in complex regenerates containing both serial duplications and supernumerary arms (Table 8.5). There is still some measure of uncertainty implicit in the experiments described here. The grafted blastemata were always incomplete, lacking the basal region in the interests of purity, while the implantation sites usually contained some basal remnants of a previous blastema. It would be more precise to regard these operations as exchanging apical and basal portions of different blastemata, in proportions which must have changed for the different stages of regeneration tested. Placing grafts on freshly amputated stumps produces very similar results, however, presumably because stump cell dedifferentiate promptly (cf. Iten and Bryant, 1975), but all duplicated structures necessarily arise some way distal to the original site of amputation.

The practical restrictions on blastemal transplantation do not apply to mature arm tissues which can readily be rotated or shifted with respect to the PD axis of the rest of the arm. Most operations of this kind have been performed on the skin. Selective reversal of the skin's PD axis by transferring it to the contralateral arm without rotating it seems to have no influence on regeneration of arms amputated through the graft, whether or not the internal tissues had been irradiated (Tables 8.6 and 8.7). A 90° rotation of the skin to make its PD axis parallel to the plane of amputation, however, increases the frequency of both arrested and duplicated regenerates (Table 8.8). A single piece of skin rotated in this way before being grafted onto an irradiated arm at best forms a hypomorphic regenerate and is commonly unable to regenerate at all. When skin is transposed along the PD axis without reorientation, amputation through the graft produces results identical to the corresponding blastemal transpositions mentioned previously. Lheureux (1975a, b) provided the most recent and complete set of examples using *Pleurodeles* larvae. Amputation through fore-arm skin transplanted to the upper arm tends to yield a short regenerate and, conversely, upper arm skin on the amputated fore-arm often produces an abnormally elongated regenerate. This difference is intensified by prior irradiation of the internal tissues: fore-arm skin only produces a hand, whereas upper arm skin produces a fore-arm as well (Figure 8.4). Earlier transpositions of muscle or skeleton along the PD axis of axolotl and *T. cristatus* limbs gave essentially

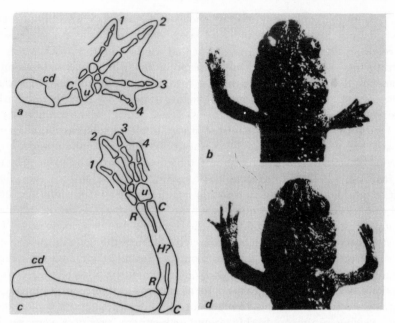

Figure 8.4. Regenerates produced on larval *Pleurodeles* by grafting skin with PD transposition followed by amputation through the graft (from Lheureux, 1975b). Fore-arm skin on an irradiated upper arm gives a foreshortened regenerate (*a*, *b*). Upper arm skin on an irradiated fore-arm gives an elongated regenerate with serial duplications (*c*, *d*) *H?*, possible humerus; *R*, radius; *C*, cubitus (ulna); *u*, ulnare. (Reproduced by permission of Spinger-Verlag.)

similar results (Vorontsova, 1937; Liosner, 1939; Juge, 1940). Also grafts of nerve tracts (Wallace, 1972) and cartilage taken from the upper arm and inserted into the irradiated forearm of axolotls yielded several regenerates with serial duplications of the skeleton. These last experiments involved quite small grafts, no residue of which were detected in the final regenerate and limb stump. They probably dedifferentiated completely in order to establish a blastema. Unlike the previous skin grafts which could be interpreted as contributing to a permanent stump influence on blastemal morphogenesis, therefore, these small grafts indicate that blastemal cells themselves remember their former PD position and are able to develop into all structures which normally lie distal to that position. Since the larger skin grafts on irradiated arms do not perturb the regenerate structure to a noticeably greater extent, their influence may also be the indirect one of providing determined blastemal cells. Similarly, regulation within a chimaeric blastema could easily account for the more normal regenerates obtained in the absence of irradiation. A conventional view of this evidence would portray the blastema as showing considerable determination in maintaining its regional specificity, PD position and orientation largely in defiance of a postulated stump influence. The more extreme view adopted here denies that the stump has any influence at all once dedifferentiation has established a blastema. If axial or positional incongruities still occur, then the stump may produce a second blastema and some regulation between the two blastemata may follow. The only evident difference between the PD axis and the transverse axes is that reversing the former has so little effect on morphogenesis. This axis seems to lack polarity, a proposition examined in the following section.

8.5 Reversed limbs

Contrary to Swett's (1927, 1928) conclusion that the formerly proximal end of an inverted limb bud was incapable of growth, proximal wound surfaces of older limbs have been observed repeatedly to form blastemata which regenerate into a normal succession of more distal limb parts. In fact Gräper (1922b) had already demonstrated such regeneration in the advanced limb buds of *Rana esculenta* and *Bufo viridis* tadpoles by first removing the tip of the limb, then excising and reimplanting the remainder in a reversed orientation. The former tarsal region was thus embedded in a pelvic wound, and a new leg of uncertain derivation sometimes arose from this complex. The cut end of the projecting thigh also grew at least part of a leg. Some of these regenerates consisted only of part of the thigh and all the more distal leg regions, regarded as a distal transformation during morphogenesis of a typical blastema, but other regenerates included a complete leg and a rudimentary pelvic cartilage. The latter would indicate proximal regeneration (cf. Harrison, 1925) if it were certain that the excision had not impinged on the true pelvic region. The existence of proximal transformation was not well documented in this study and has only found support elsewhere as a product of regulation among massed blastemata (see section 9.1).

In contrast to this scanty evidence, distal transformation from a proximal wound surface has been established beyond question. Several investigators have transplanted arms with reversed PD orientation onto the back or flank of adult European newts (Kurz, 1912; 1922; Milojevic and Grbic, 1925; Juge, 1940). The few regenerates which occurred always formed as mirror images of the grafted limb stump and thus represented distal transformation from the wound surface. The majority of reversed limbs were probably unable to regenerate because of an inadequate blood supply or innervation. Efimov (1933) attempted to overcome this problem by inserting a skinned arm segment through the tail, so as to leave both proximal and distal amputation surfaces exposed. Whenever regenerates formed on both surfaces, they were mirror images of each other. Monroy (1942) devised an ingenious means of delaying amputation until the arm was firmly attached at the elbow with alternative vascular and nervous connections. His operation involved skinning the elbow of adult *T. cristatus* and removing a piece of the humerus to render the arm immobile. He then sutured the elbow to a flank wound and allowed it to heal there for 20–30 days before amputating it both at the shoulder and through the fore-arm. He thus obtained a low proportion of 'bipolar regenerates' which invariably showed distal transformation and were usually more or less mirror images of each other. All subsequent limb reversals may be seen as derivatives of this technique. Butler (1951, 1955) simplified the operation in adapting it to *A. maculatum* larvae. He first amputated the hand and inserted the wrist into a fairly deep wound in the flank, to attain the posture of a hand in one's pocket or the cartoon figure of Napoleon. After allowing 3–4 days for healing and some recovery of circulation, Butler amputated the arm close to the shoulder. This resulted in a short shoulder stump which regenerated without difficulty, and a reversed arm stump which regressed to a variable extent before healing permanently or regenerating the normal distal succession of structures. More detailed analyses of such reversed arms and legs (Deck, 1955; Deck and Riley, 1958), although awkwardly expressed in terms of reversed laterality, revealed that the limb's AP and DV axes were faithfully retained in the regenerate but the two PD axes were mirror images of each other. Dent (1954) reached the same conclusion for the arms of adult *N. viridescens*, after further improving the operation by diverting brachial nerves to the implanted wrist and by allowing a recovery period of 2 weeks before the final amputation. The virtual absence of regression in these adequately supported adult arms allowed Dent to vary the amputation site and thus demonstrate that the most proximal region of the regenerate corresponded in structure to the end of the stump where the blastema arose. Within the limits of observation, therefore, regeneration had occurred exclusively as distal transformation. Oberheim and Luther (1958) also improved the innervation of grafted limbs, using a novel combination of Efimov's and Monroy's techniques. They grafted the arm of one *S. salamandra* larva transversely across the back muscles of another and diverted the latter's sciatic nerves into the wound area. After leaving the linked larvae under light narcosis for 3–4 days, they amputated the connecting arm at the shoulder and wrist to allow regeneration of both ends at the same time. This operation sometimes resulted in virtually perfect bipolar regenerates (Figure 8.5). All the operations described here probably produced poorly innervated limbs. Reversed larval limbs usually regressed considerably, often eliminating the entire upper arm segment and only regenerating a mirror image fore-arm. Some cases retained and duplicated part of the upper arm, but none regenerated structures more proximal than those remaining in the stump: they provided no evidence of proximal transformation.

Another instructive version of this 'Napoleon experiment' consists of removing both hands and suturing the wrists together end to end, in a mandarin posture. The stumps heal together and denervation of one arm allows nerves from the other to enter it and extend much beyond their normal length (Goshgarian, 1976, 1979). Delayed amputation of the formerly denervated arm then results in typical regeneration from a reversed limb

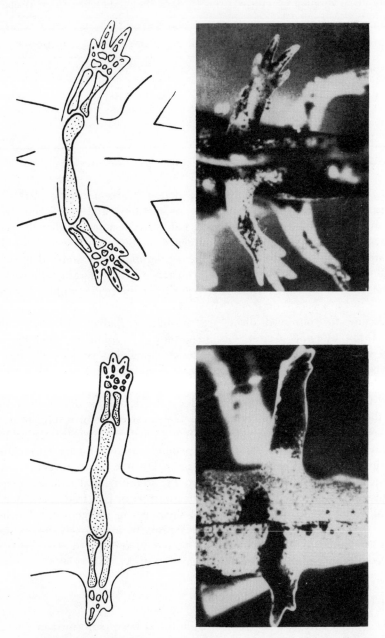

Figure 8.5. Bipolar regenerates on *S. salamandra* larvae (from Oberheim and Luther, 1958). The reversed arm is the upper one in both cases. (Reproduced by permission of Spinger-Verlag.)

Table 8.12   Classification of reversed arms 25 days after amputation (from Wallace, 1980).

Type of arm	Innervated	Denervated	Aneurogenic
Total amputated	34	32	26
Completely resorbed	4	5	2
Free end regenerated	7 (23%)	12 (44%)	10* (42%)
Implanted wrist regenerated	26 (87%)	18 (67%)	14 (58%)
Expected coincidence	20%	30%	24%
Bipolar regeneration	7 (23%)	7 (26%)	5 (21%)

*Only one of these cases had no detectable nerves, the other nine were sparsely innervated.

segment, although not yet analysed in detail (Carlson *et al.*, 1974). A comparison between these mandarin regenerates and the Napoleon regenerates innervated by relatively short flank nerves indicates that the type of innervation can have little or no influence on the morphogenesis of the regenerate. All these studies shared the assumption that some degree of innervation was indispensible for the occurrence of regeneration, however, and none of them had altered the PD polarity of the nerve supply. Indeed, there have been several comments suggesting the direction of nerve growth might dictate the observed distal transformation. Table 8.12 shows the results of my attempt to test this suggestion by producing aneurogenic reversed limbs on axolotls. A control series employing Butler's technique showed how difficult it is to implant a wrist deeply in the flank, for normal swimming or arm movements tended to pull it out again. The arm movements were eliminated in a second control series by denervating the arm at the time its wrist was implanted. The most common result for both control series and for aneurogenic arms was a partly implanted wrist which regenerated 1–4 digits from its exposed wound surface (Figure 8.6). After amputation close to the shoulder a week later, 20–40% of the reversed arms regenerated a mirror-image fore-arm from the free proximal end. Although aneurogenic arms behaved exactly like the controls in this test, a few nerve fibres had managed to penetrate most of them during the month following reversal. Distal transformation occurs as regularly in sparsely innervated reversed arms as in well innervated ones, therefore, and it also occurred in a few reversed arms which contained no nerves at all (Wallace, 1980).

8.6   Partial amputation surfaces

The operation of cutting away part of a limb has been performed in several ways, for diverse objectives and the rather variable results have been interpreted in quite contradictory fashions. Della Valle (1913) cut most of the way through an upper arm of *T. cristatus*, ligatured the incision to prevent it healing and completely severed the arm close to the elbow (Figure 8.7). The loosely attached distal segment produced in this way regenerated

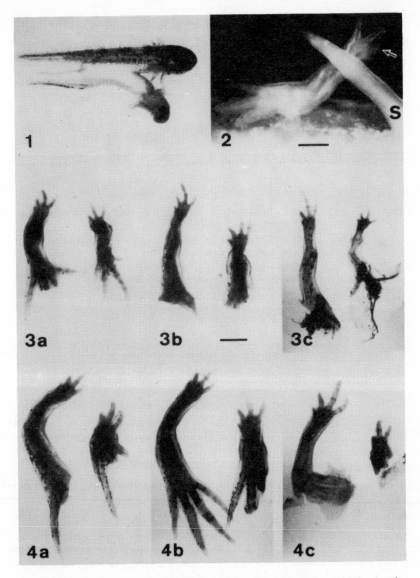

Figure 8.6. Reversal of aneurogenic arms (from Wallace, 1980). 1, the operated larva lacking the hind brain and trunk spinal cord is supported by a 12 mm normal host larva. 2, a control larva with three regenerates: a normal right arm from the shoulder (S); three right digits from the implant wrist; and a left arm from the free end of the implant (arrow). 3, largest and smallest reversed arm regenerates from each series: a, innervated control; b, denervated control; c, aneurogenic. 4, reversed arms which only regenerated following a second amputation, arranged identically to the photographs above them. Scale bars = 1 mm.

186

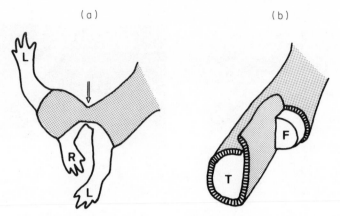

(a)   (b)

Figure 8.7. (a) Three regenerates obtained by ligating a left
upper arm of *T. cristatus* as performed by Della Valle (1913).
The ligature (arrow) partially isolated a stub and produces
two partial wound surfaces which regenerate complete
fore-arms, distinguished by left (L) or right (R) hands. The
amputated end of the arm also regenerates a left hand.
   (b) A stepped amputation on the lower leg of *T. cristatus*
(after Weiss, 1926), with a skin flap clipped over the lateral
wound surface. The major nerves lie between the tibia (T)
and fibula (F). Either half-limb surface may produce the
more complete regenerate, perhaps depending on where these
nerves were severed.

at both ends, providing yet another example of distal transformation from a
proximally facing wound surface. This operation has been repeated by
Nassonov (1930) on axolotls and by Eiland (1975) on *N. viridescens*. When the
adjacent incomplete wound surfaces are kept apart, either or both of them
may produce a complete regenerate. Eiland related the success of
regeneration to the degree of local innervation in terms of a threshold density
of nerves about one-third that thought necessary by Singer (1952).

Weiss (1926) invented a stepped amputation on the fore-arms and lower
legs of *T. cristatus* to produce a terminal amputation surface comprising half
the cross-sectional area of internal tissues but completely surrounded by skin.
The remaining tissue was cut back to provide another more proximal half
amputation surface (Figure 8.7). One of the two amputation surfaces usually
regenerated a complete hand or foot with normal structure and orientation,
while the other surface either healed or formed an incomplete regenerate
with one or two digits. Complete regeneration seemed to occur equally
frequently on the anterior or posterior half-surface, irrespective of which
happened to be the proximal and which the distal half. Gräper (1926)
obtained equivalent results from axolotls. Since either half of the limb stump
could produce a complete regenerate, the mosaic concept of each part of the
stump furnishing or organizing the corresponding part of the blastema

became quite untenable. Weiss consequently considered the blastema as indifferent building material subject to the stump's morphogenetic field which, by definition, would represent the whole limb in each half-section. Goss (1957a) repeated this type of operation on the arms and legs of *N. viridescens*, always removing the posterior half of the lower limb and amputating 2 weeks later through the middle of the remaining radius or tibia. He found that the absence of the ulna or fibula and their associated muscles usually did not prejudice normal regeneration provided the posterior skin was still present at the amputation surface. Removal of this skin as well as the posterior internal tissues, however, commonly resulted in defective regenerates which Goss characterized as anterior halves. Localized irradiation of the posterior tissues had much the same effect on regeneration as excising them (Goss, 1957b). These results are commonly taken to mean that each part of a wound surface specifies a corresponding part of the regenerate, with some allowance for a correction by skin of adjacent internal defects and vice versa. Such regulatory correction must be considerable, for either the skin or the internal tissues may be almost entirely eliminated or irradiated without preventing normal regeneration. In fact, the interpretation given by Goss was biased by concentrating on the average number of digits present on successful regenerates, although even his most severe operations sometimes yielded the full complement of digits (Table 8.13). Probably all of the half-limb surfaces were capable of forming a complete regenerate, but the operation (especially when it involved local irradiation) enhanced the frequency of hypomorphic regeneration. This interpretation seems preferable. in that it also fits the previous results, despite the arguments which Goss (1957a) mounted against it. It may merely indicate a relationship between the number of progenitor blastemal cells and the size of the blastema when morphogenesis sets in. Furthermore, the experimental defects inflicted by Goss were always on the posterior side of the limb which has been claimed to provide the majority of blastemal cells (Weiss, 1926; Singer *et al.*, 1964; Maden, 1976) and which may also contain a dominant morphogenetic centre. When Maden (1979c) repeated the partial irradiation experiment on axolotls, he also obtained defective regenerates from shielded anterior half-limbs but a shielded posterior half often formed a relatively complete regenerate. That refutes the general conclusions drawn by Goss (1957, 1961) whose 'half-limb field' was a rather mischievous subversion of the original field concept and quite unwarranted by the evidence at hand.

The decidely mosaic point of view adopted by Goss clearly influenced Stinson's investigations into the regenerative capability of anterior and posterior half fore-arms isolated by a more elaborate procedure. Stinson (1964a, b) irradiated one arm of *N. viridescens* with 1–4 krad X-rays, amputated it above the elbow and extirpated part or all of its humerus. He then inserted a fore-arm graft into this cavity and reamputated the arm through the graft 3 or 4 weeks later. The grafts were skinned 60 day fore-arm regenerates which had been trimmed down to possess only two

Table 8.13   Defect regulation in regenerating forearms of *N. viridescens*.

Amputation surface	Total cases	Percentage of cases classified by number of digits					Average no. of digits*
		0	1	2	3	4	
Exposed by excision of other parts (Goss, 1956a, 1957a)							
Anterior half	30	0	27	27	20	27	2.5
same plus posterior skin	35	0	0	6	9	85	3.8
Complete (control)	11	0	0	0	18	82	3.8
same less posterior skin	11	0	0	0	18	82	3.8
Skinned fore-arm graft in X-rayed site (Stinson, 1963, 1964a, b)							
Anterior half (digits 1 and 2)	23	13	30	48	4	4	1.8
Posterior half (digits 3 and 4)	24	21	29	46	0	4	1.7
Central part (digits 2 and 3)	30	7	20	20	30	23	2.6
Anterior half (digits 1 and 2)	40	22	38	40	0	0	1.5
Central part (digits 2 and 3)	40	30	15	17	8	30	2.7

*Excluding cases where regeneration failed completely (classified as 0 digits).

digits: either central portions bearing the second and third digit, or anterior of posterior halves. They were all autografts obtained from the contralateral shielded arms and were mostly inserted with reversed AP orientation. Stinson's main concern was to demonstrate whether or not the irradiated host limb had any influence on regeneration. Unable to detect any such influence, he adopted the provisional assumption that the grafts regenerated autonomously on an inert base. Subsequent evidence allows us to accept this assumption without reservation. The regenerates contained only fore-arm skeletal elements, representing distal transformation to complement the remaining portion of the graft. They apparently maintained the orientation, asymmetry and incompleteness of the graft. Most of these points confirm conclusions reached previously from less elaborate experiments and do not need to be argued here. The claim that anterior and posterior half grafts produced the corresponding half hand regenerates, however, requires a more scrupulous assessment. Stinston compared these grafts to what he called complete unirradiated control grafts, but these controls had been skinned and trimmed down to a central portion and all the grafts had been mildly irradiated. Stinston considered these central control grafts should be able to regenerate complete hands. Less than a third of them did so (Table 8.13) but all those which produced three or four digits might reasonably be taken as evidence for defect regulation. Alternatively, accepting Stinson's contention that they were controls for normal regeneration, the majority of hypomorphic regenerates from these central grafts must reduce the expectation of what can regenerate from the half grafts: those which regenerated two digits under such inimical conditions may well be thought to have compensated for some of their defects. Stinson placed considerable emphasis on the average number of regenerated digits in reaching his conclusion that partial grafts formed partial regenerates, and his analysis is clearly even more vulnerable than that of Goss to the criticism that any incomplete regenerate may be considered hypomorphic. Yet both these sets of experiments include results which vindicate the earlier discovery that a partial amputation surface can regulate to form a more complete distal regenerate.

8.7 Increased amputation surfaces

Weiss (1924) created a virtually double amputation surface by flexing and suturing together the fore-arm and upper arm of *T. cristatus*, then amputating through both segments close to the elbow. The variable amount of regulation towards a single regenerate which he recorded can be rationalized quite easily in terms of the preceding sections of this chapter. Providing they became sufficiently innervated, both amputation surfaces should regenerate by distal transformation and could do so jointly by accomodating the slight initial discrepancy in PD position. The regenerated structures should maintain the common DV orientation of the two stump surfaces, some structures being present as duplicates and others united by a central fusion. Monroy (1946) repeated this operation with similar results, which he compared to those he had obtained earlier by suturing both arms together behind the newt's back. In the latter case, the fore-arm stumps were attached together along their anterior surfaces and thus tended to regenerate expanded hands with up to seven digits when the stumps lay almost parallel to each other. As the angle between the arm stumps increased, however, he obtained a series of cases with smaller regenerates and fewer digits: an obtuse inter-arm angle resulted in extremely hypomorphic hands, for example with only a single finger. Monroy (1946) suggested that an angle approaching 90° might yield a single normal regenerate, but in doing so he ignored the obvious bilateral symmetry which the results displayed. All the larger regenerates could be classified as double posterior structures with a variable suppression of the intervening 'anterior' skeletal elements. He illustrated the central suppression by representing the limb's morphogenetic field as lines of force along its PD axis and projected from the amputation surface (Figure 8.8). The lines extrapolated in this way from double stumps interfered with each other to a degree which corresponded to the observed reduction of regenerate structure. The interference is only another geometrical expression of the inter-arm angle, of course, and Monroy admitted that mechanical interference between two converging blastemata could equally well account for the occurrence of hypomorphic regenerates.

These observations have been confirmed twice following similar operations on *N. viridescens*. Goss (1956b) sutured arms together along their ventral sides and then amputated them to obtain some completely duplicated regenerates whose hands were arranged as though clapping, palm facing palm.

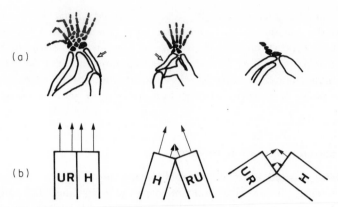

Figure 8.8. (a) A range of regenerate skeletons obtained by suturing an arm of *T. cristatus* in a fully flexed posture and then cutting off the tip of the elbow (adapted from Monroy, 1946). (b) Explanatory diagrams showing the angle between arm segments and 'lines of force' (arrows) projecting from each segment. Defective regeneration could be attributed to interference between lines of force, or to reduced growth from a restricted wound surface, H, humerus; R, radius; U, ulna; white arrows mark extra regenerated ulnae.

Oberpriller (1968) attached the posterior sides of the two fore-arms together, and obtained bilaterally symmetrical regenerates with anterior structures on both sides and a variable amount of 'posterior' structures occupying the centre of the hand. Both investigators concurred with Monroy's finding that even minor inter-arm angles reduced the number of regenerated digits from the expected maximum of eight, but their interpretations of this relationship differed considerably. Goss (1956b, 1961) favoured the more pragmatic explanation of a mechanical interference between the two potential blastemata, while stressing the duplicated base present in all regenerates as evidence of stump influence. He described the more distal unification as mutual suppression of homologous parts but barely considered it as an instance of regulation. Oberpriller (1968) adopted Monroy's speculation concerning lines of force projecting from each stump surface and attempted to demonstrate it by joining a normal left arm to an irradiated right one. The vast majority of regenerates from this operation were obviously left hands with 3–5 digits, even at obtuse inter-arm angles (Table 8.14). The simple mechanical explanation of that result is that the irradiated stump does not produce a blastema and so cannot interfere with regeneration from the normal stump, apart from providing an inert extension of the amputation surface and thus allowing the single blastema to spread laterally. Oberpriller, however, invoked the postulated longitudinal control by the stump regeneration fields and the further premise that irradiation blocks communication along the PD axis. These are mutually consistent fancies, of course, and consequently could account for her results even though neither of

Table 8.14 Regeneration from a common amputation surface on left and right arms sutured together (data of Oberpriller, 1968).

Inter-arm angle (degrees)	No irradiation			Right arm irradiated		
		Number of digits			Number of digits	
	Cases	Mode	Range	Cases	Mode	Range
ca. 0	5	6	5–8	2	3–4	3–4
30–60	13	5	3–5	8	4	2–6
ca. 90	7	1	0–3	8	5	0–5
100–150	2	0–1	0–1	10	4	1–5

them was necessary or even likely (see section 4.5). The experiment thus conveniently supports whatever assumptions are built into its interpretation. The unirradiated control specimens, however, furnished particularly clear evidence of regulation between two convergent blastemata by the suppression of their common central elements. In the absence of any evidence of a morphogenetic control exerted by the limb stumps, this should be ascribed to internal blastemal regulation.

8.8 Conclusions

The mosaic concept of strict part-for-part regeneration prevalent at the beginning of this century has been gradually replaced by a more regulative view. Regeneration can occur following stimulatory damage within a regenerative territory such as the appendage and its immediate surroundings. Although territories map as a mosaic pattern over the body surface, they resemble the morphogenetic fields of embryos and retain some of the same capacity for internal regulation. The wound surface of an amputated limb should thus be considered as part of a field and its regeneration seen as essentially a regulation to complete that field: covering the surface with a corresponding blastema or one from a more proximal position completes the field artificially and suppresses normal regeneration. Furthermore, partial and increased amputation surfaces both show a strong tendency to form a complete distal regenerate. Embryonic fields are expressions of regional determination, as the arm field has a different developmental fate from the leg or tail field, and that is equally true of the corresponding regenerative territories. When an arm field or territory is transplanted to a neutral region it develops or regenerates an arm regardless of its new position; when combined with another field or territory it produces a structural chimaera.

The embryonic limb field undergoes a successive determination of cardinal axes while still part of the flank mesoderm, even though considerable regulative ability persists in the later limb bud. Regulation is equally apparent in the growing blastema, whose axes must be either inherited on a cellular basis from adjacent stump tissues or imposed by an inductive action of the limb field. The term 'stump influence' is currently used to include both possibilities and thus perpetuates the ambiguity. Experimental evidence concerning regional and axial determination is always found to be consistent with, and usually in favour of, the conclusion adopted here: once it has provided the dedifferentiated mesenchymal cells which are already imprinted by their former axial position within a specific territory, the limb stump has no further influence on blastemal morphogenesis. The blastema grows as an autonomous unit so that its regulative and morphogenetic alterations are not susceptible to stump inductions. Depending on the validity of this conclusion, there are no known differences between developing limb buds and regenerating blastemata: morphogenesis should involve the same mechanisms in both.

Conversely, the blastema has a lasting influence on neighbouring stump tissues, normally acting to suppress their further dedifferentiation. When a blastema is shifted to a more proximal amputation surface, however, the latter continues to dedifferentiate and eventually regenerates an intercalary limb segment. Any discrepancy between the AP and DV axes of the stump and those of a transplanted blastema may also allow the stump to escape from this suppression and to produce a supernumerary regenerate. Reorientation of limb tissues prior to amputation also results in multiple regenerates, presumably by identical axial conflicts within the blastema. Since displacements of dermis or muscle are found to be the major or exclusive causes of such conflicts, just as they are the major sources of mesenchymal cells, it follows that axial determination is a property of the blastemal mesenchyme. Regional specificity is also encoded in the blastemal mesenchyme and limb bud mesenchyme possesses identical properties. The epidermis covering blastemata or limb buds does not display such determination when subjected to the equivalent tests, although it may guide the direction of growth.

Morphogenesis is thus envisaged as an autonomous process within the blastema, aided by regulation for defects or excesses of mesenchymal cells and perhaps by a mutual affinity of similar cells which causes them to aggregate. Heterotopic grafts prior to amputation or later blastemal transplantations presumably lead to multiple subpopulations of blastemal cells differing in their regional or axial instructions. Such differences are often sufficient to prevent them from merging into a coherent unit as they normally would before redifferentiating. Based on this type of study, blastemal cells seem strongly committed to their regional origin but rather less firmly to their axial position within that particular territory, while merely evincing a slight nostalgic tendency to revert to their previous form of histological differentiation. Some such order of priorities may well be required for regulation to occur at all. Contrary to the earlier definition of regeneration in the teleological terms of completing a field, studies of reversed limb stumps have revealed the general mechanism of morphogenesis to be one of distal transformation. The base of the blastema retains virtually the same specification as the amputation surface, the apex of the blastema corresponds to the tips of digits, and intervening blastemal regions match those of a normal limb between these specified points. In other words, more distal regions of the growing blastema display an altered response to the predetermined AP and DV axes to produce a progressive broadening and flattening with separate skeletal condensations through the fore-arm, wrist and hand regions; perhaps culminating in restricted growth centres corresponding to each digit. This description of morphogenesis is merely a speculative outline which may serve to indicate the sort of detailed mechanisms involved in the process. Other equally speculative treatments will be considered in the following chapter.

Chapter 9

Blastemal Morphogenesis

The morphogenesis of a limb is such a complex process that it has only been considered in greatly simplified and abstract terms. The major simplifying assumption consists of stating that detailed anatomical features not only mask an underlying basic pattern but are in fact derived from such a pattern during development or regeneration. The basic pattern is usually related to the three cardinal axes or spatial dimensions by which the limb is conventionally described, as used in earlier chapters. Each of these axes has at one time or another been considered as an expression of a gradient, of a morphogenetic field or of positional information—and all three concepts overlap considerably.

The vogue for discovering physiological and metabolic gradients in eggs, embryos or growing and regenerating systems generally has been amply documented by Child (1941), Needham (1931), and Huxley and de Beer (1934). The best examples are probably the physiological polarity which can be imposed on *Fucus* eggs by several external agents such as light or pH gradients (Jaffe, 1968, 1970), and the longitudinal polarities in regenerating planaria and hydroids which can be reversed by metabolic inhibitors or by applying electric potentials (reviewed by Bronsted, 1969; Rose, 1970b). Nothing of this kind has been established in the limb bud or blastema, whose basic patterns are understood as hypothetical gradients but usually conceived as the concentration gradient of some diffusible molecule between a steady source of supply and a 'sink' where it escapes or is destroyed. Any such gradient which happens to be discovered might in principle specify the polarity of an axis or of course, be merely a subsidiary effect of the polarity.

The morphogenetic field (expounded by Weiss, 1939) represents an attempt to adapt theoretical gradients into a two- or three-dimensional pattern and at the same time accommodate the regulative aspects of development. The embryonic limb field, for instance, includes all the flank area which can be stimulated to produce a limb. The intact field normally forms only one limb and determines its orientation, yet an isolated part can also form an entire limb and two parts can be combined into a single field. If the postulated gradients are significant factors in determining this reaction, then the partial gradients present in each part must be able to reconstitute a single complete one. Two gradients corresponding to the determined AP and DV axes should suffice to specify the embryonic limb field, with outgrowth of the limb bud occurring at a particular intersection. The proximodistal axis of outgrowth, however, is usually considered to have an equally determined polarity represented by yet a third gradient. It is this axis which is principally affected by amputation, and limb regeneration can be envisaged as regulation to re-establish its normal dimensions. Unlike regulation of the embryonic limb field, this involves obvious growth. Morgan (1901) initially distinguished between morphallactic regeneration, a remodelling of part to reconstitute the whole organ or organism in the absence of growth (as happens in *Hydra* and embryonic regulation), and epimorphic regeneration by means of blastemal growth. It is not profitable to pursue this distinction too far, however, for the main axial pattern of the limb regenerate seems to be present already in the earliest blastema. Growth merely allows that basic pattern to be expressed.

Basic patterns involving gradients or fields greatly influenced contemporary thought for 20 years or more but, as Weiss (1939) disarmingly observed, were more successful in systemizing previous knowledge than in predicting future discoveries. They were largely forgotten for a period in the enthusiasm over biochemical and molecular approaches to development which some now claim to be equally disappointing, only to emerge again disguised as positional information. This concept, formulated by Wolpert (1969, 1971) and applied primarily to regeneration of *Hydra* and chick limb development, mainly differs from its predecessors by concentrating more on individual cells than on the entire field. According to Wolpert, cells need to be aware of their position in terms of reference coordinates: all cells using the same coordinates will constitute a field and a disturbed field must re-establish its boundary values in orders to regulate. I shall follow the general practice of assuming that these boundary values are identical to the reference points used by all cells within the field, although that is not theoretically necessary. The boundary values then set up a simple pattern such as gradients across the field which individual cells can interpret and modify their behaviour accordingly (by proliferating, migrating, differentiating, setting up subsidiary centres of influence, or even dying). The consequences of fixed boundary values for both diminished and expanded fields are that the gradients alter but remain complete. Morphallaxis would then involve each cell reassessing its position according to an altered gradient. Epimorphosis is more complicated in that the exposed edge of the previous field retains its positional value, while generating a blastema whose apex may adopt the missing boundary value either initially or finally. In addition to that, according to evidence cited earlier, blastemal cells have an excellent memory of their previous positional value and may match that against the position supplied by new reference points for their future navigation.

This rudimentary account of the main concepts of morphogenesis is only intended to facilitate the following description of how they have been applied to regeneration. It will be assumed that morphogenesis is autonomously controlled within the blastema, based on arguments presented in the two previous chapters and reinforced by other experiments mentioned later. It is convenient to deal first with the proximodistal axis of the blastema, about which there is a measure of agreement, before comparing diverse views about the transverse axes.

9.1 Proximodistal axis of the blastema

As recounted earlier, there was general agreement until quite recently that the blastema was initially very much subservient to the limb stump. In particular, the PD limb axis was considered to be projected on to the blastema as an extension of the morphogenetic field. Attempts to transplant blastemata to other sites initially seemed to support this view, for an early blastema was found incapable of maintaining its regional or axial determination when confronted with heterotopic host tissues but similar transplantations of a blastema supported by a short stump segment were not affected by the host site (Giorgi, 1924). Such results are attributed nowadays to regression of the grafted blastema following dedifferentiation and erosion of its basal cells. Much of the recent debate about blastemal morphogenesis also hinges on the extent of basal erosion after transplantation. Mettetal (1939) grafted limb blastemata to the head or back of the same animal, using *T. cristatus* and larval *S. salamandra*, and observed the grafts often formed defective limbs but evidently retained a considerable degree of determination in these neutral territories. All the grafts seemed able to grow digits and successively older blastemata also formed more proximal parts of the hand or fore-arm. Mettetal accordingly suggested the blastema grows as a fountain of mesenchyme cells emerging from the end of the limb stump. Those arriving first to constitute the early blastema are determined to produce the most distal region of the regenerate and still do so after isolation to a neutral site. Later arrivals are either predetermined to form more proximal structures or are more subject to induction by the limb stump (Mettetal, 1952).

Faber (1960) effectively refuted most of this interpretation by showing that later blastemal growth in axolotls occurs by internal cellular proliferation instead of basal addition. By tracing carbon marks inserted near the tip of an early blastema to the fore-arm of the later regenerate, Faber deduced that most proliferation and growth occurred in the distal region of the blastema which he designated an apical proliferation centre. It is surely true that intense cell division continues in the apical parts of the blastema when the basal region shows fewer divisions among differentiating tissues, but there is

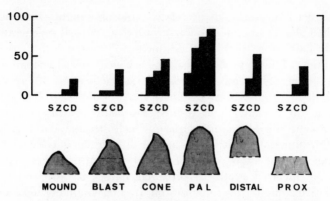

Figure 9.1. Histograms showing the principal results obtained by Faber (1960) by grafting upper arm blastemata into the flank. The ordinate is a rough measure of morphogenesis: the mean number of skeletal elements detected in each region (S, Z, C, D) of successful cases expressed as a percentage of the number in a normal limb. S, stylopodium (humerus); Z, zeugopodium (radius and ulna); C, 8 carpals; D, 13 metacarpals and phalanges. The blastemal stage or part grafted is shown underneath each histogram. Completely resorbed grafts were ignored. Note the more complete development achieved by older stages, and the failure of half-palettes to form more than hand skeleton.

no evidence of a discrete apical centre with well-defined limits. Even the carbon marks employed by Faber were widely dispersed when observed in sections of the regenerates, being merely more concentrated and thus more visible in the forearm. After repeating and extending Mettetal's series of blastemal grafts, Faber (1960) reached the conclusion that his apical proliferation centre was also an autonomous organization centre determined to produce distal structures. This was conceived as antagonistic to a gradient of influence from the limb stump which tended to induce neighbouring blastemal mesenchyme to become more proximal in structure. When grafted to the mid-flank, whole upper arm blastemata produced a spectrum of results represented by the histograms (Figure 9.1). Young blastemata often regressed completely and most of the survivors only produced digits. Cone and palette stages usually survived the operation better to form carpal elements in addition to digits, with several cases of fore-arm skeleton and an occasional humerus. Faber recognized that the operation caused renewed dedifferentiation and considerable erosion from the base of the blastema but doubted if that could entirely explain the relative deficiency of proximal skeletal elements. By transplanting early blastemata which contained carbon marks in their distal mesenchyme, he showed that apical proliferation still occurred after grafting in addition to basal regression. Faber then transplanted proximal and distal halves of palette stages side by side to the flank in order to assess their relative morphogenetic tendencies. The proximal half should provide the fore-arm during normal regeneration, while the distal half was normally destined to become the hand. After transplantation, however, both halves behaved in a virtually identical manner to form digits and some carpals as inefficiently as grafted whole early blastemata (Figure 9.1; cf. Faber, 1962). Basal regression and apical growth can still explain these results, especially as the proximal halves carried an extra distal wound surface which became covered by a new wound epithelium and apical cap and dedifferentiated completely (Faber, 1965). That would allow regulation into an entire small blastema, as described for inverted grafts of proximal halves by Michael and Faber (1961). Since the site of most intense cell division gradually moves out along the growing blastema (Figure 1.3), followed by waves of tissue demarcation and differentiation, it is reasonable to envisage Faber's centre of proliferation and organization as comprising the whole blastema. In that case, its observed autonomy stands as evidence against any induction from the limb stump. Faber (1971, 1976) has partly conceded this argument when revising his interpretation in the light of subsequent investigations.

The autonomy of the blastema was also successfully tested by Stocum (1968a) who demonstrated that the mesenchyme of cone and later stages from the upper arms of *A. maculatum* larvae could differentiate cartilage and muscle when explanted into a suitable culture medium. In fact, the cultured mesenchyme differentiated better in complete isolation than when attached to a piece of the limb stump. He was unable to assay morphogenesis by this technique because a considerable amount of the mesenchyme spread over the surface of the culture vessel. Stocum (1968b) therefore also transplanted the blastemata as autografts in the dorsal fin, and generally confirmed many of Faber's observations as shown in Figure 9.2. Cone and palette stages produced relatively complete skeletons if segments of the stump were transplanted with them but developed considerably less well if transplanted alone. The defects of proximal structure were quite pronounced, but less so than in Faber's experiments. Grafted distal halves produced hands and digits, while some proximal halves also produced fore-arm skeleton, despite being impeded by regression at both wound surfaces. Stocum reduced the regression by placing the proximal halves on their sides partly buried in the fin. These grafts developed much more completely to show high yields of fore-arm and carpals (Figure 9.2), and those which managed to protrude from the fin also formed digits. These results amply confirm a cone blastema's capability for autonomous morphogenesis when isolated from its stump, and also provide strong evidence of some tendency for self-determination of proximal and distal parts—with the provision that any part which dedifferentiates can then grow a complete series of distal structures. Stocum and Dearlove (1972) examined this last aspect by stripping the epidermis from young blastemata and implanting them into tunnels excavated in the fin. Those which lay superficially enough to come

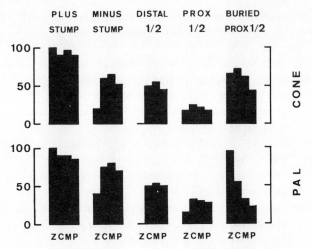

Figure 9.2. Histograms of the results obtained by Stocum (1968b) by grafting upper arm blastemata to the dorsal fin, using cone stages (upper row) and palette stages (lower row). The ordinate is identical to that in Figure 9.1 but the four metacarpals (M) and nine phalanges (P) are considered separately here, and no humerus was recorded in the regenerates. Note the fairly complete development in the absence of stump tissue, and that protected proximal-half grafts form mainly proximal structures.

into contact with the host wound epithelium eventually grew out to form more metacarpal and phalangeal cartilages than did deeply implanted blastemata, by extra growth rather than at the expense of losing proximal structures.

The importance of graft regression has been pursued by de Both (1970b), principally by transplanting the mesenchyme from several blastemata into a flank wound or the enucleated orbit of axolotls. The latter site yielded superior differentiation but de Both found it easier to insert multiple implants into a cavity in the flank musculature. Grafting a complete early upper arm blastema over the mesenchyme of two or three others in the flank, de Both found they regulated to produce a single regenerate with a more complete skeleton than he could obtain from a single grafted blastema. More surprisingly, a multiple graft of four hand plates from upper arm palette stages sometimes regulated into a regenerate containing fore-arm skeleton and once with a humerus, in addition to the expected carpal and digital cartilages. Suspecting that an enlarged composite blastema might be able to show proximal transformation, de Both (1970b) implanted a palette stage carpal blastema on top of the mesenchyme of three others in the orbit. He identified a radius or ulna (which such blastemata would not form normally) in two regenerates out of a total of 44 such operations. These two cases stand as unique examples of proximal transformation, for the other cases mentioned were derived from the distal portions of upper arm blastemata, as pointed out in a perceptive commentary by Stocum (1975b) who was unable to confirm these last results. The identification of the radius or ulna in these two cases apparently depended on the size and position of the cartilage as much as its shape: it was large and embedded in host tissue below the regenerate's carpals, according to Faber (1976). Those criteria would usually be accepted without question were it not for the theoretical importance attached to the results. Michael and Faber (1971) subsequently reported that a row of fore-arm blastemata completely embedded in the flank could combine to differentiate into a single large piece of cartilage, but refrained from identifying it as a humerus.

Perhaps blastemal mesenchyme tends to differentiate into excess cartilage when its distal growth is impeded, in which case even the skeleton would provide misleading evidence of morphogenesis. Even so, de Both's results confirm the ability of mesenchyme to regulate into an autonomous mass which controls its own development into a variety of limb regions, in the absence of any stump induction. That conclusion conforms perfectly to the results of shifting blastemata to more proximal levels in the arm (section 8.4), where the graft suffers some delay but is not diverted from its expected course of development, while defect regulation occurs as intercalary regeneration by the host arm stump. Maden (1980a) has produced the most recent and convincing demonstration of this by exchanging blastemata between dark and white axolotls and by irradiating either the host arm stump or the graft. He found no sign of proximal transformation even in the most favourable circumstance when the proximal stump had been irradiated and so was unable to remedy the proximodistal defect. This study also revealed a novel form of intercalation: a proximal blastema placed on a distal stump could sometimes produce an intercalaray regenerate, probably of reversed polarity and thus by distal transformation from the base of the blastema.

Among many experiments which could have revealed the effect, only one has been interpreted as supporting the possibility of proximal transformation. Dinsmore (1974) confirmed de Both's observation that a carpal blastema only regenerated into hand and digits when transplanted into the orbit of an *A. maculatum* larva. When transplanted on to a bed of minced muscle in the orbit, however, such a blastema quite often formed parts of the radius and ulna as well. Dinsmore suggested the blastema not only incorporated dedifferentiated cells released from the muscle and organized them to suit its needs, but also inferred the increased size achieved by incorporating extra cells was the only important feature in the blastema's ability to form more proximal skeleton. There is no need to take his argument seriously, for the minced muscle had been obtained from an upper arm. Assuming dedifferentiated muscle cells were incorporated into the blastema, as seems likely, they presumably retained their normal propensity to regenerate into upper arm structures or to produce the observed radius and ulna by distal transformation.

9.2 Models of the proximodistal axis

The investigators mentioned above were rarely able to resist the opportunity offered by their results to speculate about the nature of determination. We may limit the variety of theories to be discussed by excluding those which involve stump induction, an apical organization centre or proximal transformation—because there is insufficient evidence of these phenomena. Growth of the blastema occurs by mesenchymal proliferation guided by an apical epidermal cap; it is followed by a proximodistal sequence of determination and tissue demarcation, and then by a similar sequence of cellular differentiation. Blastemal mesenchyme may reform the structures from which it arose or any more distal structures, under autonomous internal regulation of the whole blastema. Accordingly, the blastema is an independent self-differentiating system or field and its proximodistal polarity can be represented as an accentuated section of the gradient which depicts

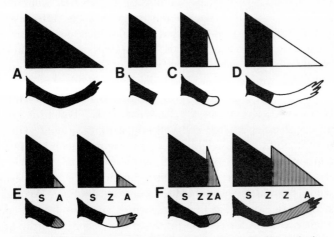

Figure 9.3. The proximodistal axis of a limb and its postulated gradient. A, an intact limb; B, after amputation; C, during regeneration; D, completely regrown. No curves are included in the graphs as they could imply particular diffusion gradients. E and F, application of the gradient to proximal and distal transposition of blastemata as executed by Stocum (1975a). Stump dedifferentiation occurs when confronted with more distal tissue (in E), allowing intercalary regeneration. Confrontation with more proximal tissue (in F) does not elicit any comparable reaction from the stump, but the blastema may sometimes grow back as a reversed polarity intercalary regenerate (Maden, 1980a). S, stylopodium; Z, zeugopodium; A, autopodium (explained in Figure 1.1).

the PD polarity of the whole limb (Figure 9.3). The simplest basic concept of the limb as a morphogenetic field containing a gradient of some diffusible or propagated substance, emanating from the base of the limb or boundary of the regenerative territory, can be dismissed immediately. The development of transplanted blastemata is normal enough to convince us they contain a complete set of morphogenetic instructions. They must be aware of their position along the PD axis, and thus contain at least one reference point and the ability to exhibit distal transformation during apical growth. That leaves three categories of model which have all been espoused during the last few years: distinguished by invoking basal or apical boundaries of the blastema as reference points, or both.

Faber (1976) abandoned the apical organisation centre of his previous articles in favour of the idea that the apical epithelium creates a distal boundary or reference point in the immediately underlying mesenchyme (which may not be a very different viewpoint). The 'progress zone' theory of chick limb development applied to regeneration by Smith *et al.* (1974) involves an initial basal boundary, but growth occurs in the progress zone of the apical mesenchyme, where proliferating cells autonomously take on a progressively more distal value, by monitoring the growth or number of cell divisions. When cells are left behind the progress zone and cease dividing, their positional value is fixed—so that initial relicts form more proximal structures than later ones. The mechanisms operating in these models seem to be a classical morphogenetic gradient on one hand and an internal egg-timer on the other. The model involving both basal and apical boundaries invented by Maden (1977b), is curious in that the two reference points only affect other cells in the field indirectly. The principle invoked here is that any dedifferentiated cell will adjust its positional value to adopt the average of its immediate neighbours, a procedure which is repeated when the neighbours also change their status as the blastemal PD axis elongates (Figure 9.4). Such a scheme could operate by surface contact between cells and so avoids postulating a diffusible chemical gradient. Wolpert (1969) pointed out that one advantage of models which stipulate both boundaries of an axis is their provision for cells to interpret the steepness of the gradient as an instruction for linear growth—an essential requirement for epimorphic regeneration. Maden's model adapts this idea by assigning integral positional values to the cells and stating that cells will continue to divide until a complete series has been created by intercalation or adjustment between the boundary values of the blastemal base and the apical epithelium.

All three models necessarily have much in common, being attempts to describe normal regeneration in abstract terms, besides incorporating those experimental results which most impressed their designers. The progress zone invokes a proximal boundary and does not seem to require a distal one. Faber (1976) reverses this scheme by arguing that mesenchyme cells can retain their original value or even attain a more proximal one unless influenced by his postulated distal boundary. Maden's model stipulates both

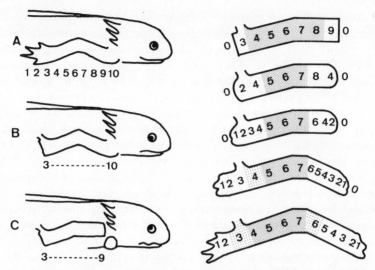

Figure 9.4. Bipolar regeneration interpreted according to the numerical model proposed by Maden (1976, 1977b). Cartoons on the left show the numbers assigned to the PD axis of a right arm (A), the first operation of partly implanting its wrist into the flank (B) and later amputation at the shoulder (C). Diagrams on the right show successive stages of regeneration. Wound epithelium with a value 0 induces underlying tissue to dedifferentiate and drop to an intermediate value, thus inducing adjacent tissue to follow suit. Cell division in these regions allows intercalation of missing values and growth. Redifferentiation then spreads from the unaffected central region (shaded), resulting in a completed right hand and left fore-arm dictated by the reversed sequence of numbers (see Figure 8.6).

boundaries and would also allow overall proximal transformation. These discrepancies expose our ignorance and emphasise the fundamental decision which hangs upon whether or not de Both's (1970b) claim of proximal transformation can be confirmed. A decision between the three models could also depend upon the identification of a discrete apical centre of proliferation which is fundamental to the progress zone theory but less important or incidental to the others. Chalkley (1954) found that cell division was not localized at any stage of regeneration in *N. viridescens* and Maden (1976) confirmed the general distribution of dividing cells in axolotl blastemata. Smith and Crawley (1977) claimed to have detected an apical localization of cells which had entered the S-phase, but only after the onset of basal redifferentiation and only by repeatedly smoothing their raw data from blastemata of *T. cristatus*. I take that to mean no distinct progress zone can be detected in any of the three species routinely employed for studies of regeneration.

A further difference lies in whether distal transformation occurs by apical growth (exclusively in a progress zone or more diffusely perhaps in Faber's model) or is merely an example of intercalary regulation which can occur between any two reference points. An experiment performed by Shuraleff and Thornton (1967) is often cited in this connection and this is perhaps the best place to describe it. They amputated legs of large axolotls in the mid-thigh region, sutured a skin flap over two-thirds of the wound surface and grafted a foot on to the remaining wound surface. After allowing 10 days for a recovery of circulation and innervation, they sliced off the sutured skin to expose the former wound which, as expected, formed the basis of a 'supernumerary' leg. The experiment did not achieve its intended purpose of testing if the presence of a foot would prevent regeneration (cf. Carlson, 1977), but it yielded an intriguing side-effect: the grafted foot was displaced distally until it reached a definitive position attached to the tarsus of the regenerated leg. Thigh segments grafted in an indentical manner remained at the original site of amputation, and it was assumed that no growth would have occurred if the wound had not been re-opened (although I have not discovered a demonstration to that effect). It certainly seems as though a defect along the PD axis can be remedied by a genuine intercalation involving limited distal transformation of the stump tissue which ceases once the missing positional values have been restored. Presumably this response could only occur following local dedifferentiation involving either an open wound or the equivalent of a wound epithelium, as required for normal regeneration (Thornton and Kraemer, 1951). There are several analogous observations of relatively modest amounts of intercalary regeneration after transplanting blastemata on to limb stumps. The base of the blastema is probably always eroded to some extent, so that it tends to form a regenerate with proximal defects which can be replaced by distal transformation of the stump. Any supernumerary limbs caused by conflicts on the transverse axes usually emerge somewhat distal to the site of transplantation—as digits after fore-arm grafts or fore-arms after operations on the upper arm. Gross intercalary regeneration occurs typically when a fore-arm blastema has been grafted on to an upper arm stump, except when the stump has been irradiated (Maden, 1980a).

These examples appear to favour the model formulated by Maden (1977b) and could be accommodated by Faber's (1976) latest proposals, but seem quite incompatible with the progress zone theory. We must be content to leave the two most plausible models at present, for several other versions could be formulated. This year's model (Slack, 1980), for instance, treats the proximodistal axis in much the way advocated by Maden but dispenses with his intercalatory averaging process. It employs a series of intercalatory steps down from the blastemal base to an apical zero value, which resemble the successive steps of a progress zone but are controlled by a gradient of decremental stump induction. Stump induction was dismissed as a fallacy (section 9.1), but the rest of the model seems no worse than any other. All these models should be treated as fairy stories. They may provide a psychological insight into the way their authors like to consider morphogenesis, but they are too divorced from reality to help us understand the process itself.

9.3   Transverse axes of the blastema

The determination of these blastemal axes has usually been tested by reversing their polarity with respect to the corresponding stump axes. Replacing a blastema without reorientation on its former site provides a suitable control and usually allows perfect regeneration to continue, after some delay while a new blood supply and innervation are established. If the blastema is rotated 180° about its PD axis before replacement, however, both the AP and DV axes are reversed in polarity relative to the stump axes. Exchanging blastemata between contralateral limbs can similarly create a reversal of either the AP or the DV axis. Such aligned axes with reversed polarities are said to be in conflict with each other. Milojevic (1924) employed this test on regenerates of *T. cristatus* and concluded that very early blastemata were undetermined because they grew into regenerates which conformed to the host axes, but the blastemal axes became defined about 12 days after amputation. Studies of graft resorption cited in section 9.1 support the widespread criticism of these results: the early blastemata may have been completely replaced by a host regenerate, so their axial determination remained unknown.

Milojevic's article was intended as a résumé of a broad investigation and provides little detail of his results. It is more profitable to examine the very similar study performed by Schwidefsky (1935) on *T. vulgaris* which provides enough data to us to follow his interpretation. Schwidefsky transplanted 8–15 day blastemata and 16–20 day cone stages on to contralateral limb stumps, to reverse either their AP or DV axes. He also exchanged them ipsilaterally between arms and legs without reorientation, as a test of their regional determination. The results are set out in Table 9.1 to show the basis for Schwidefsky's conclusion that AP and DV axes become determined in 16–17 day cone stages, whereas regional determination is only established at about 18 days. He considered the younger blastemal grafts formed harmonic (ortsgemass) regenerates under directions from the host limb stump, and so must have been undetermined at the time of transplantation. Most of the cone grafts evidently had rigidly determined axes and so formed disharmonic (herkunftsgemass) regenerates despite the presumed host influence. The criticism levelled against this conclusion is essentially that there is no independent evidence that young blastemal grafts survived. Erosion of the graft and its replacement by a host blastema would give the observed results and, in that case, there is no evidence of any host influence on a graft. Schwidefsky's data actually provide some support for the occurrence of graft erosion. The number of regenerates which developed well enough to be fully analysed was little more than half of the cases available for classification as

Table 9.1 Progressive determination of blastemal axes and regional specificity in *T. vulgaris* (according to Schwidefsky, 1935).

Numbers of regenerates from three separate operations*

Age of graft (days at 19 °C)		Reversed AP axis		Inverted DV axis		Exchanged arm–leg	
		Harmonic	Disharmonic	Harmonic	Disharmonic	Conform	Differ
8–14	Blastema	25	0	31	0	27	0
15	Blastema	5†	0	8†	0	3	0
16	Cone	4	7	3	10	14†	1
17	Cone	1	5	1	8	6	1
18–20	Cone	1	16	2	16	2	13
		Single	Duplicated	Single	Duplicated	Single	Duplicated
8–15	Blastema	76	0	87	0	46	0
16–20	Cone	27	33	32	31	64	0

*Contralateral grafts to create AP or DV axial conflicts—regenerates scored as harmonic (conforming to host axis) or disharmonic (retaining graft polarity); grafts exchanged between arms and legs without any axial conflict are scored as conforming to host structure (e.g. a hand on an arm) or different (e.g. a foot on an arm).

†These numbers include unusually high proportions (ca. 50%) of distorted, defective or hypomorphic regenerates, perhaps the first signs of conflict or graft survival.

single or duplicated (Table 9.1). The remainder were probably hypomorphic growths from reduced grafts. Schwidefsky supposed the duplicated regenerates were entirely derived from the graft (which is certainly incorrect), and even distinguished a category of grafts with partial or labile determination responsible for duplications which were mainly harmonic. Table 9.1 shows that duplications only occurred when determined graft axes were in conflict with those of the host, and only in 50% of such cases. Other species which regenerate more consistently than *T. vulgaris* provide much higher yields of duplications in precisely the same conditions of axial conflict.

The best example of such results was provided by Lodyzenskaja (1930). She obtained 2–35 day-old foot regenerates by amputating juvenile axolotls through the ankle, and applied them as autografts to the excavated wound surfaces of freshly amputated fore-arms. This operation allowed the host skin to contract over the edge of the graft and hold it in place. A control series of operations maintaining the normal orientation of the graft usually resulted in simple regenerates with four or five digits. She could not decide if these were entirely feet or partially hands, so they only provided a standard for judging the effects of similar operations which created axial conflicts. The regenerates produced by reversing the AP and DV blastemal axes differed from the control series in two respects. The grafts rarely grew enough digits to reveal their final orientation (classed as harmonic or disharmonic to the host axes in Table 9.2) but were usually supplemented by extra digits whose orientation conformed to the host arm and which arose from separate host blastemata next to the base of the graft. A blastema transplanted with reversed AP orientation usually provoked the formation of one or more extra digits both in front of and behind it, resulting in duplicated fan-shaped hand. A conflict on the DV axis sometimes produced identical duplications, but almost as frequently a normal host regenerate grew out below the graft, which then appeared as hypomorphic hand (or foot) back to back with that of the host. Conflicts along both axes produced both these types of duplications, besides palm facing palm forms resulting from a dorsal host regenerate, and others with intermediate arrangements of digits. Lodyzenskaja concluded from a comparison of the four series that axial conflicts provoke duplications, rather more frequently from the anterior than the posterior side of the host arm and much more commonly from the ventral than from the dorsal side. In other words, most duplicate structures arose on the sides of the arm which show most rapid initial growth during normal regeneration. Since even 2 day-old blastemata provoked these duplications and the occasional failure to do so could be attributed to resorption of the small graft, it follows that both AP and DV axes were already determined when tested virtually at the onset of regeneration. The cases considered in the last chapter where limb tissues were rotated prior to amputation suggest that axial determination exists even earlier: it is not effaced during dedifferentiation and thus precedes regeneration. The validity of this deduction is largely independent of how one envisages the mechanism underlying the production of duplicated hands. Whether a harmonic blastemal graft prevents stump tissues from dedifferentiating, as suggested by Schotté (Schotté et al., 1941; Schotté and Harland, 1943a), or if it can assimilate dedifferentiated host cells with similar axial properties, the occurrence of duplications still indicates a discordance between determined axes. The reoriented grafts permit partial or hypomorphic regeneration by the limb stump, often at sites along the conflicting axis.

Iten and Bryant (1975) performed a similar experiment by exchanging blastemata and later regenerates between left and right arms of *N. viridescens* to produce an AP axial conflict. They obtained examples of duplicated hands and supernumerary arms, but failed to recognize these were alternative expressions of the same fundamental response. Their results have been summarized earlier (Table 8.5), together with a criticism of their belief that the stump influences the graft's AP polarity to yield intermediate patterns of regenerated hand. The supposed intermediates obviously correspond to the fan duplications described above and are reclassified accordingly in Table 9.2, with additional data for the other forms of axial conflict on newt upper arms. When some allowance is made for imposing a common format on them, the data obtained by different investigators using different species show a remarkable similarity in the frequency, position and orientation of duplications. The latter two features will be considered later in more detail following an attempt to justify such an analysis. Iten and Bryant (1975) placed their grafts on to limb stumps which still carried the remnants of previous regenerates, but stated they obtained virtually identical results using freshly amputated host arm stumps, as Lodyzenskaja had done. In the latter case, the appearance of lateral blastemata testifies to the dedifferentiation of host tissue. If there is any necessity to postulate an interaction between host and graft, as Iten and Bryant argued, then this interaction is probably restricted to adjacent blastemal mesenchyme of host and graft origin or to basal and apical portions of a composite blastema. There is no evidence that the host influences the axial determination of the graft, which may form a defective hand of uncertain orientation merely because it suffered some resorption after being transplanted. The graft obviously blocks or distorts the normal regeneration expected of the amputated host arm, but the nature of this perturbation can only be deduced from the characteristics of the duplications.

The occurrence of duplications on more than 80% of the cases of axial conflict shown in Table 9.2 indicates the presence of determined axes in the blastemata used for grafting. Either one or two accessory structures could be recognized on each arm and they were usually aligned on the conflicting axis. The elliptical cross-section of the fore-arm perhaps distorts this positional relationship for accessory digits which are most easily ascribed to the elongated AP axis, so it is preferable to concentrate on the upper arm whose AP and DV axes are nearly equal (Tank and Holder, 1979). As Bryant and Iten (1976) pointed out, the concept of axial conflicts easily

Table 9.2 Comparison of experiments testing the determination of blastemal axes by transplanting them on to arm stumps of (*a*) juvenile axolotls (Lodyzenskaja, 1930), (*b*) adult newts (Iten and Bryant, 1975; Bryant and Iten, 1976).

Host site	Axial conflict	Total cases	Simple hands (%)			Duplications (%)		Duplications present at following positions on host arm (%)						
			Harmonic	Disharmonic	Uncertain	Extra digits	Extra arms	A	A + P	P	V	V + D	D	Uncertain
Fore-arm			*(a) Early blastemal to early digit foot regenerates*					*(groups of extra digits)*						
	AP	40	0	0	2	98	0	15	75	0	5	0	0	2
	DV	39	0	5	13	82	0	8	20	13	23	0	0	18
	APDV	69	1	1	9	88	0	4	51	3	14	0	4	12
	None	69	92	0	1	7†	0	0	0	0	0	0	0	7
			(b) Blastemal to early digit fore-arm regenerates					*(extra hands considered only)*						
	AP	30	3	30	30	36	52	16	20	16	0	0	0	0
Upper arm			*(b) Blastemal to early digit upper arm regenerates*					*(extra arms considered only)*						
	AP	36	3	72	8	17	61	8	42	11	0	0	0	0
	DV	29	0	69	14	17	69	0	0	0	28	34	7	0
											AV	AV + PD	PD	
	APDV	24	0	75	25	0	79	8†	0	4†	29	38	12	0
	None	20	100	0	0	0	0	0	—	—	—	—	—	—

*Consisting of a single extra digit in each case; similar minor duplications were classed as simple hands in (*b*). Duplications occurred at frequencies of 80–100% when axes conflicted but rarely otherwise. Most duplications arose on the conflicting axis, all of them except those listed as PD conformed to the host arm's orientation.

†Present in addition to either an AV or PD supernumerary arm.

accounts for the predominant position of both supernumerary arms when only one axis has been reversed, but simultaneous conflicts on both AP and DV axes might then be expected to produce up to four extra arms, one at each of the poles. Instead, they still found a maximum of two supernumerary arms which were mostly located on the AV–PD diagonal. These dominant positions constituted a novel discovery, apparently unknown even after analogous experiments on limb buds, with a theoretical importance which warrants an independent confirmation. No confirmation has been obtained from two separate investigations on axolotls (Maden and Turner, 1978; Wallace and Watson, 1979), perhaps because the grafts tended to revert to a normal orientation and thus altered the position of the supernumeray arms. A more fundamental contradiction stems from the occasional appearance of three accessory structures, best documented by Rahmani (1960) in *T. cristatus* legs with APDV skin rotations.

Supernumerary arms and accessory digits also exhibit a predictable orientation, although it cannot be discerned so easily as the other features mentioned. After transplantations causing a conflict on either the AP or DV axis, the duplications show the same orientation as the host limb. It would be easy to suppose that the host not only dictates this orientation but does so by exclusively providing the blastemal cells from which they grow. Conversely, any graft contribution to these duplications would arise from its basal tissue and thus automatically tend to grow from a proximally-facing amputation surface (or reversed limb) to become a mirror image of the graft. That is in fact an equally accurate description of the orientation exhibited by these duplications, because the graft and host are mirror images of each other. It follows that the orientation of duplications caused by AP or DV axial conflicts provides completely ambiguous evidence of their origin. APDV dual axial conflicts are rather more revealing, although such grafts show a distressing tendency to undergo correctional rotation after which the expected duplications often fail to appear or grow so poorly that their orientation cannot be ascertained. Bryant and Iten (1976) managed to classify about half of the supernumerary arms formed after APDV axial reversal and found those occupying an AV position on the host arm conformed to its orientation. They were presumably derived from the host stump because they were not mirror images of the graft. The PD located supernumeraries, on the other hand, were mirror images of the grafts but lacked any simple relation to the host axes, so they should provisionally be considered as produced or controlled by the adjacent AV quadrant of the graft. It is of some interest to note that the AV quadrants of both host and graft seem to act as dominant centres of influence on the local formation of a duplicate regenerate. Bryant and Iten (1976) certainly appreciated this point but had lost confidence in the ability of conventional axial conflicts to account for these features of supernumerary arms. They claimed the following model offered a complete explanation of their own results as well as similar duplications encountered after skin rotation, besides accommodating experiments on the regeneration of insect legs and imaginal discs and the development of limb buds. The model is formally defined as one to describe pattern regulation in epimorphic fields by means of polar coordinates (French *et al.*, 1976). In preference to such a ponderous title it became popularly known as the clockface model.

9.4 The clockface model

This model describes the pattern of a morphogenetic field as a series of concentric circles. When applied to the limb's embryonic field or its regenerative territory, the tip of the limb is represented by the centre and the outermost circle indicates the field boundary, hence all radii follow the proximodistal axis of the limb. Positional information can then be expressed in terms of radial and circumferential coordinates. The latter, which concern us here, were initially assigned arbitrary values of 1–12 (i.e. a regular clockface, Figure 9.5a) with the implication that no hiatus existed between adjacent numbers. The operations considered previously have all involved local defects, excesses or transpositions of tissue which created a confrontation between regions of different positional value. The model stipulates that such confrontations must be reconciled by intercalary growth according to the following two rules.

(1) Intercalation between numbers in the circular sequence occurs by the shorter of the two available paths (e.g. adjacent positions 11 and 3 always intercalate 12, 1 and 2 as intermediates, never employing the longer sequence 4–10). This implies the circle has no vectorial properties, being equally satisfied by clockwise or anticlockwise sequences.

(2) The formation of a complete circle with all 12 positional values present at least once is a prerequisite for distal outgrowth and the formation of a more apical limb pattern (distal transformation).

Thus, simple amputation is followed by regression or lateral repair until a complete circle is present at the end of the stump, which can then extend distally. Amputation is equivalent to reaming out the centre of the diagram; regeneration is depicted as filling in the centre again. A sector less than half the limb diameter should form a lateral mirror image of itself according to the first rule and be unable to grow distally according to the second. An attempt to test this prediction will be described shortly. Duplications are only formed when diametrically opposite positional values are brought into contact (e.g. axially reversed blastemal grafts), for an intercalation between these values can occur by two equally short sequences to generate a subsidiary complete circle—which also follows rule (2).

Bryant and Iten (1976) found it necessary to displace the circular positional values from the original regular intervals into a relatively dense cluster in the AV quadrant (Figure 9.5b), in order to explain the position of supernumerary arms produced after APDV blastemal conflicts. That is equivalent to the deduction mentioned earlier concerning a dominant morphogenetic region located in the AV sector, and serves as a good example for evaluating the model relative to one employing conventional axes. Although the clockface model is set out much more precisely than any alternative, this precision is achieved at the expense of considerable artificiality. It is a two-dimensional model which may serve well enough for the embryonic limb field but provides an inescapably superficial view of a developed limb and cannot easily be adapted to events which separate skeleton from muscles or the divergence of fingers. In terms of whole regenerates and duplications, however, the clockface model can be used to make rather precise predictions which are amenable to experimental tests, such as the following.

Bryant (1976) constructed 'double half' arm stumps of adult *N. viridescens* to check if a wound surface lacking a complete circle of positional values could regenerate. A typical operation consisted of removing skin and muscles from the posterior half of an upper arm, while leaving intact the humerus and the adjacent major artery and nerve, then replacing the missing tissue with muscle and skin from the anterior half of the contralateral arm—to result in a double anterior half arm. The complementary operation produced double posterior half arms. The arms used for donating tissue served as simple half arm controls and a second control involved the sham operation of regrafting the excised tissue to reconstitute a normal limb. Most of these arms were allowed to heal for at least 3–4 weeks before being amputated through the distal region of the graft, with results shown in Table 9.3. The double half limb stumps formed normal blastemata but then usually stopped growing or differentiated into grossly defective structures containing only a few small cartilages. Many of the double posterior arms produced a supernumerary arm from the proximal junction of graft and host tissue. It might even be

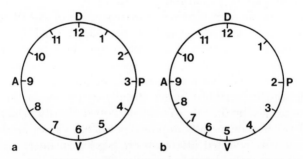

Figure 9.5. Amputation surface of a left limb stump to illustrate the regular clock-face model (a), and the amended model (b) required to explain the occurrence of supernumerary arms at only two specific sites after grafting blastemata with APDV axial reversal (Bryant and Iten, 1976). The conventional axes AP and DV are also indicated. (Reproduced by permission of Birkhäuser Verlag.)

Table 9.3   Regeneration of reconstructed arms (data of Bryant, 1976; Bryant and Baca, 1978).

| Type of arm | Numbers of arms | | | |
	Regenerate normally	Inhibited or hypomorphic	Total	With a lateral supernumerary
Double anterior	0	16	16	1
Double posterior	1	17	18	10
Half anterior	26	4	30	—
Half posterior	27	5	32	—
Sham operation	20	0	20	0

considered that their regenerative tendencies were satisified in this way or that the proximal blastema lured nerves away from the distal part of the graft. Bryant argued these experimental arms were adequately innervated, but could only demonstrate that nerve tracts were present. Sham-operated control arms regenerated normally, as did most of the half-arm stumps. On a strict reading of the clockface model (Figure 9.5b), an anterior half includes positions 6–12 and so should intercalate the missing values and thus regenerate perfectly. A posterior half with values 5–12 should intercalate a duplicate mirror image series of values and fail to regenerate, unless it manages to recruit some anterior values from higher up the arm stump. The observed regeneration need not invalidate the model's shortest intercalation rule, but does emphasize the danger of ignoring simplifications inherent in the model. Similarly, the model successfully explains the formation of supernumerary arms (Table 9.3), but incorrectly predicts they should occur as frequently on double anterior arms as on double posterior ones. Bryant and Baca (1978) extended this experiment by constructing the analogous double dorsal and double ventral arms, and found they also usually failed to regenerate more than an arrested blastema. Since the clockface model predicted the kind of results obtained, more or less, and no other current model would do that, Bryant could fairly claim to have obtained some support for it.

Stocum (1978a) repeated this experiment on the legs of 6–7 cm A. tigrinum larvae and similarly sized axolotls. Amputating experimental legs half way down the thigh, he found double anterior stumps were often arrested at a blastemal stage and established that the arrested blastemata were well innervated. Double posterior stumps, however, commonly regenerated symmetrical double posterior feet with up to eight toes (arranged as two mirror image sets, e.g. 5–4–3–2–2–3–4–5). Amputating such legs in the shank was usually followed by regeneration of symmetrical feet, either double anterior or double posterior, often with relatively few digits. In other words, regulation of the blastema occurred by suppression of the central elements to produce a hypomorphic double foot. The arrested blastemata obtained on

Table 9.4   Regeneration of reconstructed axolotl arms, from data of Tank (1978b).

Type of arm	Number amputated	Category of regeneration		
		Arrested blastema	Hypomorphic (1–3 digits)	Normal (4 digits)
Double anterior	15	15	0	0
Double posterior	15	9	6	0
Double dorsal	15	6	9	0
Double ventral	13	5	8	0
Sham operation (sum of four types)	12	0	0	12

double anterior thighs and more regularly on double newt arms could be merely more extreme expressions of hypomorphic regeneration. That would reconcile the results obtained by Bryant and Stocum, but does not explain them in terms of the clockface model. Stocum (1978b) sought to amend the model's complete circle rule to comply with his results. There is little point in describing his modifications for the model has been subjected to a more stringent test (section 6).

Tank (1978b) also tested the regeneration of double half arms by repeating Bryant's (1976) operation on the upper arms of juvenile axolotls. Although his results were quite similar to Bryant's and were interpreted as supporting the clockface model, they raise further questions. All the double half arms regenerated less well than sham-operated arms, where half was excised and replaced, but only the double anterior arms were consistently arrested, as demanded by the clockface model (Table 9.4). Even in this most favourable case, the frequency of arrested blastemata increases as amputation is delayed longer (Tank and Holder, 1978). Tank (1979) examined how muscle and dermis contributed to the arrest of regeneration. Table 9.5 summarizes most of these results by combining data from double anterior and double posterior arms. Lines 1–3 of the table show the inhibition is most pronounced when the graft has been allowed to heal for 30 days before the amputation, when all the hypomorphic regenerates were recorded from double posterior arms. Lines 4–7 suggest that the rearrangement of the dermis supplies the dominant inhibitory influence, especially as rearranged muscle does not prevent regeneration if the skin occupies its normal position. If the majority of sensory nerves penetrate the dermis, there may be a neurotrophic explanation of these results involving a withdrawal of misdirected nerves about 30 days after the operation. What is known of the redirection of misplaced nerves (see Chapter 10) does not favour this explanation, yet the only precedents of blastemal arrest have come from delayed denervation experiments (Schotté, 1962a; Singer and Craven, 1948; Schotté and Butler, 1944). Furthermore, these double arms support the growth of normal blastema grafted on to them, but their arrested blastemata do not recover when grafted on to normal arm stumps (Holder and Tank, 1979).

Table 9.5 Analysis of what inhibits the regeneration of double half arms (from data of Tank and Holder, 1978; Tank, 1978b, 1979: reproduced by permission of the Wistar Press).

Symmetrical tissue*	Delay of amputation (days)	Total cases	Percentage of arms regenerate			
			Arrested blastema	Hypomorphic (1–3 digits)	Normal (4 digits)	Multiple (>4 digits)
Skin and muscle	5–15	68	15	84	1	0
Skin and muscle	20–60	69	71	29	0	0
Skin and muscle	30	30	80	20	0	0
Skin (normal muscle)	30	23	65	17	9	9
Skin (muscle extirpated)	30	16	31	50	19	0
Muscle (normal skin)	30	46	2	2	83	13
Muscle (dermis extirpated after 30 days)	33	20	45	5	30	20
None (sham operations for above lines)	30	26	0	0	100	0

*The data comprise an almost equal number of arms made double anterior or double posterior for the tissues designated symmetrical. Double posterior arms tended to produce more complete regenerates, except where only muscle had been made symmetrical.

Slack (1976, 1977) had previously obtained double posterior arms by the totally different method of displacing the posterior polarizing zone to lie in front of the arm primordium in axolotl embryos. When such arms were amputated, they regenerated more or less the same partially duplicated structures, which Slack and Savage (1978a, b) claimed to be in contravention of the clockface model's complete circle rule. The model can actually accommodate this result by representing Slack's 'reduplicated limb' as two complete circles compressed into a figure-of-eight, in which the centrally placed values are only expressed distally as anterior digits. These reduplicated arms exemplify a much more basic law implicit in the term regeneration, that it tends to perpetuate stump abnormalities while allowing some scope for regulation (Newth, 1958a, b). Symmetrical double posterior arm regenerates have been produced by surgical reconstruction of axolotl upper arms and immediate amputation (Holder *et al.*, 1980). When these regenerates were re-amputated below the elbow, they usually formed duplicated hands with 5–7 digits, which confirmed the results of Slack and Savage (1978). Krasner and Bryant (1980) constructed double anterior and double posterior axolotl fore-arms and found that the majority of them were capable of growing hypomorphic symmetrical regenerates like those obtained previously by Stocum (1978a).

The type of regenerate formed by a double half limb stump thus seems to be quite variable and the frequency of regeneration depends on several factors. The geometry of the limb is one factor, for anterior and posterior ones differ; other factors are how long the limb was allowed to heal before amputation and whether it was amputated in the upper arm or fore-arm. We should not let the complexity of these results distract our attention from the purpose of the experiments. The operation was originally devised to test a specific prediction of the clockface model, that such arms should not regenerate at all. The various circumstances in which regeneration does occur all point to inadequacies of the model and have led to major revisions of the model by Bryant and Baca (1978) and by Krasner and Bryant (1980). Even the latest revision only adapts the model to fit some of the peculiarities of double half limbs and notably evades the issue, considered in section 6, that the model had been refuted 2 years earlier.

9.5   Conventional models

All the uncontested features of duplications mentioned so far can be incorporated into models which employ the conventional transverse axes, where axial conflicts between host and graft tissues promote duplications. Several versions of such a model have been envisaged but rarely stated precisely, partly because the axes have always been realized to be quite arbitrary creations. Slack (1980) deliberately converted the clockface dial values into series of numbers arranged on two transverse axes. That is an engaging idea but results in a most cumbersome model. Having tried and failed to follow his instructions for manipulating the numbers, I am content to accept Slack's assertion that he can account for the duplications produced by all the blastemal rotations considered in the following section. The only other version of the conventional model to be presented in any detail was developed by Lheureux (1972–77) on the basis of results considered in chapter 8. He described the transverse axes as gradients between opposite poles (faces or qualities) which characterize both the skin and internal tissues of the limb. According to this model, outgrowth of a regenerate or supernumerary arm requires a contact between two opposite poles of any transverse axis, normally provided by tissues closing over an amputation surface. A single quality, as provided by anterior skin grafted onto an irradiated limb stump, only leads to abortive or hypomorphic regeneration. In the absence of irradiation, however, the same operation not only permits normal regeneration following apical contact of opposite poles present in the internal tissues, but also creates an additional growth centre where the posterior internal tissue is confronted by anterior quality skin. Lheureux (1977) found he could depict his model as a regular clockface (Figure 9.5a, but not b) involving an infinite number of polarized diameters, although the only poles he actually tested were the familiar A, P, D, V points and his notion of confrontation between opposite poles strongly resembles the older concept of axial conflicts.

Lheureux (1977) attempted to distinguish between his model of regeneration and the prevailing view of how the transverse axes are established in developing limb buds. He pointed out that a confrontation restricted to D and V qualities produced a normal regenerate and thus must have specified all intermediate values, including A and P. He therefore denied that regeneration could involve a special morphogenetic centre such

Table 9.6 Proportion of arms bearing duplications after transplantation of blastemata in adult *T. cristatus* (deduced* from Abeloos and Lecamp, 1931).

Axial conflict	Transposition along proximo-distal axis		
	None	Distal	Proximal
AP or DV	~50%	96%	5%
APDV	78%	Many	None
None	10%	None	None

*From the available data, about 20 arms were produced in each of the listed categories using an undisclosed blastemal stage. Transpositions were from upper limb to lower limb (distal) and vice versa (proximal); untransposed blastemata may combine results from upper and lower limbs and included exchanges between arms and legs. Contralateral exchanges, some inverted and others not (listed here as AP or DV conflicts), were recorded together as giving virtually identical results. Duplications comprise the whole range from an extra digit to two supernumerary arms.

as the zone of polarizing activity (ZPA) found on the posterior side of limb buds. The embryonic ZPA itself shows field properties, however, for excising the most active zone does not alter the AP pattern of the remnant (MacCabe *et al.*, 1973). Similarly, a ZPA could be reconstituted in the experimental arms considered by Lheureux, so his argument is at least premature. I prefer to make a distinction according to the number of polarized axes stipulated by various versions of this model. Carlson (1974a, 1975b) only needed to consider a single AP axis to explain the results of his operations; chick limb buds are routinely considered in terms of AP and DV axes, while larger numbers are invoked in the clockface model and by Lheureux.

The notions of axial conflict and confrontation of opposed poles are obviously consequences of the polarity reversals executed in most of the relevant experiments. Abeloos and Lecamp (1931) performed one such experiment, by transplanting limb regenerates of *T. cristatus* to impose AP, DV or APDV axial conflicts in addition to proximal or distal transposition. They reported the experiment with such disdain for detail that their results (Table 9.6) only serve as a summary of the more painstaking analyses described previously, but they deduced two general rules.

(1) Duplications occur commonly when the axes of the stump and the graft *do not coincide*.

(2) The first rule applies only when no proximal transposition has occurred.

Whether by accident or design, the first rule clearly predicts that 90° or even smaller rotations of a regenerate may cause duplications. This 'axial discrepancy' rule has little obvious application to the events of normal regeneration, but is quite valuable in pointing out the shortcomings of those models which rely on the concept of axial conflict.

9.6 Testing the models

Each of the models outlined in the last two sections can be fitted to much of the known data about the basic pattern of limb morphogenesis and duplication. Various predictions can be made and tested for each of them, like Bryant's (1976) test of the clockface model. What is really needed, however, is a test which discriminates between the models so that some can be pursued and others abandoned. Rotating blastemata through successive 90° intervals seems to fulfill this requirement. Not only does it explicitly compare the predictions of the axial discrepancy rule with those based on axial conflict, but it also tests predictions of the clockface model and might reveal either a dominant morphogenetic centre or some vectorial (clockwise versus anticlockwise) property associated with the transverse axes. The most obvious predictions from different models are explained in Figure 9.6 and Table 9.7. Seeing this as a particularly discriminating test without any obvious technical difficultues, I decided to perform it on some *Pleurodeles* larvae which I was rearing for another purpose.

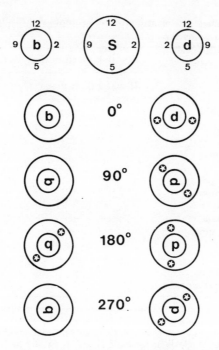

Figure 9.6. Blastemal rotations designed to test models of the limb's transverse axes. Top centre, surface of a left arm stump (S) with clock-face numbers (Figure 9.5b) to show its major axes. The orientation of this stump is maintained in the lower diagrams (outer circles). Left column, a left arm blastema (b) can be replaced on S so that all axes are aligned or rotated clockwise by 90°, 180° (APDV axial conflict, q), or 270°. Right column, a right arm blastema turned to match the PD polarity of S (converting b to d) creates an AP axial conflict, or can be rotated clockwise by 90°, 180° (DV axial conflict, p), or 270°. Stars mark the positions of duplications predicted by the clock-face model (see also Table 9.5) or on which it was based (cases of q, d and p).

Figure 9.6 shows the eight configurations produced by autografts of palette stages onto ipsilateral and contralateral fore-arms with rotations at 90° intervals. Ten examples of each configuration were obtained and maintained for 5–6 weeks, by which time they had reached a stable state. Recording the occurrence of duplications is sufficient for the present purpose, because it distinguishes between the models which have been proposed (Table 9.7). Any rotation approaching or exceeding 90° resulted in the appearance of accessory digits from the majority of these fore-arm operations (Table 9.8). A repetition of the experiment on axolotl fore-arms confirmed this result, while similar operations on the upper arms of axolotls produced multiple hands or fore-arms as alternatives to accessory digits (Figure 9.7). This incidentally vindicates the earlier assertion that extra digits or arms are equivalent responses which can both be considered together as duplications. Grafts performed just above the elbow produced high yields of duplications but

220

Table 9.7 Predicted presence (+) or absence (−) of duplications following blastemal transplantation at successive rotations shown in Figure 9.6.

Models used for the predictions	Rotation given in degrees for							
	Ipsilateral blastema				Contralateral blastema			
	0	90	180	270	0	90	180	270
(1) *Clockface*	−	−	+	−	+	+	+	+
(2) *Axial conflicts* concerning:								
(2*a*) AP axis only	−	−	+	−	+	−	−	−
(2*b*) DV axis only	−	−	+	−	−	−	+	−
(2*c*) AV–PD diagonal only	−	−	+	−	−	−	−	+
(2*d*) both AP and DV axes	−	−	+	−	+	−	+	−
(2*e*) Four or more diameters*	−	−	+	−	+	+	+	+
(3) *Axial discrepancy* concerning:								
(3*a*) AP axis only	−	+	+	+	+	+	−	+
(3*b*) DV axis only	−	+	+	+	−	+	+	+
(3*c*) AV–PD diagonal only	−	+	+	+	+	−	+	+
(3*d*) Any two or more diameters	−	+	+	+	+	+	+	+

*Corresponds to the model invoking a confrontation between diametrically opposite poles (Lheureux, 1977).

Assigning a vectorial quality to an axis is equivalent to stipulating another axis, i.e. would lead to predictions (2*d*) or (3*d*) for any single diameter and (2*e*) or (3*d*) for any pair of diameters.

grafts close to the shoulder often reverted to a more normal orientation and then failed to provoke duplications (Table 9.9). Both the clockface model and any model depending on axial conflicts predict the 90° and 270° ipsilateral rotations should not generate duplications (Table 9.7), so they are clearly invalidated by these results. The axial discrepancy rule correctly predicts these ipsilateral duplications, but only a model which stipulates two or more determined transverse axes survives the test of contralateral transplantations which produced duplications at each interval of rotation. These could well be the traditional AP and DV axes, although the experiment cannot specify their position. Incorporating the axial discrepancy rule to these axes probably removes any need to envisage additional polarized diameters (and there is certainly no evidence that more than two axes exist), allowing us to use limb development as a model for limb regeneration. That model will be considered in the following chapter.

The location and orientation of the duplications were not easily assessed, and not particularly informative either. Fore-arm grafts usually produced only 1–3 digits, with accessory groups of 1–3 digits arranged along the host AP axis and conforming to the host DV orientation for the most part.

The prevalent result was a fan-shaped hand, with other conformations occurring much less frequently, exactly as described by Lodyzenskaja (1930). Upper arm grafts usually produced four-digit hands with well-separated accessory hands or fore-arms. Many of these arose as pairs on either side of host arm, conforming to its orientation and aligned on either the AP or DV

Table 9.8 Results of autografting fore-arm palette stages with rotation at 90° intervals, *Pleurodeles waltl* larvae (data of Wallace, 1978: reproduced by permission of Birkhäuser Verlag).

	Rotatation given in degrees for							
	Ipsilateral grafts				Contralateral grafts			
	0	90	180	270	0	90	180	270
Numbers of arms bearing								
(a) simple hands (4 digits)	10	3*	1*	2*	0	2*	0	0
(b) complex hands (5–9 digits)	0	7	9	8	10	8	10	10
Mean number of digits								
Assigned to graft	4.0	2.3	2.4	1.7	2.3	2.8	2.9	2.6
Assigned to host	0	2.8	3.7	3.1	4.4	3.1	3.4	3.6

*Anatomically composite, e.g. two graft digits and two host digits.

axis. A minor category of disharmonic accessory hands also occurred after ipsilateral graft rotations, corresponding to those discovered by Bryant and Iten (1976) after APDV axial reversals. These disharmonic hands could all have grown as mirror image regenerates from the base of the graft (but did not occur in any constant position relative to the host or graft axes), unless they correspond to the double-dorsal and double-ventral hands subsequently identified by Maden (1980b).

I have described this test as a complete novelty, as it seemed to be when I

Table 9.9 Numbers of arms bearing duplications after grafting palette stages with rotation at 90° intervals, juvenile axolotls (data of Wallace and Watson, 1979).

	Rotation given in degrees for							
	Ipsilateral grafts				Contralateral grafts			
	0	90	180	270	0	90	180	270
(A) *Fore-arm grafts*								
Simple hands (4 digits)	9	0	0	0	1*	0	1*	0
Duplicated hands (5–11 digits)	1†	10	10	10	9	10	9	10
(B) *Elbow level grafts*								
Simple hands (4 digits)	8	1*	0	1*	0	1*	0	0
Duplicated hands or fore-arms (totalling 5–14 digits)	2†	9	10	9	10	9	10	10
(C) *Shoulder level grafts*								
Simple hands (4 digits)	10	10	5	8	0	2	5	4
Duplicated hands or fore-arms (totalling 5–12 digits)	0	0	5	2	10	8	5	6

*Anatomically composite; all other simple hands conformed to the host AP axis.
†Normal hands except for a single extra digit.

performed it. In fact, Nicholas (1924) discovered that limb buds rotated through 90° or 270° provoked duplications except when they reverted to a normal orientation. Schwidefsky (1935) also reported the occurrence of duplications following the 90° rotation of cone stage blastemata on the limbs of *T. alpestris*. Despite these early accounts, there is considerably more information to be gained from the operation. Maden and Turner (1978), for instance, have already demonstrated that blastemal rotations considerably less than 90° provoke duplications of axolotl arms. According to their preliminary account and a report by Tank (1978a), grafts exchanged

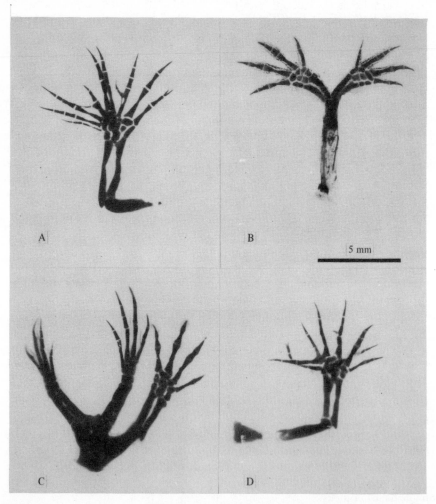

Figure 9.7. Examples of duplications caused by rotated blastemal grafts (from Wallace and Watson, 1979). A–C are all contralateral 270° rotations on the forearm, above the elbow, and near the shoulder respectively. D is an ipsilateral 90° rotation above the elbow which has produced a partial duplication of the radius and hand, contrary to the prediction of the clock-face model.

between dark and white axolotls show that supernumerary arms are commonly composite structures, derived from graft and host cells, yet structurally normal—indicating the regulation of chimaeric blastema into a coherent unit. The rules governing regulation may only be the converse of those specifying the production of duplications, but an appropriate reformulation would probably be more pertinent to normal regeneration.

9.7 Conclusions

Traditionally, blastemal morphogenesis has been viewed in terms of a simple basic pattern involving gradients, and the derivative concepts of the morphogenetic field and positional information. Furthermore, the proximodistal and transverse axes have usually been treated in isolation. Despite these simplifications, the analysis has not progressed far: the following conclusions are as certain to contain errors of ignorance and misjudgement as they are sure to be contested.

The proximodistal axis is labile with a reversible polarity, as suggested in the last chapter. Distal transformation is encountered almost universally, in contrast to the meagre and questionable evidence of proximal transformation. Apical growth is stimulated by a wound epidermis, but intercalary growth can also be demonstrated as the distal transformation of stump mesenchyme under a grafted blastema. The best available model derived from these considerations is that distal transformation always occurs as intercalary growth between reference points provided by the most proximal and distal dedifferentiated cells. That strikes me as a pretentious statement, for I have been unable to translate it into concrete terms or convert it into a testable proposition.

The transverse axes of the blastema are predetermined, being present before amputation and maintained throughout regeneration. They are not susceptible to stump inductions but can be modified by a process of regulation within a composite blastema. Duplications are attributable to axial discrepancies between blastemal subpopulations. This means a host limb stump can only influence a transplanted blastema by contributing cells to it. There is some evidence that an accurately aligned blastema inhibits dedifferentiation in adjacent stump tissues, while a misaligned graft may lack that ability but can divert host mesenchyme away to the side. Determination seems to reside in the conventional AP and DV axes, which permits blastemal morphogenesis to be equated to that of a developing limb bud. Both limb buds and blastemata have to be considered as autonomous self-regulative units or fields which might contain identical morphogenetic centres.

Chapter 10

Comments and Speculations

Each of the previous chapters ends with a summary or set of conclusions based upon a restrained and sceptical assessment of the available evidence. I expect the reader to treat my conclusions in an equally critical spirit, for otherwise there would be little point in presenting the evidence and arguing about it. Lack of evidence also places a constraint on the conclusions which can be drawn at present, but that need not prevent me from expressing a few opinions on the main themes which have appeared earlier. The following commentary on fields and territories is intended to justify my belief that limb regeneration is strictly a repetition of limb development. Even if it were an unjustified speculation, it would still be interesting enough to merit further testing and an enquiry into its implications. The main implication, which is considered finally, is the prospect of turning our knowledge to medical advantage.

10.1 Fields and territories

As a necessary part of depicting current views of morphogenesis, the last two chapters took on a somewhat conceptual tone which requires additional explanation. The idea of morphogenetic fields has pervaded embryology so thoroughly that it cannot be shaken off, any more than a religious upbringing. Most investigators have accepted the idea as a useful one even if it did not lead anywhere, probably because it subsumed the concepts of metabolic gradients and determinative interactions without paying them much attention. Of course this attitude infuriated Child (1941) who castigated the morphogenetic field as 'a concept without definite content . . . a verbalistic refuge . . . almost mystical in character'. That is quite true: the field is merely a useful shorthand term for a self-regulatory area without definite boundaries (in contrast to a mosaic territory), and should not be supposed to hold any deeper meaning. Positional information is largely a rephrasing of the field concept but also postulates definite boundary values to the field, apparently alluding more to agriculture than to magnetism.

224

Adopting a Childish view of the underlying gradients, where a particular metabolic rate decreases as a function of distance from a centre of activity, a field is defined by this centre and fades out into adjacent fields without either detectable or theoretical boundaries. That strikes me as the more accurate description of the embryonic limb primordium and a quite plausible description of the blastema, even if we can't get to grips with it.

To some extent, at least, we have to take on trust the assertion that a regenerative territory is equivalent to the earlier embryonic field. They are both defined in the same way, as areas where the growth of a limb can be provoked, and both seem to fade away at their margins enough to deter precise mapping. The arm and leg fields of the embryo are held to overlap or be confluent, for limbs may be induced in the middle of the flank–but they are only induced rarely and it is hard to decide whether they are arms or legs. Similarly, nerves deviated to the margins of the limb territory tend to induce smaller and less well-formed supernumerary limbs, and perhaps do so less frequently, than nerves diverted to the base of the limb. Either this field property or individual variability prevented Guyénot and his colleagues from actually preparing a definitive map of the regeneration territories that could be compared to a standard butcher's chart. There is no known structural means of identifying the boundaries of regeneration territories or of relative positions within them, then, but growing nerves can recognize specific regions under certain circumstances.

The embryonic limb bud may fairly be claimed to attract its nerve supply, although nerves have reached the primordium before it projects from the flank. Both the urodele arm and its innervation normally grow from the area of the third to the fifth somites. The nerves from this region are deflected further back to enter a posteriorly transplanted limb bud, but then somites 6–11 can provide an additional or alternative innervation (Detwiler, 1920). Limbs transplanted even later into the larval flank can gain a nerve supply, even if they are usually unable to move. Whether developing nerves seek out particular regions within the limb or grow anywhere and later form appropriate connections is not completely settled (cf. Gaze, 1970; Horder, 1978; McGrath and Bennett, 1979), but regrowing nerves seem to have a strong sense of identity or purpose. When Grimm (1971) exchanged the extensor and flexor nerves of axolotl fore-arms, an invisible fraction of each tract rediscovered its appropriate muscles and regained control of them. Similarly, Cass and Mark (1975) first mapped the regions of the axolotl leg served by motor nerves 15–17 and then showed the maps were not appreciably altered when the pre-axial and post-axial nerve trunks had been exchanged. In 6 weeks or so after rerouting the nerves, some of the axons had homed in to their old muscles. Correct re-innervation probably involved trial and error plus an element of competition, for the territory left vacant by one nerve failing to regrow became innervated by nerves from an adjacent territory. Cutaneous sensory nerves do not share this ability to return to their original territory, but permanently innervate uncharacteristic areas of skin to which they have been redirected (Johnston *et al.*, 1975). Simply cutting a nerve trunk can cause it to re-innervate a chaotic pattern of skin patches, although crushed axons usually regrow along their former sheaths to re-establish a fairly normal receptive field. Cutaneous nerves are also capable of serving an adjacent territory whose normal nerve supply has been destroyed. Their collateral sprouting is restricted to the immediately adjacent territory, however, so that nerve 15 can take over the area of nerve 16 but not that of nerve 17 (Diamond *et al.*, 1976). Even these cutaneous territories are apparently defined by internal tissues, for they are not altered when the limb skin is displaced. Deispite this last conclusion, the skin may also have regional properties which influence either the peripheral or central connections made by sensory nerves. Some such influence is probably needed to account for the misdirected reflexes elicited by stimulating displaced skin grafts (Jacobson and Baker, 1969; Heidemann, 1978). Furthermore, the underlying territory or the local nerve supply seems to have a marked influence on regenerating skin (Whimster, 1965, 1971, 1979).

The admittedly perplexing features of the interactions mentioned above need not concern us here. For the present purpose, they simply reveal territories which escape traditional anatomical observation. In addition to discrete tissues, these territories subdivide a gross region such as the limb. They must form gradually as the limb develops from part of its embryonic field and they must reform within a regenerating blastema, presumably by elaborating on an axial pattern present in the undifferentiated mesenchyme. The consistent expression of that pattern suggests it is not easily perturbed, being a buffered or self-regulatory system. It is subject to genetic influence, however, as shown by the abnormal limbs of several mutants and is thus exposed as a metabolic pattern.

10.2 The limb bud paradigm

At some time the dominant growth centre of a field must be influenced by other centres, each characterized by their own specific metabolic activity. They may be subsidiary centres of the main field or lie entirely outside it, and they may only exert a relatively brief influence or induction which has a permanent effect. Limb development provides a good example of the postulated influence. A zone of polarizing activity (ZPA) exists within the limb field but placed somewhat caudally to the main growth centre, so that it occupies the posterior part of the chick wing bud or an area of the flank immediately behind the arm primordium of an amphibian embryo. Assuming the action of the ZPA is to determine the anteroposterior axis of the future limb (Summerbell, 1979), its inductive period seems to be quite variable. Determination of the AP axis occurs from shortly after gastrulation to tail bud stages of different newts and salamanders, perhaps even later in the chick. The ZPA also exhibits field properties, for removing its active centre does not prevent the rest of the limb from becoming polarized along an AP axis, but transplanting it in front of the growth centre induces adjacent parts of the bud to form posterior structures of a duplicated limb (MacCabe et al., 1973–9; Slack, 1977). A supernumerary arm induced in the flank of an amphibian embryo is typically found to have a reversed AP axis, because it originates behind the ZPA active centre. We thus envisage the ZPA as a subsidiary field whose gradients are imposed on limb-forming mesodermal cells so that the nearest ones become determined to form the most posterior structures. The manner in which a dorsoventral axis is imposed on the limb primordium has not been ascertained so clearly, although Swett (1927–37) has established the existence of a polarizing influence and defined its time of action: a dorsal analogue of the ZPA awaits rediscovery.

I believe a dorsal polarizing centre has been discovered already, first in adult newts and later in embryos, in the following way. Whenever Guyénot and his colleagues (1926–48) diverted a brachial nerve to emerge at the base of the arm of T. cristatus, they noted the induced extra arms always copied the orientation of the primary arm. A nerve diverted dorsally to the shoulder, however, tended to induce the growth of either a misoriented or a duplicated arm. At the risk of breaking a Geneva convention by converting the axes employed by Guyénot to the standard ones used here, these arms can be identified as genuine DV inversions and their mirror-image duplications. Kiortsis (1953) pursued this discovery to demonstrate that the orientation of the induced arm depended upon that of the local territory where the nerve emerged, and to show the larval shoulder territory produced duplicate limbs even when transplanted to the pelvic region. Finally he recorded the appearance of inverted and duplicated arms when shoulder skin was shifted down to the centre of the arm primordium in tail bud embryos. In his own terms, Kiortsis had anticipated the official discovery of the ZPA by some 20 years. The topography of the region and the occurrence of palm to palm duplications, however, imply he had identified a dorsal centre with the equivalent property of polarising the limb's DV axis (Swett, 1938). Here is an occasion where investigating limb regeneration has provided a successful analogy for development. It deserves wider recognition.

The AP and DV axes of the limb are obviously either maintained or re-established in the growing blastema. Diverse cases of regeneration from partial or increased amputation surfaces, or from small grafts in irradiated limbs, all show enough regulation toward normal structure to demonstrate that no rigid point-for-point determination of their axial arrangement would suffice. Even differential migration and affinity could not explain regeneration from small grafts of cartilage, muscle or connective tissue: there must be considerable respecification in a positional sense as well as histological transformation. Furthermore, experiments on transplanted blastemata show they already possess regional and axial characteristics very early in regeneration and they are probably not susceptible to later influence from the limb stump. The only way to reconcile these properties (or so it seems to me) is to regard mesodermal cells as being imprinted with a sense of their position within the limb, a sense which they retain during dedifferentiation and proliferation but which can be over-ridden to some extent by regulation within the blastema. Based on this argument, it is quite reasonable to suspect the existence of dominant centres which specify the AP and DV axes of the blastema and to propose, as a working hypothesis, that such centres are identical to those detected during limb development. The potential to form such centres would have to exist all along the limb, but they need only become active at the end of an amputated stump to extend into the growing blastema.

Limb development is surely a defective model for regeneration, if only because it is imperfectly understood and our ideas about it might change markedly over the next decade. Yet the chick wing bud in particular has been investigated so successfully during the past 30 years that the best prospect for an advance in understanding limb regeneration would be afforded by simply testing to what extent a blastema resembles a wing bud. This course of action has the practical advantage of postponing detailed consideration of the various elaborate models which have been devised for the limb bud itself. Several features which merit investigation in the blastema emerge from Saunders' (1977) review of wing development and from other articles in the same volume. The analogy between the apical ectodermal ridge and the blastemal apical cap has been pursued by Thornton (1954–68) but still deserves further exploration, either by direct surgery or by searching for morphogenetic mutants in amphibians. As just mentioned, there is a good excuse to look for a ZPA on the posterior side of the blastema and to note any inhibitory effects which might revolutionize the analysis of blastemal rotations along the lines proposed by Fallon and Crosby (1975). It would be worthwhile to confront the prevalent idea that epidermis is completely non-specific in regeneration with the considerable evidence of an ectodermal influence on the DV axis of chick limb buds (Pautou and Kieny, 1973; MacCabe *et al.*, 1973, 1974). Even our current models of the blastemal proximodistal axis would benefit from establishing if proximal transformation can occur as clearly as found in the chick limb bud by Kieny (1977).

Although analogies to the developing limb will surely help to solve the problems of blastemal morphogenesis, they do not provide much additional insight to questions raised in earlier chapters. Fortunately, our understanding of regeneration is more self-sufficient in these respects. There is already a coherent outline of knowledge (despite some obvious gaps) concerning the major events of regeneration which have been identified and adequately described in terms of gross structure and histology. What is known of the fine structure and metabolism of the blastema generally confirms this description. If electrical or immunological events intervene at the onset of regeneration, that would only add an extra phase without fundamentally changing our present view of the process. Similarly, discovery of some systematic change of the cell cycle during blastemal growth would have interesting repercussions but would not revolutionize our concept of regeneration as organized growth by means of cellular proliferation. It is surely no coincidence that the three best means of impeding regeneration (irradiation, denervation and hypophysectomy) all attack the blastemal growth phase by reducing or preventing the normal increase of mesenchymal cells. Once enough cells have accumulated, they seem to form a more self-sufficient unit which can produce a miniature or hypomorphic regnerate even in the absence of further growth. It is even possible the ability to regenerate may simply depend on the outcome of a race between cellular proliferation to attain a self-sufficient blastema and a typical wound-healing response which forces the cells to differentiate into scar tissue. The concept of a critical blastemal mass, as envisaged by Guyénot and Schotté (1923), seems as good an answer as any to the final perplexing question of why some organisms can regenerate but others cannot.

10.3 Towards human limb regeneration

Ever since Morgan (1901) remarked on the advantage of not having a wooden leg, there has been an extra motivation for exploring the conditions and mechanism of limb regeneration. Pure science might be of practical use to humanity here, and relegate plastic surgeons to more lucrative cosmetic operations. The whole matter has been the subject of a desultory philosophical debate for most of this century, without any noticeable connection to the occasional attempted induction of regeneration in frogs or higher vertebrates. These attempts were made in open defiance of the once popular assertion that amniotes, mammals and especially man have evolved in some aspect which compelled them to lose the primitive ability to regenerate. Vorontsova and Liosner (1960) sought to establish a homology between regeneration and asexual reproduction, which they regarded as two aspects of a fundamental property of living matter more regularly expressed among primitive organisms than in highly specialized ones. Such an argument has a certain appeal when illustrated by asexually reproducing flatworms or colonial coelenterates, yet these cases can be matched by contradictory examples. It is impossible to assert that frogs are an evolutionary advance on tadpoles, for instance, but metamorphosis certainly restricts their regnerative capacity. Similarly, the ability to regenerate a new lens from the dorsal iris is found in many true newts (Salamandridae) but not among allied salamanders, an example elaborated by Reyer (1977): 'Although the newt eye has a structural and functional complexity equal to that of the mammalian eye, it has the amazing capacity for the replacement of the neural retina, optic nerve, and lens with subsequent restoration of visual function. This shows that regeneration is not necessarily limited to the restoration of systems with a simple organization'.

The sporadic occurrence of regenerative ability among multicellular animals persuades me that searching for an evolutionary explanation is essentially a vain enterprise. Starting from the same premise, that regenerative capacity reflects a fundamental property of living matter, Polezhaev (1972) simply concluded that meant it could be evoked even in such a recalcitrant creature as man. That is a much more encouraging philosophy, but not much help in finding the means of doing so.

The variable capacity for regeneration encountered in evolutionary surveys strikes me as having a rather different significance. From the viewpoint of research on urodeles, it is easy to adopt high standards of anatomical and functional normality in a regrown limb. We should be even more fastidious about human limb regeneration, of course, yet we cannot ignore the many reports of regeneration according to lower criteria. Some facets of regeneration or very similar processes occur regularly in the amputated limbs of frogs and toads, reptiles and mammals. Easily the most striking example concerns young children who can regenerate crushed or lost fingertips, including the nail, in about 3 months (Figure 10.1), provided treatment is restricted to cleaning and dressing the wound (Illingworth, 1974). Any attempt to suture a skin flap over the wound impedes regeneration, just as it does in amphibians. There is a further parallel between this example and regeneration in frogs. In both cases, the best regeneration occurs in young, rapidly growing individuals where some regenerative ability persists at the tip even when it seems to have been totally lost from the rest of the limb. Despite our very limited prowess at limb regeneration, we and other mammals are quite adept at healing surface wounds and internal injuries, such as bone fractures or torn muscles and ligaments. The similarities between tissue repair and regeneration have been considered by Needham (1952), Carlson (1978) and in such detail by Schmidt (1968) that it is sufficient to say they are overwhelming. All the processes implicated in amphibian limb regeneration have also been recorded in mammalian wound-healing, internal tissue repair or the regrown finger tips. In an admirably succinct review of this similarity, Hay (1974) concluded that it must be some higher control which prevents human limbs from regenerating. I think we can identify the control more precisely. Healing and repair processes involve only a limited amount of cellular proliferation, barely causing a perceptible extension of the limb stump in most frogs or amniotes. The missing higher control, therefore, is probably one which stimulates or prolongs the period of cellular division until enough cells have accumulated to form a self-sufficient blastema. If higher vertebrates could be persuaded to form a blastema, there is a reasonable expectation that it would resemble a limb bud in requiring no further encouragement to develop according to its own internal instructions. We have to assume a blastema would be self-sufficient anyway, because we are still far too ignorant of morphogenetic prescriptions to have any hope of supplying artificial ones.

The grand strategy advocated here is to produce a blastema and trust it will know what to do next. There are three obvious tactics which should help to produce a blastema: exaggerating dedifferentiation, stimulating cell division

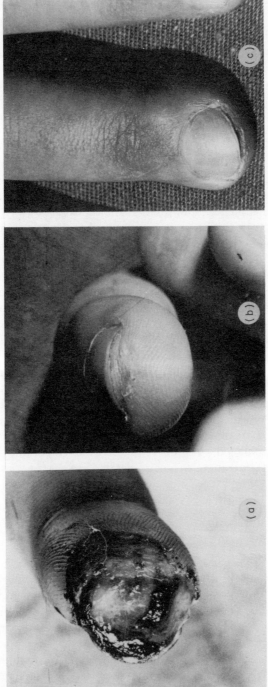

Figure 10.1. Regeneration of a child's finger tip (From Illingworth, 1974: reproduced by permission of Grune & Stratton Inc.)

and delaying redifferentiation. The means at our disposal only allow a minor manipulation of any one process at present, but we might well need to link consecutive stimulations to cause a marked enhancment. Traumatic attempts to increase dedifferentiation, by lacerating the limb stump, excavating its skeleton, or even temporarily interrupting its blood supply, have only produced slight or undetectable improvements of regeneration in frogs. Excess vitamin A or electrical stimulation seem only slightly more effective but there must be a considerable range of treatments, such as depolarizing cell membranes (Holley, 1975), which might work. A stimulation of cell division has only been observed as a partial relief from inhibitory treatments in urodeles but, at least, shows growth hormone and prolactin to be appropriate stimulants. Their usefulness remains in doubt, for they would need to be applied at a high local concentration while avoiding any adverse systemic response. The neurotrophic agent initially seemed at an advantage here, for the trick of augmenting the limb's innervation must act locally. Strengthening the stimulus enough to yield uncontestibly enhanced regeneration, however, requires isolation of the neurotrophic agent and application with the same risk of diffusion. A fibroblast growth factor which resembles the neurotrophic agent in several respects (Mescher and Gospodarowicz, 1979) is the most promising candidate for this purpose, but has not yet pushed cell proliferation above the level found in normally regenerating newt arms. Despite sharing the current scepticism about chalones (Houck, 1973) it would be unwise to ignore the concept that tissue growth may be susceptible to external inhibition (Birch, 1976) or to forget the stimulatory effect of foetal serum and embryo extracts on the proliferation of cells in tissue culture. Hormonal manipulations, especially of thyroxine and corticosteroids, which once seemed the most promising means of delaying redifferentiation, have been most disappointing. As with the other tactics mentioned here, however, there are surely alternative treatments and ideas from allied fields of research which deserve to be explored. I propose to consider the influence of thyroid hormones on regeneration in more detail, just as one example of the exploration still needed.

10.4 Thyroid antagonism

First of all, it is necessary to point out that thyroid hormones do not have the same effect on all vertebrates. They control moulting in newts, for instance, but anurans moult in response to corticosteroids. Although the thyroid controls metamorphosis in both these groups, the delayed growth of tadpole (but not newt) legs seems specifically adapted to increasing amounts of circulating thyroxine. This last feature is surely irrelevant to amniotes which do not metamorphose and probably maintain a fairly constant level of thyroxine after hatching or birth, apart from circadian rhythms and seasonal changes. Furthermore, the available measurements of thyroxine in circulation mainly reflect the amount tightly bound to plasma albumins and globulins, whereas there is a growing suspicion that the active hormone is either free

Table 10.1 Thyroxine (T_4) concentration in the blood of adult vertebrates.

Species	$T_4(\mu g/l$ serum)	Comments
Lamprey	5	Falls at metamorphosis from 73 in larva
Hagfish	34	
Salmon	20	Rising to 61–95 in spring smolt
Trout	45	About 0.5% as free T_4
Spotted newt	6.3	Rising to 10–15 in spring; 0.2 μg T_3
Crested newt	Undetected[*]	
Ribbed newt	Undetected[*]	
Axolotl	Undetected[*]	<0.1 in juvenile (Schultheis, 1979)
Tiger salamander	2.2	Rising to 5.5 in spring
Leopard frog	17	When starved
Bullfrog	Undetected	3–6 at metamorphosis; 1.5 μg T_3
Common toad	3.5	In autumn
Green toad	5	In autumn, rising to 11–14 in spring
Cobra	10–20	With seasonal variation
Chicken	8–16	With diurnal rythm; 2–4 μg T_3
Mammals (nine species)	10–40	About 0.05% as free T_4
Human (normal range)	10–120	Higher during pregnancy and just after birth

[*]No measurable reaction in standard clinical tests sensitive to 8 $\mu g/l$ and to 0.06 $\mu g/l$ of T_3. Most of the other data may be traced from Liversage and Korneluk (1978) or subsequent issues of *General and Comparative Endocrinology*.

thyroxine or triiodothyronine. With this caveat that the data may be quite misleading, Table 10.1 indicates the presence of lower thyroxine levels in amphibians than in most other vertebrates. In fact many of the amphibian measurements are too low to be at all reliable, and it is possible there is some threshold level of thyroid homone which prevents limb regeneration in anurans and higher vertebrates but which urodeles do not normally reach.

This tenuous hypothesis is really based on the way metamorphosis influences regeneration (see Chapter 3). Most tadpoles lose their capacity for limb regeneration as they approach metamorphic climax, although perhaps incompletely in the clawed toad (Beetschen, 1952; Dent, 1962) or very gradually in the midwife toad (Lecamp, 1950, 1952). Limb regeneration is typically retarded at the onset of metamorphosis in urodeles (Figure 3.3) and may become defective or hypomorphic then, either temporarily in Iberian newts or permanently in common newts (Schwidefsky, 1935) and fire salamanders (Roguski, 1961). The common factor underlying these examples could be an elevated thyroid activity causing premature redifferentiation in mesenchymal cells. The comparative survey shown in Table 10.1 might have invalidated such a simple view, but actually provides mild encouragement for it. Some further support comes from the slight acceleration of limb regeneration noticed by Schmidt (1958a) in newts several weeks after he had extirpated their thyroids, and from the postmetamorphic recovery of regenerative ability in *Pleurodeles* (Table 3.1).

Where thyroxine seemed most obviously implicated as an inhibitor of limb regeneration, in anuran metamorphosis, there is now a mass of contradictory evidence which at least complicates the issue. Tadpole legs do not grow appreciably after thyroidectomy, so presumably a regenerate would not grow either, while thyroid antagonists also inhibit regeneration. Tadpole limb buds begin to lose their regenerative ability quite early in their development, when still small with poorly formed digits. Between this time and metamorphic climax, their capacity for regeneration is progressively restricted to more distal parts of the leg, and later amputation at any particular level yields more defective regenerates (Figure 3.4). Marcucci (1916), who discovered these gradual changes, related them to a proximodistal wave of differentiation down the leg of frog and toad tadpoles. Forsyth (1946) provided histological evidence for the same conclusion in *Rana sylvatica*. He stated the formation of perichondrium and striated muscle in any region corresponded to a critical stage when amputation there only elicited the formation of a blastema which was incapable of further growth (Table 10.2). The proximodistal recession of regenerative capacity has also been confirmed in *Rana clamitans* tadpoles, although without evident histological changes, by Schotté and Harland (1943c), who endorsed Marcucci's argument that an elevated thyroxine content in the blood should affect all parts of the leg simultaneously. Thyroxine can only exert an indirect control.

One way in which thyroxine could act as an indirect antagonist of regeneration here is through its effect in the limb's innervation. There is a dramatic loss of sensory and motor neurons which supply leg nerves, during the period when leg regeneration becomes increasingly defective in *Xenopus* tadpoles. Hughes (1961, 1968) and Prestige (1965) have estimated that only about one-tenth of the motor neurons and about one-third of the sensory neurons survive this destruction to provide the definitive leg innervation. Amputation shortly before metamorphosis postpones the degeneration of some neurons (Prestige, 1967a, b). If the leg innervation is elevated or maintained to a corresponding degree, it could explain Marcucci's (1916) contention that reamputating a young regenerate causes renewed regeneration even beyond the critical period (when amputating the normal contralateral leg of the same

Table 10.2 Loss of regenerative capacity in the thigh of *Rana sylvatica* tadpoles (data of Forsyth, 1946: reproduced by permission of The Wistar Institute).

Stage[*]	Total amputated	Regenerated	Arrested blastema	Healed stump	Uncertain
V–IX (leg bud)	17	17	0	0	0
X	21	12	6	2	1
XI	16	1	3	10	2
XII (five toes)	30	0	0	30	0
XIII–XVIII	43	0	0	43	0

[*]Converted to the stages of Taylor and Kollros (1946).

tadpole did not cause it to regenerate). Neuronal degeneration is a normal feature of metamorphosis and subject to control by thyroxine, for Prestige (1965) recorded that it does not occur in tadpoles kept in thiourea.

Other examples cited in chapter 3 also suggest the basic idea of a thyroid control over regeneration must be qualified. The most telling arguments there are that tadpole legs which have already lost the power to regenerate do not regain it when grafted on to younger larvae. We could accommodate this failure by invoking the development of receptors which bind thyroxine tightly

Figure 10.2. Spikes regenerated by *Xenopus laevis*, amputated shortly after metamorphosis. Upper row, long and average spikes grown by controls; lower row, retarded spikes grown during recovery from treatment with thiourea.

enough to retard its depletion for several weeks after grafting, by which time the host might be supplying enough hormone to saturate the receptors. Assuming inhibition depended upon an increased number of saturated receptors would also go some way to explaining the proximodistal progression of skin gland maturation (Verma, 1965) which parallels the loss of ability to regenerate. Receptor sites with a high affinity for thyroxine have been identified as a component of chromatin and of the mitochondrial membrane in responsive mammalian tissue (Baxter *et al.*, 1979; Sterling *et al.*, 1978). Their involvement introduces quite a drastic complication for anyone wishing to test how thyroxine affects regeneration. The standard antithyroid agents such as thiourea, thiouracil or perchlorate only interfere with hormone production in the thyroid gland. They cannot affect tissue-bound hormone until the stored thyroglobulin and circulating hormone have been substantially depleted. No wonder such agents do not facilitate regeneration in late tadpoles, especially as they are suspected to have a more direct inhibitory action.

My only attempt at this kind of manipulation involved keeping postmetamorphic *Xenopus* in 0.4% thiourea from 1 month before until 2 months after amputation, by which time controls in tapwater had produced long spikes (Figure 10.2). The treatment virtually arrested regeneration but no scar tissue formed over the limb stumps, which were able to grow modest spikes after the treatment had been terminated. Like so many experiments considered in these pages, this one yielded enigmatic results. I still think the analysis and general approach were reasonable enough to justify trying milder treatments, and venture to hope this account may direct someone towards a correct solution.

References

The italicized numbers in brackets following each item provide a partial author index by indicating the main text pages where each article is cited.

Some items have been shortened in one of the following ways:

1. Where more than three authors claim to have written an article, only the first one is named here.
2. A few excessively long titles have been truncated, either by not repeating the main title of a serialized study or by omitting pointless subtitles.
3. The most commonly cited journals are identified by the acronyms listed below instead of the conventional abbreviations:

DAN   *Doklady Akademiya Nauk S.S.S.R.*
DB   *Developmental Biology*
JEEM *Journal of Embryology and experimental Morphology*
JEZ   *Journal of experimental Zoology*
JM   *Journal of Morphology*
RA   *(Wilhelm) Roux' Archiv für Entwicklungsmechanik der Organismen* (including all previous versions of the title)
RSZ   *Revue Suisse de Zoologie.*

Abe, T., T. Haga, and M. Kurokawa (1973). Rapid transport of phosphatidylcholine occurring simultaneously with protein transport in the frog sciatic nerve. *Biochem. J.* **136**, 731–740. (*41*)

Abeloos, M., and M. Lecamp (1931). Sur la production de formations anormales et multiples dans les membres du triton par la transplantation de régénérats. *C. R. Acad. Sci., Paris*, **192**, 639–641. (*178, 217–18*)

Allen, B. M. (1918). The relation of the thyroid gland to regeneration in *Rana pipiens. Anat. Rec.*, **14**, 85–86. (*67*)

Allen, B. M. (1925). The effects of extirpation of the thyroid and pituitary glands upon the limb development of anurans. *JEZ*, **42**, 13–30. (*67*)

Allen, B. M., and L. M. Ewell (1959). Resistance to X-irradiation by embryonic cells of the limb-buds of tadpoles. *JEZ*, **142**, 309–335. (*86*)

Anton, H-J. (1956). Das Auftreten von Unverträglichkeitsreaktion bei heteroplastichen Transplantationen zwischen verschiedenen *Triturus*-Arten. *RA*, **149**, 26–44. (*126*)

Anton, H-J. (1961). Zur Frage der Aktivierung der Gewebe im Extremitatenstumpf bei Urodelen vor der Blastembildung. *RA*, **153**, 363–369. (*101*)

Anton, H-J. (1965). The origin of blastemal cells and protein synthesis during forelimb regeneration in *Triturus*. In *Regeneration in Aminals and Related Problems* (Eds. V. Kiortsis and H. A. L. Trampusch), pp. 377–395, North-Holland, Amsterdam. (*101, 126*)

Babich, G. L., and J. E. Foret (1973). Effects of dibutyryl cyclic AMP and related compounds on newt limb regeneration blastemas *in vitro*. II. ^{14}C-leucine incorporation. *Oncology*, **28**, 89–95. (*101*)

Bacq, Z. M., and P. Alexander (1961). *Fundamentals of Radiobiology*, 2nd edition, Pergamon, Oxford. (*72*)

Balinsky, B. I. (1925). Transplantation des Ohrbläschens bei Triton. *RA*, **105**, 718–731. (*159*)

Balinsky, B. I. (1957). New experiments on the mode of action of the limb inductor. *JEZ*, **134**, 239–273. (*159*)

Bantle, J. A., and R. A. Tassava (1974). The neurotrophic influence on RNA precursor incorporation into polyribosomes of regenerating adult newt forelimbs. *JEZ*, **189**, 101–114. (*100*)

Bardeen, C. P., and F. H. Baetjer (1904). The inhibitive action of the Roentgen rays on regeneration in planarians. *JEZ*, **1**, 191–195. (*71*)

Barr, H. J. (1964). The fate of epidermal cells during limb regeneration in larval *Xenopus*. *Anat. Rec.*, **148**, 358. (*121*)

Basleer, R., P. Collognon, and F. Matagne-Dhoossche (1963). Effets cytologiques du Myleran sur le muscle strié de l'embryon de poulet *in ovo* et de la queue de l'axolotl en régénération. *Arch. Biol. Liège*, **74**, 79–94. (*106*)

Bast, R. E., M. Singer, and J. Ilan (1979). Nerve-dependant changes in content of ribosomes, polysomes, and nascent peptides in newt limb regenerates. *DB*, **70**, 13–26. (*42*)

Baxter, J. D., and 11 others (1979). Thyroid hormone receptors and responses. *Recent Prog. Hormone Res.*, **35**, 97–153. (*236*)

Bazzoli, A. S., and 3 others (1977). The effects of thalidomide and two analogues on the regenerating forelimb of the newt. *JEEM*, **41**, 125–135. (*107*)

Beck, C. F., and G. J. Howlett (1977). The nature of miscoding caused by growth in the presence of 2-thiouracil. *J. mol. Biol.*, **111**, 1–17. (*105*)

Becker, R. C. (1960). The bioelectric field pattern in the salamander and its stimulation by an electronic analog. *Inst. Radio-Engineers Transactions on Medical Electronics*, **7**, 202–207. (*113*)

Becker, R. O. (1961a). The bioelectric factors in amphibian-limb regeneration. *J. Bone & Joint Surgery*, **43A**, 643–656. (*112–13*)

Becker, R. O. (1961b). Search for evidence of axial current flow in peripheral nerves of salamander. *Science*, **134**, 101–102. (*112–13*)

Becker, R. O. (1972). Stimulation of partial limb regeneration in rats. *Nature, London*, **235**, 109–111. (*114*)

Becker, R. O., and J. A. Spadaro (1972). Electrical stimulation of partial limb regeneration in mammals. *Bull. N.Y. Acad. Med.*, **48**, 627–641. (*114*)

Beetschen, J. C. (1952). Extension et limites du pouvoir régénérateur des membres après la métamorphose chez *Xenopus laevis* Daudin. *Bull. Biol.*, **86**, 88–100. (*2, 33*)

Belkin, R. (1934). Régénération des extrémités de l'axolotl pendant la métamorphose. *C.R. Soc. Biol.*, **115**, 1162–1163. (*59*)

Belkin, R. I. (1950). Influence of thyreoiodine on the regeneration of limbs in the axolotl and *Amblystoma* (in Russian). *DAN*, **70**, 327–329. (*60*)

Bergonié, J. and L. Tridondeau (1906). Interprétation de quelques résultats de la radiothérapie et essai de fixation d'une technique rationnelle. *C.R. Acad. Sci., Paris*, **143**, 983–985. (*71*)

Bewley, T. A., and C. H. Li (1971). Sequence comparison of human pituitary hormone, human chorionic somatomammotropin and ovine pituitary lactogenic hormone. *Experientia*, **27**, 1368–1371. (*56*)

Bieber, S., R. Bieber, and G. H. Hitchings (1954). Activities of 6-mercaptopurine and related compounds on embryonic and regenerating tissues of *Rana pipiens*. *Ann. N.Y. Acad. Sci.*, **60**, 207–211. (*106–8*)

Bieber, S., and G. H. Hitchings (1959). Effects of growth-inhibitors on amphibian tail blastema. *Cancer Res.*, **19**, 112–115. (*106, 110*)

Birch, P. R. J. (1976). *The Biology of Cancer: A New Approach*. MTP Press, Lancaster, England. (*232*)

Bischler, V. (1926). L'influence du squelette dans la régénération, et les potentialités des divers territoires du membre chez *Triton cristatus*. *RSZ*, **33**, 431–560. (*85, 126, 139, 156*)

Bodemer, C. W. (1958). The development of nerve-induced supernumerary limbs in the adult newt, *Triturus viridescens*. *JM*, **102**, 555–581. (*126*)

Bodemer, C. W. (1959). Observations on the mechanism of induction of supernumerary limbs in adult *Triturus viridescens*. *JEZ*, **140**, 79–99. (*126*)

Bodemer, C. W. (1960). The importance of quantity of nerve fibers in development of nerve-induced supernumerary limbs in *Triturus* and enhancement of the nervous influence by tissue implants. *JM*, **107**, 47–59. (*28, 45*)

Bodemer, C. W. (1962a). The effects of micro-quantities of ribonuclease on the regenerating amphibian limb. *Anat. Rec.*, **142**, 105–110. (*107–9*)

Bodemer, C. W. (1962b). Distribution of ribonucleic acid in the regenerating urodele limb as determined by autoradiographic localization of uridine-H^3. *Anat. Rec.*, **142**, 457–468. (*100*)

Bodemer, C. W. (1964). Evocation of regrowth phenomena in anuran limbs by electrical stimulation of the nerve supply. *Anat. Rec.*, **148**, 441–457. (*114*)

Bodemer, C. W., and N. B. Everett (1959). Localization of newly synthesized proteins in regenerating newt limbs as determined by radio-autographic localization of injected methionine-S^{35}. *DB*, **1**, 327–342. (*101*)

Borgens, R. B., J. W. Vanable, and L. F. Jaffe (1977a). Bioelectricity and regeneration. I. Initiation of frog limb regeneration by minute currents. *JEZ*, **200**, 403–416. (*115–16*)

Borgens, R. B., J. W. Vanable, and L. F. Jaffe (1977b). Bioelectricity and regeneration: Large currents leave the stumps of regenerating newt limbs. *Proc. Natl. Acad. Sci., Wash.*, **74**, 4528–4532. (*112*)

Borgens, R. B., J. W. Vanable, and L. F. Jaffe (1979a). Small artificial currents enhance *Xenopus* limb regeneration. *JEZ*, **207**, 217–225. (*115*)

Borgens, R. B., J. W. Vanable, and L. F. Jaffe (1979b). Role of subdermal current shunts in the failure of frogs to regenerate. *JEZ*, **209**, 49–55. (*115*)

Borgens, R. B., J. W. Vanable, and L. F. Jaffe (1979c). Reduction of sodium dependent stump currents disturbs urodele limb regeneration. *JEZ*, **209**, 377–386. (*115*)

Both, N. J. de (1970a). Transplantation immunity in the axolotl (*Ambystoma mexicanum*) studied by blastemal grafts. *JEZ*, **173**, 147–158. (*122*)

Both, N. J. de (1970b). The developmental potencies of the regeneration blastema of the axolotl limb. *RA*, **165**, 242–276. (*146, 199*)

Both, N. J. de (1971). The regeneration territories – a critical note. *Folia Morph.*, **19**, 177–181. (*160*)

Bovet, D. (1930). Les territoires de régénération; leurs propriétés étudiées par la méthode de déviation du nerf. *RSZ*, **37**, 83–145. (*157*)

Brandt, W. (1924). Extremitätentransplantionen an *Triton taeniatus*. *RA*, **103**, 517–554. (*164*)

Bromley, S. C., and C. S. Thornton (1974). Effect of highly purified growth hormone on limb regeneration in the hypophystectomised newt, *Notophthalmus viridescens. JEZ*, **190**, 143–153. *(55)*

Bronsted, H. V. (1969). *Planarian Regeneration*, Pergamon, Oxford. *(194)*

Brunst, V. V. (1950a). Influence of X-rays on limb regeneration in urodele amphibians. *Quart. Rev. Biol.*, **25**, 1–29. *(77–82)*

Brunst, V. V. (1950b). The effects of local X-ray irradiation on the tail development of young axolotls. *JM*, **86**, 115–151.

Brunst, V. V. (1952). Technics of local low voltage roentgen ray irradiation of experimental animals. *Lab. Invest.*, **1**, 432–438. *(80)*

Brunst, V. V. (1960). Reaction of limb regenerates of adult axolotl (*Siredon mexicanum*) to X-irradiation. *Radiation Res.*, **12**, 642–656. *(79–81, 114, 128)*

Brunst, V. V. (1961). Some problems of regeneration. *Quart. Rev. Biol.*, **36**, 178–206.

Brunst, V. V., and E. A. Scheremetieva (1933). Untersuchung des Einfluss von Röntgenstrahlen auf die Regeneration der Extremitäten beim Triton. *RA*, **128**, 181–215. *(75)*

Brunst, V. V., and E. A. Scheremetieva (1936). Sur la perte local du pouvoir régénérateur chez le triton et l'axolotl causée par l'irradiation avec les rayons X. *Archs Zool. exp. gén.*, **78**, 57–67.

Bryant, S. V. (1976). Regenerative failure of double half limbs in *Notophthalmus viridescens. Nature, London*, **263**, 676–679.

Bryant, S. V. (1977). Pattern regulation in amphibian limbs, in *Vertebrate Limb and Somite Morphogenesis* (Eds. D. A. Ede *et al.*), pp. 311–327, Cambridge U.P., Cambridge.

Bryant, S. V. (1978). Pattern regulation and cell commitment in amphibian limbs, in *The Clonal Basis of Development* (Eds. S. Subtelny and I. M. Sussex), pp. 63–82, Academic Press, New York.

Bryant, S. V., and B. A. Baca (1978). Regenerative ability of double-half and half upper arms in the newt, *Notophthalmus viridescens. JEZ*, **204**, 307–323 *(213—16)*

Bryant, S. V., and L. E. Iten (1974). The regulative ability of the limb regeneration blastema of *Notophthalmus viridescens. RA*, **174**, 90–101. *(146)*

Bryant, S. V., and L. E. Iten (1976). Supernumerary limbs in amphibians: experimental production in *Notophthalmus viridescens* and a new interpretation of their formation. *DB*, **50**, 212–234. *(208–11, 222)*

Bryant, S. V., and L. E. Iten (1977). Intercalary and supernumerary regeneration in regenerating and mature limbs of *Notophthalmus viridescens. JEZ*, **202**, 1–15.

Burgess, A. M. C. (1967). The developmental potentialities of regeneration blastema cell nuclei as determined by nuclear transplantation. *JEEM*, **18**, 27–41. *(121)*

Burnett, A. L. (1968). The acquisition, maintenance, and lability of the differentiated state in *Hydra*, in *The Stability of the Differentiated State* (Ed. H. Ursprung), pp. 109–127, Springer-Verlag, Berlin. *(140)*

Burnett, A. L., C. E. Kary, and A. M. Lagorio (1971). Induction of growth in newt and frog limbs after perfusion with extracts from newt blastemas. *Nature, London*, **234**, 98–99. *(37–8)*

Burnett, B. M., and R. A. Liversage (1964). The influence of chloramphenicol on the regenerating forelimb in adult *Triturus viridescens. Am. Zool.*, **4**, 427. *(107)*

Butler, E. G. (1933). The effects of X-radiation on the regeneration of the fore limb of *Amblystoma* larvae. *JEZ*, **65**, 271–316. *(73, 84)*

Butler, E. G. (1935). Studies on limb regeneration in X-rayed *Amblystoma* larvae. *Anat. Rec.*, **62**, 295–307. *(75, 80)*

Butler, E. G. (1951). The mechanics of blastema formation and regeneration in urodele limbs of reversed polarity. *Trans. N.Y. Acad. Sci.*, **13**, 164–167. *(182)*

Butler, E. G. (1955). Regeneration of the urodele forelimb after reversal of its proximo-distal axis. *JM*, **96**, 265–281. *(182)*

Butler, E. G., and H. F. Blum (1963). Supernumerary limbs of urodele larvae resulting from localised ultraviolet light. *DB*, **7**, 218–233. *(98)*

Butler, E. G., and J. P. O'Brien (1942). Effects of localized X-radiation on regeneration of the urodele limb. *Anat. Rec.*, **84**, 407–413. *(76, 80)*

Butler, E. G., and W. O. Puckett (1940). Studies on cellular interactions during limb regeneration in *Amblystoma*. *JEZ*, **84**, 223–239 *(144)*

Butler, E. G., and O. E. Schotté (1941). Histological alterations in denervated non-regenerating limbs of urodele larvae. *JEZ*, **88**, 307–341. *(23)*

Butler, E. G., and O. E. Schotté (1949). Effects of delayed denervation on regenerative activity in limbs of urodele larvae. *JEZ*, **112**, 361–392. *(23, 147)*

Callan, H. G., and J. H. Taylor (1968). A radioautographic study of the time course of male meiosis in the newt *Triturus vulgaris*. *J. Cell Sci.*, **3**, 615–626. *(15, 20)*

Carlone, R. L., and J. E. Foret (1979). Stimulation of mitosis in cultured limb blastemata of the newt, *Notophthalmus viridescens*. *JEZ*, **210**, 245–252. *(111)*

Carlson, B. M. (1967a). The histology of inhibition of limb regeneration in the newt, *Triturus*, by Actinomycin D. *JM*, **122**, 249–263. *(107)*

Carlson, B. M. (1967b). Studies on the mechanism of implant-induced supernumerary limb formation. I. Histology *JEZ*, **164**, 227–241. *(126)*

Carlson, B. M. (1969). Inhibition of limb regeneration in the axolotl after treatment of the skin with Actinomycin D. *Anat. Rec.*, **163**, 389–401. *(107, 151)*

Carlson, B. M. (1972). Muscle morphogenesis in axolotl limb regenerates after removal of the stump musculature. *DB*, **28**, 487–497.

Carlson, B. M. (1974a). Morphogenetic interactions between rotated skin cuffs and underlying stump tissues in regenerating axolotl forelimbs. *DB*, **39**, 263–285. *(81, 168–9, 218)*

Carlson, B. M. (1974b). Factors controlling the initiation and cessation of early events in the regenerative process, in *Neoplasia and Cell Differentiation* (Ed. G. V. Sherbet), pp. 60–105, Karger, Basel.

Carlson, B. M. (1975a). Multiple regeneration from axolotl limb stumps bearing cross-transplanted minced muscle regenerates. *DB*, **45**, 203–208 *(175)*

Carlson, B. M. (1975b) The effects of rotation and positional change of stump tissues upon morphogenesis of the regenerating axolotl limb. *DB*, **47**, 269–291. *(173–5, 218)*

Carlson, B. M. (1977). Inhibition and axial deviation of limb regeneration in the newt by means of a digit implanted into the amputated limb. *JM*, **154**, 223–241. *(204)*

Carlson, B. M. (1978). Types of morphogenetic phenomena in vertebrate regenerating systems. *Am. Zool.*, **18**, 869–882. *(230)*

Carlson, B. M., S. F. Civiletto, and H. G. Goshgarian (1974). Nerve interactions and regenerative processes occurring in newt limbs fused end-to-end. *DB*, **37**, 248–262. *(184)*

Carlson, B. M., and C. F. Morgan (1967). Studies on the mechanism of implant-induced supernumerary limb formation. II. The effect of heat treatment, lyophilization and homogenization on the inductive capacity of frog kidney. *JEZ*, **164**, 243–249. *(126)*

Cass, D. T., and R. F. Mark (1975). Re-innervation of axolotl limbs. I. Motor nerves. *Proc. R. Soc., London*, **B190**, 45–58. *(35, 225)*

Chalkley, D. T. (1954). A quantitative histological analysis of forelimb regeneration in *Triturus viridescens*. *JM*, **94**, 21–70. *(8–12, 145, 203)*

Chalkley, D. T. (1959). The cellular basis of limb regeneration, in *Regeneration in Vertebrates* (Ed. C. S. Thornton), pp. 34–58, U. Chicago Press, Chicago. *(11)*

Chapron, C. (1974). Mise en évidence du rôle, dans la régéneration des amphibiens, d'une glycoprotéine sécrétée par la cape apicale. *JEEM*, **32**, 133–145. *(104)*.

Charlemagne, J., and A. Tournefier (1974). Obtention of histocompatible strains in the urodele amphibian *Pleurodeles waltlii* Michah (Salamandridae). *J. Immunogenetics*, **1**, 125–129. *(124)*

Child, C. M. (1941). *Patterns and Problems of Development*, U. Chicago Press, Chicago.

Ching, Y-C., and P. J. Wedgewood (1967). Immunologic responses in the axolotl *Siredon mexicanum. J. Immunol.*, **99**, 191–200. *(122)*

Choo, A. F., D. M. Logan, M. P. Rathbone (1978). Nerve trophic effects: an in vitro assay for factors involved in regulation of protein synthesis in regenerating amphibian limbs. *JEZ*, **206**, 347–354. *(111)*

Cohen, N. (1966a) Tissue transplantation immunity in the adult newt, *Diemictylus viridescens*. I. The latent phase: healing, restoration of circulation and pigment cell changes in autografts and allografts. *JEZ*, **163**, 157–172. *(122)*

Cohen, N. (1966b). II. The rejection phase: first- and second-set allograft reactions and lack of sexual dimorphism. *JEZ*, **163**, 173–190. *(122)*

Cohen, N. (1966c). III. The effects of X-irradiation and temperature on the allograft reaction. *JEZ*, **163**, 231–240. *(79, 122)*

Cohen, N. (1967). The maturation of transplantation immunity in the salamander, *Ambystoma tigrinum. Am. Zool.*, **7**, 763. *(122)*

Cohen, N. (1968). Chronic skin graft rejection in the Urodela. I. A comparative study of first- and second-set allograft reactions. *JEZ*, **167**, 37–47. *(122)*

Cohen, N. (1969). Immunogenetic and developmental aspects of tissue transplantation immunity in urodele amphibians, in *Biology of Amphibian Tumors* (Ed. M. Mizell), pp. 153–168, Springer-Verlag, New York. *(122)*

Cohen, N. (1971). Amphibian transplantation reactions: a review. *Am. Zool.*, **11**, 193–205. *(122)*

Cohen, N., and W. H. Hildemann (1968). Population studies of allograft rejection in the newt, *Diemictylus viridescens. Transplantation*, **6**, 208–217. *(122)*

Conn, H., S. Wessels, and H. Wallace (1971). Regeneration of locally irradiated salamander limbs following superficial skin incisions or nerve transection. *JEZ*, **178**, 359–368. *(80, 91, 171)*

Connelly, T. G., R. A. Tassava, and C. S. Thornton (1968). Limb regeneration and survival of prolactin treated hypophysectomized adult newts. *JM*, **126**, 365–371. *(54–5)*

Conninck, L. de, and 3 others (1955). Acides ribonucléiques, hormone somatotrope et régénération chez *Triturus* (Urodèles). *Soc. R. Zool. Belg. Annales*, **86**, 191–234. *(2, 54)*

Copeland, D. E. (1943). Cytology of the pituitary gland in developing and adult *Triturus viridescens. JM*, **72**, 379–409. *(58)*

Curtis, W. C. (1936). Effects of X-rays and radium on regeneration, in *The Biological Effects of Radiation* (Ed. B. M. Duggar), pp. 437–457, McGraw-Hill, New York. *(73)*

Dalton, H. C., and Z. P. Krassner (1956). Pituitary influence on pigment pattern development in the white axolotl. *JEZ*, **133**, 241–257. *(66)*

Dasgupta, S. (1970). Developmental potentialities of blastema cell nuclei of the Mexican axolotl. *JEZ*, **175**, 141–148. *(121)*

David, L. (1932). Das verhalten von extremitätenregeneraten des weissen und pigmentieren azolotl bei heteroplasticher, heterotoper und orthotoper transplantation und sukzessiver regeneration. *RA*, **126**, 457–511. *(78)*

David, L. (1934). La contribution du matériel cartilagineux et osseaux au blastème de régénération des membres chez les amphibiens urodèles. *Arch. d'Anat. Microsc.*, **30**, 217–234. *(144)*

Davies, D. R., and H. J. Evans (1966). The role of genetic damage in radiation induced cell-lethality. *Adv. Radiation Biol.*, **2**, 243–353. *(72)*

Dearlove, G. E., and M. H. Dresden (1976). Regenerative abnormalities in *Notophthalmus viridescens* induced by repeated amputations. *JEZ*, **196**, 251–261. *(147)*

Dearlove, G. E., and D. L. Stocum (1974). Denervation-induced changes in soluble protein content during forelimb regeneration in the adult newt, *Notophthalmus viridescens*. *JEZ*, **190**, 317–327. *(101)*

Dearlove, G. E., S. Lopez, and M. H. Dresden (1975). Disruption of normal forelimb regeneration in adult *Notophthalmus viridescens* by sublethal concentrations of Trypan Blue. *JEZ*, **194**, 413–428. *(107–9)*

Deck, J. D. (1955). The innervation of urodele limbs of reversed proximo-distal polarity. *JM*, **96**, 301–331. *(182)*

Deck, J. D. (1961a). Morphological effects of partial denervation on regeneration of the larval salamander forelimb. *JEZ*, **146**, 299–307. *(30, 50)*

Deck, J. D. (1961b). The histological effects of partial denervation and amputation in larval salamander forelimbs. *JEZ*, **148**, 69–79. *(50)*

Deck, J. D. (1971). The effects of infused materials upon the regeneration of newt limbs. II. Extracts from newt brain and spinal cord. *Acta anat.*, **79**, 321–332. *(38)*

Deck, J. D., and C. B. Futch (1969). I. Blastemal extracts on denervated limb stumps. *DB*, **20**, 332–348. *(38)*

Deck, J. D., and H. L. Riley (1958). Regeneration on hindlimbs with reversed proximo-distal polarity in larval and metamorphosing urodeles. *JEZ*, **138**, 493–504. *(182)*

Deck, J. D., and S. L. Shapiro (1963). Interference with newt limb regeneration by an inhibitor of histamine formation. *Anat. Rec.*, **145**, 316. *(107)*

Dafazio, J. V. (1968). Failure of growth hormone to support regeneration of denervated limbs in adult newts. *Anat. Rec.*, **160**, 338. *(57)*

DeHaan, R. L. (1956). The serological determination of developing muscle protein in the regenerating limb of *Amblystoma mexicanum*. *JEZ*, **133**, 73–85. *(103, 132)*

DeLanney, L. E. (1961). Homografting of sexually mature Mexican axolotls. *Am. Zool.*, **1**, 349. *(122)*

DeLanney, L. E. (1978). Immunogenetic profile of the axolotl: 1977. *Am. Zool.*, **18**, 289–299. *(122)*

DeLanney, L. E., M. K. Blackler, and K. V. Prahlad (1967). The relationship of age to allograft and tumor rejection in Mexican axolotls. *Am. Zool.*, **7**, 733. *(122)*

Della Valle, P. (1913). La doppia regenerazione inversa nelle fratture delle zampe di Triton. *Boll. Soc. nat. Napoli*, **25**, 95–160. *(184–6)*

Dent, J. N. (1954). A study of regenerates emanating from limb transplants with reversed proximo-distal polarity in the adult newt. *Anat. Rec.*, **118**, 841–856. *(182)*

Dent, J. N. (1962). Limb regeneration in larvae and metamorphosing individuals of the South African clawed toad. *JM*, **110**, 61–77. *(2, 32, 68)*

Dent, J. N. (1966). Maintenance of thyroid function in newts with transplanted pituitary glands. *Gen. comp. Endocrinol.*, **6**, 401–408. *(54–5, 122)*

Dent, J. N. (1967). Survival and function in hypophyseal homografts on the spotted newt. *Am. Zool.*, **7**, 714. *(122)*

Dertinger, H., and H. Jung (1970). *Molecular Radiation Biology*. Springer-Verlag, Berlin. *(83)*

Descotils-Heernu, F., J. Quertier, and J. Brachet (1961). Quelques effets du B-mercaptoéthanol et du dithioglycol sur la régénération. *DB*, **3**, 277–296. *(106)*

Desha, D. L. (1974). Irradiated cells and blastema formation in the adult newt, *Notophthalmus viridescens*. *JEEM*, **32**, 405–416. *(81, 89, 94)*

Desselle, J-C. (1968). Restauration de la régénération par des implants de cartilage dans les membres irradiés de *Triturus cristatus*. *C.R. Acad. Sci., Paris*, **D267**, 1642–1645. *(81, 88, 141)*

Desselle, J-C. (1974). Cytophotometrie des acides nucléiques dans les cellules musculaires de membres en régénération, de membres irradiés aux rayons X, de membres irradiés reactivés par implants de cartilage chez *Triturus cristatus*. *Acta Emb. exper.* **1974**, 207–235. *(84, 88, 99)*

Desselle, J.-C., and M. Gontcharoff (1978). Cytophotometric detection of the participation of cartilage grafts in regeneration of X-rayed urodele limbs. *Biologie cellulaire*, **33**, 45–53. (*90, 126, 138*)

Detwiler, S. R. (1920). Experiments on the transplantation of limbs in *Amblystoma*. The formation of nerve plexuses and the function of the limbs. *JEZ*, **31**, 117–169. (*225*)

Detwiler, S. R. (1930). Observations upon the growth, function, and nerve supply of limbs when grafted to the head of salamander embryos. *JEZ*, **55**, 319–379. (*162*)

Diamond, J., and 3 others (1976). Trophic regulation of nerve sprouting. *Science*, **193**, 371–377. (*226*)

Dinsmore, C. E. (1974). Morphogenetic interactions between minced limb muscle and transplanted blastemas in the axolotl. *JEZ*, **187**, 223–231. (*200*)

Donaldson, D. J., J. M. Mason, and B. R. Jennings (1974). Protein patterns during regeneration and wound healing in the adult newt: a comparative study using gel electrophoresis. *Oncology*, **30**, 334–346. (*101*)

Drachman, D. B., and M. Singer (1971). Regeneration in botulinum-poisoned forelimbs of the newt, *Triturus*. *Exp. Neurol.*, **32**, 1–11. (*36, 39*)

Dresden, M. H. (1969). Denervation effects on newt limb regeneration: DNA, RNA and protein synthesis. *DB*, **19**, 311–320. (*41*)

Dretchen, K. L., and 3 others (1976). Evidence for a prejunctional role of cyclic nucleotides in neuromuscular transmission. *Nature, London*, **264**, 79–81. (*109*)

Droin, A. (1959). Potentialités morphogènes dans la peau du triton en régénération. *RSZ*, **66**, 641–709. (*168, 173*)

Dubois, P. M., and 3 others (1975). Evidence for immunoreactive somatostatin in the endocrine cells of human foetal pancreas. *Nature, London*, **256**, 731–732. (*56*)

Dulcibella, T. (1974a). The occurrence of biochemical metamorphic events without anatomical metamorphosis in the axolotl. *DB*, **38**, 175–186. (*66*)

Dulcibella, T. (1974b). The influence of L-thyroxine on the change in red blood cell type in the axolotl. *DB*, **38**, 187–194. (*66*)

Dumont, A. E., and N. Sohn (1963). Effect of 5-fluorouracil on regeneration in tadpoles. *Nature, London*, **199**, 617–618. (*106*)

Dunis, D. A., and M. Namenwirth (1977). The role of grafted skin in the regeneration of X-irradiated axolotl limbs. *DB*, **56**, 97–109. (*89—90, 140*)

Efimov, M. (1933a). Ueber den Mechanismus des Regenerationsprozesses. II. Mitteil. Die Rolle der Haut im Prozess der Regeneratiuon eines Organs beim Axolotl. *Biol. Zhurn.*, **2**, 214–219. (*148*)

Efimov, M. (1933b). III Mitteil. Bleibt die Polarität der Extremität beim Regenerationsprozess erhalten? Die Rolle der innerem Teile des Organs in diesem Prozess. *Biol. Zhurn.*, **2**, 220–228. (*181*)

Egar, M., and M. Singer (1971). A quantitative electron microscope analysis of peripheral nerve in the urodele amphibian in relation to limb regenerative capacity. *JM*, **133**, 387–397. (*24, 29*)

Egar, M., C. L. Yntema, and M. Singer (1973). The nerve fiber content of *Amblystoma* aneurogenic limbs. *JEZ*, **186**, 91–95. (*48*)

Eggert, R. C. (1966). The responses of X-irradiated limbs of adult urodeles to autografts of normal cartilage. *Jez*, **161**, 369–390. (*81, 134, 137*)

Eiland, L. C. (1975). The relation of nerves to multiple regeneration in a single newt limb. *JEZ*, **194**, 359–371. (*29, 186*)

Evans, G. M., and H. Rees (1971). Mitotic cycles in dicotyledons and monocotyledons. *Nature, London*, **233**, 350–351. (*17*)

Faber, J. (1960). An experimental analysis of regional organization in the regenerating fore limb of the axolotl (*Ambystoma mexicanum*). *Arch. Biol.*, **71**, 1–72. (*145–8, 196–8*)

Faber, J. (1962a). Additional experiments on the self-differentiation of transplanted whole and half forelimb regenerates of *Ambystoma mexicanum*. *Arch. Biol.*, **73**, 369–378. (*198*)

Faber, J. (1962b). A threshold effect in the morphogenetic realisation of transplanted limb regenerates of *Ambystoma mexicanum*. *Arch. Biol.*, **73**, 379–403 (*198*)

Faber, J. (1965). Autonomous morphogenetic activities of the amphibian regeneration blastema, in *Regeneration in Animals and Related Problems* (Eds. H. A. L. Trampusch and V. Kiortsis), pp. 404–419, North-Holland, Amsterdam. (*198*)

Faber, J. (1971). Vertebrate limb ontogeny and limb regeneration: morphogenetic parallels. *Adv. Morphogenesis*, **9**, 127–147. (*148, 198*)

Faber, J. (1976). Positional information in the amphibian limb. *Acta Biotheoretica*, **25**, 44–65. (*198–204*)

Fallon, J. F., and G. M. Crosby (1975). The relationship of the zone of polarizing activity to supernumerary limb formation (twinning) in the chick wing bud. *DB*, **43**, 24–34. (*228*)

Faustov, V. S., and B. M. Carlson (1967). Changes in the levels of ATP and respiration in regenerating axolotl extremities (in Russian). *DAN*, **176**, 728–730. (*103*)

Filatow, D. (1927). Activierung des Mesenchyms durch eine Ohrblase und einen Fremdkörper bei Amphibien. *RA*, **110**, 1–32. (*159*)

Finch, R. A. (1969). The influence of the nerve on lower jaw regeneration in the adult newt, *Triturus viridescens*. *JM*, **129**, 401–413. (*94*)

Fischman, D. A., and E. D. Hay (1962). Origin of osteoclasts from mononuclear leucocytes in regenerating newt limbs. *Anat. Rec.*, **143**, 329–337. (*14*)

Flickinger, R. A. (1967). Biochemical aspects of regeneration, in *The Biochemistry of Animal Development* (Ed. R. Weber), vol. 2, pp. 303–337, Academic Press, New York.

Foret, J. E. (1970). Regeneration of larval urodele limbs containing homoplastic transplants. *JEZ*, **175**, 297–322. (*139–40, 147*)

Foret, J. E. (1973). Stimulation and retardation of limb regeneration by Adenosine 3′-5′ monophosphate and related compounds. *Oncology*, **27**, 153–159. (*38, 101*)

Foret, J. E., and G. L. Babich (1973a). Inhibition by acetylcholine of protein synthesis in cultured limb regenerates. *Oncology*, **27**, 281–285. (*111*)

Foret, J. E., and G. L. Babich (1973b). Effects of dibutyryl cyclic AMP and related compounds on newt limb regeneration blastemas *in vitro*. I. H^3-thymidine incorporation. *Oncology*, **28**, 83–88. (*101, 111*)

Forsyth, J. W. (1946). The histology of anuran limb regeneration. *JM*, **79**, 287–321. (*67, 234*)

Francis, M. J. O., and D. J. Hill (1975). Prolactin-stimulated production of somatomedin by rat liver. *Nature, London*, **255**, 167–168. (*56*)

Francoeur, R. T. (1968). General and selective inhibition of amphibian regeneration by vinblastine and dactinomycin. *Oncology*, **22**, 218–226. (*106*)

Francoeur, R. T., and C. G. Wilber (1968). Amphibian regeneration and the teratogenic effects of vinblastine. *Oncology*, **22**, 302–331. (*106*)

Franklin, T. T., and G. A. Snow (1975). *Biochemistry of Antimicrobial Action*, 2nd edn., Chapman & Hall, London. (*108*)

French, V., P. J. Bryant, and S. V. Bryant (1976). Pattern regulation in epimorphic fields. *Science*, **193**, 969–981. (*211*)

Fritsch, C. (1911). Experimentelle Studien über Regenerationsvorgange des Gliedenassenskelettes der Amphibien. *Zool. Jarb.*, **30**, 376–472. (*156*)

Gaze, R. M. (1970). *The Formation of Nerve Connections*, Academic Press, New York. (*225*)

Gebhardt, D. O. E., and J. Faber (1966). The influence of aminopterin on limb regeneration in *Ambystoma mexicanum*. *JEEM*, **16**, 143–158. (*105*)

Geraudie, J., and M. Singer (1978). Nerve dependent macromolecular synthesis in the epidermis and blastema of the adult newt regenerate. *JEZ*, **203**, 455–460. (*43*)

246

Gidge, N. M., and S. M. Rose (1944). The role of larval skin in promoting limb regeneration in adult anura. *JEZ*, **97**, 71–93. (*39*)

Giorgi, P. de (1924). Les potentialités des régénérats chez *Salamandra maculosa*. Croissance et différentiation. *RSZ*, **31**, 1–52. (*152, 159, 196*)

Glade, R. W. (1957). The effects of tail tissue on limb regeneration in *Triturus viridescens*. *JM*, **101**, 477–521. (*157*)

Glade, R. W. (1963). Effects of tail skin, epidermis, and dermis on limb regeneration in *Triturus viridescens* and *Siredon mexicanum*. *JEZ*, **152**, 169–193. (*122–5, 157*)

Glick, B. (1931). The induction of supernumerary limbs in *Amblystoma*. *Anat. Rec.*, **48**, 407–414. (*159*)

Globus, M., and S. V. Globus (1977). Transfilter mitogenic effect of dorsal root ganglia on cultured regeneration blastemata in the newt, *Notophthalmus viridescens*. *DB*, **56**, 316–328. (*111*)

Godlewski, E. (1928). Untersuchungen über Auslösung und Hemmung der Regeneration beim Axolotl. *RA*, **114**, 108–143. (*11, 149*)

Goldfarb, A. J. (1909), The influence of the nervous system in regeneration. *JEZ*, **7**, 643–722. (*22*)

Goldfarb, A. J. (1911). The central nervous system in its relation to the phenomenon of regeneration. *RA*, **32**, 617–635. (*22*)

Goode, R. P. (1967). The regeneration of limbs of adult anurans. *JEEM*, **18**, 259–267. (*4*)

Goodwin, P. A. (1946). A comparison of regeneration rates and metamorphosis in *Triturus* and *Amblystoma*. *Growth*, **10**, 75–87. (*60*)

Goshgarian, H. G. (1976). The regeneration of brachial nerves of contralateral origin into denervated fused newt forelimbs. *JEZ*, **197**, 347–356. (*183*)

Goshgarian, H. G. (1979). The interaction of nerves of opposite regenerating polarity in fused newt forelimbs. *JEZ*, **209**, 297–307. (*183*)

Goss, R. J. (1956a). The relation of bone to the histogenesis of cartilage in regenerating forelimbs and tails of adult *Triturus viridescens*. *JM*, **98**, 89–123. (*147, 152, 188*)

Goss, R. J. (1956b). The unification of regenerates from symmetrically duplicated forelimbs. *JEZ*, **133**, 191–209. (*189*)

Goss, R. J. (1956c). Regenerative inhibition following limb amputation and immediate insertion into the body cavity. *Anat. Rec.*, **126**, 15–27. (*149*)

Goss, R. J. (1956d). The regenerative responses of amputated limbs to delayed insertion into the body cavity. *Anat. Rec.*, **126**, 283–297. (*149*)

Goss, R. J. (1957a). The relation of skin to defect regulation in regenerating half limbs. *JM*, **100**, 547–563. (*187–8*)

Goss, R. J. (1957b). The effect of partial irradiation on the morphogenesis of limb regenerates. *JM*, **101**, 131–148. (*81, 187*)

Goss, R. J. (1961). Regeneration of vertebrate appendages. *Adv. Morphogenesis*, **1**, 103–152. (*158–9, 187*)

Goss, R. J. (1969). *Principles of Regeneration*. Academic Press, New York.

Gräper, L. (1922a). Extremitätentransplantationen an Anuren. I. *RA*, **51**, 284–309. (*164*)

Gräper, L. (1922b). II. Mitteilung: Reverse Transplantationen. *RA*, **51**, 587–609. (*181*)

Gräper, L. (1926). Zur Genese der Polydactylie. *RA*, *107*, 154–161. (*186*).

Grillo, H. C., and 3 others (1968). Collagenolytic activity in regenerating forelimbs of the adult newt (*Triturus viridescens*). *DB*, **17**, 571–583. (*104*)

Grillo, R. S. (1971). Changes in mitotic activity during limb regeneration in *Triturus*. *Oncology*, **25**, 347–355. (*18*)

Grillo, R. S., and P. Urso (1968). An autoradiographic evaluation of the cell reproduction cycle in the newt, *Triturus viridescens*. *Oncology*, **22**, 208–217. (*18*)

Grim, M., and B. M. Carslon (1974). A comparison of morphogenesis of muscles of the forearm and hand during ontogenesis and regeneration in the axolotl (*Ambystoma mexicanum*). (in 2 parts). *Z. Anat. Entwick-Gesch.*, **145**, 137–167. (*154*)

Grim, M., and B. M. Carslon (1979). The formation of muscles in regenerating limbs of the newt after denervation of the blastema. *JEEM*, **54**, 99–111. (*23, 154*)

Grimm, L. M. (1971). An evaluation of myotypic respecification in axolotls. *JEZ*, **178**, 479–496. (*35, 225*)

Grobstein, C. (1959). Differentiation of vertebrate cells, in *The Cell* (Eds. J. Brachet and A. E. Mirsky), vol. I, pp. 437–496, Academic Press, New York. (*139*)

Gurdon, J. B. (1974). *The Control of Gene Expression in Animal Development*, Oxford U.P., London. (*144*)

Gurdon, J. B., and H. R. Wodland (1975). *Xenopus*, in *Handbook of Genetics* (Ed. R. C. King), vol. 4, pp. 35–50, Plenum, New York. (*120*)

Guth, L. (1968). 'Trophic' influences of nerve on muscle. *Physiol. Rev.*, **48**, 645–687. (*36*)

Guyénot, E. (1926). La perte du pouvoir régénérateur chez les anoures, étudiée par la méthode des hétérogreffes. *C.R. Soc. Biol.*, **94**, 437–439. (*58*)

Guyénot, E. (1927a). La perte du pouvoir régénérateur des anoures, étudiée par les hétérogreffes, at la notion de territoires. *RSZ*, **34**, 1–53. (*58*)

Guyénot, E. (1927b). Le problème morphogénétique dans la régénération des urodèles: détermination et potentialités des régénérats. *RSZ*, **34**, 127–154. (*148, 151, 211*)

Guyénot, E., J. Dinichert-Favarger, and M. Galland (1948). L'exploration du territoire de la patte antérieure du triton. *RSZ*, **55**, suppl. 2, 1–120. (*147, 157, 164*)

Guyénot, E., and O. Schotté (1923). Relation entre la masse du bourgeon de régénération et la morphologie du régénérat. *C.R. Soc. Biol.*, **89**, 491–493. (*229*)

Guyénot, E., and O. E. Schotté (1926a). Le rôle du système nerveux dans l'édification des régénérats de pattes chez les urodèles. *C.R. Soc. Phys. Hist. nat. Genève*, **43**, 32–36 (*23*)

Guyénot, E., and O. E. Schotté (1926b). Démonstration de l'existence de territoires specifiques de régénération par la méthode de la déviation des troncs nerveux. *C.R. Soc. Biol.*, **94**, 1050–1052. (*23, 156*)

Hahn, H. P. von and F. E. Lehman (1960). Verschiedenartige synergistiche Effekte zweier SH-substituierter Morphostatika. *RSZ*, **67**, 353–371. (*106, 110*)

Hall, A. B., and O. E. Schotté (1951). Effects of hypophysectomies upon the initiation of regenerative processes in the limbs of *Triturus viridescens*. *JEZ*, **118**, 363–387. (*54, 65*)

Harris, J. B., and S. Thesleff (1972). Nerve stump length and membrane changes in denervated skeletal muscle. *Nature, London,* **236**, 60–61. (*41*)

Harrison, R. G. (1918). Experiments on the development of the fore limb of *Amblystoma*, a self-differentiating equipotential system. *JEZ*, **25**, 413–461. (*146–8*)

Harrison, R. G. (1921). On relations of symmetry in transplanted limbs. *JEZ*, **32**, 1–136. (*161–3*)

Harrison, R. G. (1925). The effect of reversing the mediolateral or transverse axis of the fore-limb bud in the salamander embryo (*Amblystoma punctatum*). *RA*, **106**, 469–502. (*160–3, 181*)

Hay, E. D. (1952). The role of epithelium in amphibian limb regeneration, studied by haploid and triploid transplants. *Am. J. Anat.*, **91**, 447–481. (*133*)

Hay, E. D. (1956). Effects of thyroxine on limb regeneration in the newt, *Triturus viridescens*. *Bull. Johns Hopkins Hospital*, **99**, 262–285. (*61*)

Hay, E. D. (1959). Electron microscopic observation of muscle dedifferentiation in regenerating *Amblystoma* limbs. *DB*, **1**, 555–585. (*132*)

Hay, E. D. (1962). Cytological studies of dedifferentiation and differentiation in regenerating amphibian limbs, in *Regeneration* (Ed. D. Rudnick), pp. 177–210, Ronald Press, New York. (*132*)

248

Hay, E. D. (1966). *Regeneration*, Holt, Rinehart & Winston, New York.

Hay, E. D. (1968). Dedifferentiation and metaplasia in vertebrate and invertebrate regeneration, in *The Stability of the Differentiated State* (Ed. H. Ursprung), pp. 85–108, Springer-Verlag, Berlin. *(140)*

Hay, E. D. (1970). Regeneration of muscle in the amputated amphibian limb, in *Regeneration of Striated Muscle and Myogenesis* (Eds. A. Mauro, S. A. Shafiq, and A. T. Milhorat), pp. 3–24, Excerpta Medica, Amsterdam. *(103)*

Hay, E. D. (1974). Cellular basis of regeneration, in *Concepts of Development* (Eds. J. Lash and J. R. Whittaker), pp. 404–428, Sinauer Ass., Stamford, Conn. *(230)*

Hay, E. D., and D. A. Fischmann (1961). Origin of the blastema in regenerating limbs of the newt *Triturus viridescens*. *DB*, **3**, 26–59. *(12–16)*

Hayashida, T., P. Licht, C. S. Nicoll (1973). Amphibian pituitary growth hormone and prolactin: immunochemical relatedness to rat growth hormone. *Science*, **182**, 169–171. *(56)*

Heath, H. D. (1953). Regeneration and growth of chimaeric amphibian limbs. *JEZ*, **122**, 339–366. *(126)*

Heidemann, M. K. (1978). Quantitative behavioral studies of misdirected responses in skin-grafted *Rana pipiens*. *JEZ*, **204**, 249–257. *(226)*

Hertwig, G. (1927). Beitrage zum Determinations- und Regenerations- problem mittels der Transplantation haploid-kerniger Zellen. *RA*, **111**, 292–316. *(133)*

Hofmann, D. K., D. Klein, and W. Luther (1978). Limb regeneration from X-irradiated tails of *Ambystoma mexicanum* following transplantation of flank skin from region adjacent to hindlimb. *RA*, **185**, 227–234. *(159)*

Holder, N., S. V. Bryant, and P. W. Tank (1979). Interactions between irradiated and unirradiated tissues during supernumerary limb formation in the newt. *JEZ*, **208**, 303–309. *(169)*

Holder, N., and P. W. Tank (1979). Morphogenetic interactions occurring between blastemas and stumps after exchanging blastemas between normal and double-half forelimbs in the axolotl, *Ambystoma mexicanum*. *DB* **68**, 271–279. *(214)*

Holder, N., P. W. Tank, and S. V. Bryant (1980). Regeneration of symmetrical forelimbs in the axolotl, *Ambystoma mexicanum*. *DB*, **74**, 302–314. *(216)*

Holley, R. W. (1975). Control of growth of mammalian cells in culture. *Nature, London*, **258**, 487–490. *(232)*

Holtzer, H. (1959). The development of mesodermal axial structures in regeneration and embryogenesis, in *Regeneration in Vertebrates* (Ed. C. S. Thornton), pp. 15–33, U. Chicago Press, Chicago *(141)*

Holtzer, H. (1961). Aspects of chondrogenesis and myogenesis, in *Synthesis of Molecular and Cellular Structure* (Ed. D. Rudnick), pp. 35–87, Ronald Press, New York. *(103)*

Holtzer, S. W. (1956). The inductive activity of the spinal cord in urodele tail regeneration *JM*, **99**, 1–39. *(9)*

Hoperskaya, O. A. (1975). The development of animals homozygous for a mutation causing periodic albinism (aᵖ) in *Xenopus laevis*. *JEEM*, **34**, 253–264. *(121)*

Horak, A. (1939). L'influence de l'ablation et de la greffe de l'hypophyse sur la régénération de la queue chez l'axolotl. *Bull. Int. Acad. Polonaise Sci. Lettres ser. B*, **2**, 499–517. *(57)*

Horder, T. J. (1978). Functional adaptability and morphogenetic opportunism, the only rules for limb development? *Zoon*, **6**, 181–192. *(225)*

Horn, E. C. (1942). An analysis of neutron and X-ray effects on regeneration of the forelimb of larval *Amblystoma*. *JM*, **71**, 185–219. *(85, 97)*

Horton, J. D. (1969). Ontogeny of the immune response to skin allografts in relation to lymphoid organ development in the amphibian *Xenopus laevis* Daudin. *JEZ*, **170**, 449–465. *(122)*

Houck, J. C. (ed.) (1973). Chalones: concepts and current researches. *Nat. Cancer Inst. Mongr.*, **38**, 1–233. *(232)*

Hughes, A. (1961). Cell degeneration in the larval ventral horn of *Xenopus laevis* (Daudin). *JEEM*, **9**, 269–284. (*234*)

Hughes, A. F. W. (1968). *Aspects of Neural Ontogeny*. Academic Press, London. (*234*)

Hui, F., A. Smith (1970). Regeneration of the amputated amphibian limb: retardation by hemicholinium-3. *Science*, **170**, 1313–1314. (*39*)

Hui, F. W., and A. A. Smith (1976). Degeneration of Leydig cells in the skin of the salamander treated with cholinolytic drugs or surgical denervation. *Exp. Neurol.*, **53**, 610–619. (*154*)

Humphrey, R. R. (1966). A recessive factor (o, for ova deficient) determining a complex of abnormalities in the Mexican axolotl (*Ambystoma mexicanum*). *DB*, **13**, 57–76. (*120*)

Humphrey, R. R. (1967). Genetic and experimental studies on a lethal trait ('short toes') in the Mexican axolotl (*Ambystoma mexicanum*). *JEZ*, **164**, 281–295.

Humphrey, R. R. (1975). The axolotl, *Ambystoma mexicanum*, in *Handbook of Genetics* (Ed. R. C. King), vol. 4, pp. 3–17, Plenum Press, New York. (*119*)

Huxley, J. S., and G. R. de Beer (1934). *The Elements of Experimental Embryology*, Cambridge U.P., Cambridge. (*194*)

Illingworth, C. M. (1974). Trapped fingers and amputated finger tips in Children. *J. Pediatric Surgery*, **9**, 853–858. (*4, 230–1*)

Illingworth, C. M., and A. T. Barker (1980). Measurement of electrical currents emerging during the regeneration of amputated finger tips in children. *Clin Phys. Physiol. Meas.*, **1**, 87–89. (*113*)

Iten, L. E., and S. V. Bryant (1973). Forelimb regeneration from different levels of amputation in the newt, *Notophthalmus viridescens*: length, rate and stages. *RA*, **173**, 263–282. (*7–8*)

Iten, L. E., and S. V. Bryant (1975). The interaction between the blastema and stump in the establishment of the anterior–posterior and proximal–distal organization of the limb regenerate. *DB*, **44**, 119–147. (*165–8, 178, 208–9*)

Iten, L. E., and S. V. Bryant (1976a). Stages of tail regeneration in the adult newt, *Notophthalmus viridescens*. *JEZ*, **196**, 283–292. (*9*)

Iten, L. E., and S. V. Bryant (1976b). Regeneration from different levels along the tail of the newt, *Notophthalmus viridescens*. *JEZ*, **196**, 293–306. (*9*)

Jabailly, J., T. W. Rall, and M. Singer (1975). Assay of cyclic 3′-5′ monophosphate in the regenerating forelimb of the newt, *Triturus*. *JM*, **147**, 379–383. (*102*)

Jabailly, J. A., and M. Singer (1977). Neurotrophic stimulation of DNA synthesis in the regenerating forelimb of the newt, *Triturus*. *JEZ*, **199**, 251–256. (*38*)

Jacobson, M., and R. E. Baker (1969). Development of neuronal connections with skin grafts in frogs: behavioural and electrophysiological studies. *J. comp. Neurol.*, **137**, 121–142. (*226*)

Jaffe, L. F. (1968). Localization in the developing *Fucus* egg and the general role of localizing currents. *Adv. Morphogenesis*, **7**, 295–328. (*194*)

Jaffe, L. F. (1970). On the centripetal course of development, the *Fucus* egg, and self-electrophoresis. *DB*, suppl. **3**, 83–111. (*194*)

Johnston, B. T., J. E. Schrameck, and R. F. Mark (1975). Re-innervation of axolotl limbs. II. Sensory nerves. *Proc. R. Soc., London* **B190**, 59–75. (*35, 226*)

Juge, J. (1940). Les potentialités morphogénétiques des segments du membre dans la régénération du triton. *RSZ*, **47**, 65–133. (*180*)

Just, J. J. (1972). Protein-bound iodine and protein concentration in plasma and peri-cardial fluid of metamorphosing anuran tadpoles. *Physiol. Zool.*, **45**, 143–152. (*67*)

Kamrin, A. A., and M. Singer (1959). The growth influence of spinal ganglia implanted into the denervated forelimb regenerate of the newt, *Triturus*. *JM*, **104**, 415–439. (*37–8*)

Karczmar, A. G. (1946). The role of amputation and nerve resection in the regressing limbs of urodele larvae. *JEZ*, **103**, 401–427. (*29–30, 51, 94*)

Karczmar, A. G. (1948). Influence of methyl bis (B-chloroethyl) amine (nitrogen mustard), hydroquinone and thyroxine on regeneration in urodele larvae. *Anat. Rec.*, **101**, 712. (*106*)

Keller, R. (1946). Morphogenetische Untersuchungen am Skelett von *Siredon mexicanus* Shaw mit besonderer Berücksichtung des Ossifikationsmodus beim neotenen Axolotl. *RSZ*, **53**, 329–426. (*66*)

Kelly, D. J., and R. A. Tassava (1973). Cell division and ribonucleic acid synthesis during the initiation of limb regeneration in larval axolotls (*Ambystoma mexicanum*). *JEZ*, **185**, 45–53 (*100*)

Kerr, T. (1966). The development of the pituitary in *Xenopus laevis* Daudin. *Gen. comp. Endocrinol.*, **6**, 303–311. (*53*)

Kieny, M. (1977). Proximo-distal pattern formation in avian limb development, in *Vertebrate Limb and Somite Morphogenesis* (Eds. D. A. Ede *et al.*), pp. 87–103, Cambridge U.P., Cambridge. (*228*)

Kihlman, B. A. (1966). *Actions of Chemical on Dividing Cells*, Prentice-Hall, New Jersey. (*104*)

Kiortsis, V. (1953). Potentialités du territoire patte chez le triton (adultes, larves, et embryons). *RSZ*, **60**, 301–410 (*157, 160, 227*)

Koerker, D. J., and 6 others (1974). Somatostain: hypothalamic inhibitor of the endocrine pancreas. *Science*, **184**, 482–484. (*56*)

Konieczna-Marczynska, B., and A. Skowron-Cendrzak (1958). The effect of the augmented nerve supply on the regeneration in postmetamorphic *Xenopus laevis*. *Folia Biol. Krakow*, **6**, 37–46 (*33, 39*)

Krasner, G. N., and S. V. Bryant (1980). Distal transformation from double-half forearms in the axolotl, *Ambystoma mexicanum*. *DB*, **74**, 315–325. (*216*)

Kurz, O. (1912). Die Beinbildended Potenzen entwickelter Tritonen. *RA*, **34**, 588–617. (*181*)

Kurz, O. (1922). Versuche über Polaritatsumkehram Tritonenbein. *RA*, **50**, 186–191. (*181*)

Lagan, M. (1961). Regeneration from implanted dissociated cells. II. Regenerates produced by dissociated cells derived from different organs. *Folia Biol. Krakow*, **9**, 3–26. (*81*)

Lasek, R. (1970). Protein transport in neurons. *Int. Rev. Neurobiol.*, **13**, 289–324. (*41*)

Lassalle, B. (1974a). Caractéristiques des potentiels de surface des membres du triton *Pleurodeles waltlii* Michah. *C.R. Acad. Sci., Paris*, **D278**, 483–486. (*111*)

Lassalle, B. (1974b). Origine epidermique des potentiels de surface du membre de triton, *Pleurodeles watlii* Michah. *C.R. Acad. Sci., Paris*, **D278**, 1055–1058. (*111*)

Lassalle, B. (1977). Evaluation in vitro des potentiels de surface cutanés chez le triton, *Pleurodeles waltlii* Michah. par la mesure des différences de potentiales transépithéliales. *C.R. Acad. Sci., Paris*, **D285**, 183–186. (*114*)

Lassalle, B. (1979). Surface potentials and the control of amphibian limb regeneration. *JEEM*, **53**, 213–223. (*114*)

Laufer, H. (1959). Immunochemical studies of muscle proteins in mature and regenerating limbs of the adult newt, *Triturus viridescens*. *JEEM*, **7**, 431–458. (*132*)

Lazard, L. (1959). Régénération des membres chez l'axolotl adulte, aux dépens de greffes embryonnaires. *C. R. Acad. Sci., Paris*, **D245**, 2857–2859. (*125*)

Lazard, L. (1965). Régénération de membres chez l'axolotl a partir de greffes embryonnaires, in *Regeneration in Animals and Related Problems* (Eds. V. Kiortsis and H. A. L. Trampusch), pp. 396–403, North-Holland, Amsterdam. (*125*)

Lazard, L. (1967). Restauration de la régénération de membres irradiés d'axolotl par des greffes hétérotopiques d'origines diverses. *JEEM*, **18**, 321–342 (*125, 141, 159*)

Lazard, L. (1968). Inhibition de la régénération des membres par les rayons X, chez l'axolotl (*Ambystoma mexicanum*). *Ann. Embryol. Morph.*, **1**, 49–59. (*74, 77–84, 94*)

Lazard, L. (1970). Incorporation de précurseurs des acides nucléiques dans le membre d'axolotl en voie de régénération. *C.R. Acad. Sci.*, Paris, **D271**, 2031–2034. (*99*)

Lebowitz, P., and M. Singer (1970). Neurotrophic control of protein synthesis in the regenerating limb of the newt, *Triturus. Nature, London*, **225**, 824–827. (*38, 41*)

Lecamp, M. (1950). Sur les limites du pouvoir de régénération chez le crapaud accoucheur. *C.R. Acad. Sci.*, Paris, **D230**, 1110–1112. (*233*)

Lecamp, M. (1952). Régénération et metamorphose chez un batracien anoure, *Alytes obstetricans. C.R. Acad. Sci.*, Paris, **D235**, 1699–1700. (*233*)

Lehmann, F. E. (1961). Action of morphostatic substances and the role of proteases in regenerating tissues and in tumour cells. *Adv. Morphogenesis*, **1**, 153–187. (*105, 110*)

Lentz, T. L. (1972). A role of cyclic AMP in a neurotrophic process. *Nature, London*, **238**, 154–155. (*109*)

Lheureux, E. (1972). Contribution a l'étude du rôle de la peau et des tissus axiaux du membre dans la déclenchement de morphogenèses régénératrices anormales chez le triton *Pleurodeles waltlii* Michah. *Ann. Emb. Morph.*, **5**, 165–178. (*169–70*)

Lheureux, E. (1975a). Nouvelles données sur les rôles de la peau et des tissus internes dans la régénération du membre du triton *Pleurodeles waltlii* Michah. (Amphibien Urodele). *RA*, **176**, 285–301. (*171–2, 179*)

Lheureux, E. (1975b). Régénération des membres irradiés de *Pleurodeles waltlii* Michah. (Urodele). Influence des qualités et orientations des greffons non irradiés. *RA*, **176**, 303–327. (*80, 169–72, 179–80*)

Lheureux, E. (1977). Importance des associations de tissus du membre dans le développement des membres surnuméraires induits par déviation de nerf chez le triton *Pleurodeles waltlii* Michah. *JEEM*, **38**, 151–173. (*176–8, 217*)

Li, C. W., and H. A. Bern (1976). Effects of hormones on tail regeneration and regression in *Rana catesbiana* tadpoles. *Gen. comp. Endocrinol*, **29**, 376–382. (*61, 67*)

Linsenmayer, T. F., and G. N. Smith (1976). The biosynthesis of cartilage type colagen during limb regeneration in the larval salamander. *DB*, **52**, 19–30. (*103, 132*)

Liosner, L. D. (1931). Uber den Mechanismus des Verlusts der Regenerationsfähigkeit während der Entwicklung der Kaulquappen von *Rana temporaria. RA*, **124**, 571–583. (*58*)

Liosner, L. D. (1939). On the role of the skeleton in generating new forms in the course of regeneration (in Russian) *Bull. Biol. Med. exp. URSS*, **8**, 17–20. (*180*)

Liosner, L. D. (1941). On the formation of caudal skeleton as studied by transplantation (translation). *C.R. (Doklady) Acad. Sci. URSS*, **31**, 815–817. (*159*)

Liosner, L. D., and M. A. Vorontsova (1937). Recherches sur la détermination du processus régénératif chez les amphibiens. I. Régénération du membre avec les muscles de la queue transplantés. *Arch. d'Anat. Microsc.*, **33**, 313–344. (*125, 157*)

Litschko, E. J. (1934). Einwirkung der Röntgenstrahlen auf die Regeneration der Extremitäten, des Schwanzes und eines Teiles der Dorsalflosse beim Axolotl (German summary). *Trave. Lab. Zool. exp. Morph. anim.*, **3**, 101–140. (*75, 81*)

Litwiller, R. (1939). Mitotic index and size in regenerating amphibian limbs. *JEZ*, **82**, 273–286. (*17*)

Litwiller, R. (1940). Mitotic indices in regenerating urodele limbs. II. A study of the diurnal distribution of cell divisions. *Growth*, **4**, 169–172. (*21*)

Liversage, R. A. (1959). The relation of the central and autonomic nervous systems to the regeneration of limbs in adult newts. *JEZ*, **141**, 75–117. (*62, 126*)

Liversage, R. A. (1962). Regeneration of homoplastically deplanted forelimbs following spinal cord ablation in *Amblystoma opacum* larvae. *JEZ*, **151**, 1–15. (*62*)

Liversage, R. A. (1967). Hypophysectomy and forelimb regeneration in *Ambystoma opacum* larvae. *JEZ*, **165**, 57–69. (*65*)

Liversage, R. A., and B. M. Colley (1965). Effects of puromycin on the differentiation of the regenerating forelimb in *Diemictylus viridescens. Am. Zool.*, **5**, 720. (*107*)

Liversage, R. A., and K. R. S. Fisher (1974). Regeneration of the forelimb in adult hypophysectomised *Notophthalmus (Diemictylus) viridescens* given embryonic or adult chicken anterior pituitary extract. *JEZ*, **190**, 133–141. *(55)*

Liversage, R. A., and M. Globus (1977). In vitro regeneration of innervated forelimb deplants of *Ambystoma* larvae. *Can. J. Zool.*, **55**, 1195–1199. *(111)*

Liversage, R. A., and R. G. Korneluk (1978). Serum levels of thyroid hormone during forelimb regeneration in the adult newt, *Notophthalmus viridescens. JEZ*, **206**, 223–227. *(59, 233)*

Liversage, R. A., and L. Liivamagi (1971) Forelimb regeneration in hypophysectomized adult *Diemictylus viridescens* following organ culture and autoplastic implantation of the adenohypophysis. *JEEM*, **26**, 443–458. *(54)*

Liversage, R. A., and B. W. Price (1973). Adrenocorticosteroid levels in adult *Diemictylus viridescens* plasma following hypophysectomy and forelimb amputation. *JEZ*, **185**, 259–264. *(62, 69)*

Liversage, R. A., M. P. Rathbone, and H. M. McLaughlin (1977). Changes in cyclic GMP levels during forelimb regeneration in adult *Notophthalmus viridescens. JEZ*, **200**, 169–175. *(102)*

Liversage, R. A., and S. R. Scadding (1969). Re-establishment of forelimb regeneration in adult hypophysectomised *Diemictylus (Triturus) viridescens* given frog anterior pituitary extract. *JEZ*, **170**, 381–396. *(55)*

Locatelli, P. (1924). L'influenza del sistema nervoso sui processi di rigenerazione. *Arch. Sci. Biol.*, **5**, 362–376. *(22)*

Locatelli, P. (1925). Nuovi esperimenti sulla funzione della sistema nervoso sulla rigenerazione. *Arch. Sci. Biol.*, **7**, 300–312 *(22, 36, 156)*

Locatelli, P. (1929). Der Einfluss des Nervensystems auf die Regeneration. *RA*, **114**, 686–770. *(22–3)*

Lodyzenskaja, V. I. (1928). La transplantation des bourgeons de régénération des extrémités de l'axolotle. *C.R. Acad. Sci. URSS*, **A6**, 99–101. *(165)*

Lodyzenskaja, V. I. (1930). La transplantation des bourgeons de régénération des extrémités de l'axolotl. *Trav. Lab. Zool. exp. Morph. anim.*, **1**, 61–120. *(165, 207–9)*

Luscher, M. (1946). Die Wirkung des Colchicins auf die an der Regeneration beteiligten Gewbe im Schwanz der Xenopus-Larve. *RSZ*, **53**, 683–734. *(105)*

Luther, W. (1938). Die Strahlenwirkung auf Amphibienhaut vor und nach der Metamorphose. *Naturwiss.*, **27**, 713–720. *(80)*

Luther, W. (1948). Zur Frage des Determinationszustande von Regenerationsblastem. *Naturwiss.*, **35**, 30–31. *(80, 95, 159)*

Lynn, W. G., and H. E. Wachnowski (1951). The thyroid gland and its functions in cold blooded vertebrates. *Quart. Rev. Biol.*, **26**, 123–168. *(61)*

MacCabe, J. A., J. Errick, and J. W. Saunders (1974). Ectodermal control of the dorsoventral axis of the leg bud of the chick embryo. *DB*, **39**, 69–82. *(228)*

MacCabe, J. A., P. S. Lyle, and J. A. Lence (1979). The control of polarity along the anteroposterior axis in experimental chick limbs. *JEZ*, **207**, 113–119. *(227)*

MacCabe, J. A., J. W. Saunders, and M. Pickett (1973). The control of the anteroposterior and dorsoventral axes in embryonic chick limbs constructed of dissociated and reaggregated limb-bud mesoderm. *DB*, **31**, 323–335. *(218, 227)*

McCullough, W. D., and R. A. Tassava (1976). Determination of the blastemal cell cycle in regenerating limbs of the larval axolotl, *Ambystoma mexicanum. Ohio J. Science*, **76**, 63–65. *(18)*

McGrath, P. A., and M. R. Bennett (1979). The development of synaptic connections between different segmental motoneurones and striated muscles in an axolotl limb. *DB*, **69**, 133–145. *(225)*

Maden, M. (1976). Blastemal kinetics and pattern formation during amphibian limb regeneration. *JEEM*, **36**, 561–574. *(18–20, 145, 187, 203)*

Maden, M. (1977a). The role of Schwann cells in paradoxical regeneration in the axolotl. *JEEM*, **41**, 1–13. *(9, 132, 142)*

Maden, M. (1977b). The regeneration of positional information in the amphibian limb. *J. theor. Biol.*, **69**, 735–753. *(202—4)*

Maden, M. (1978). Neurotrophic control of the cell cycle during amphibian limb regeneration. *JEEM*, **48**, 169–175. *(43–4)*

Maden, M. (1979a). Neurotrophic and X-ray blocks in the blastemal cell cycle. *JEEM*, **50**, 169–173. *(42, 45, 87)*

Maden, M. (1979b). The role of irradiated tissue during pattern formation in the regenerating limb. *JEEM*, **50**, 235–242. *(90, 95, 169)*

Maden, M. (1979c). Regulation and limb regeneration: the effect of partial irradiation. *JEEM*, **52**, 183–192. *(187)*

Maden, M. (1980a). Intercalary regeneration in the amphibian limb and the law of distal transformation. *JEEM*, **56**, 201–209. *(95, 200–4)*

Maden, M. (1980b). The structure of supernumerary limbs. (unpublished). *(222)*

Maden, M., and R. N. Turner (1978). Supernumerary limbs in the axolotl. *Nature, London*, **273**, 232–235. *(210, 222)*

Maden, M., and H. Wallace (1975). The origin of limb regenerates from cartilage grafts. *Acta Emb. exper.*, **1975**, 77–86. *(80, 123–8, 137)*

Maden, M., and H. Wallace (1976). How X-rays inhibit amphibian limb regeneration. *JEZ*, **197**, 105–113. *(20, 83–6, 134)*

Maier, C. E., and M. Singer (1977). The effect of light on forelimb regeneration in the newt. *JEZ*, **202**, 241–244. *(2)*

Mailman, M. L., and M. H. Dresden (1976). Collagen metabolism in the regenerating forelimb of *Notophthalmus viridescnes*: synthesis, accumulation and maturation. *DB*, **50**, 378–394. *(104)*

Mailman, M. L., and M. H. Dresden (1979). Denervation effects on newt limb regeneration: collagen and collagenase. *DB*, **71**, 60–70. *(104)*

Malacinski, G. M., and A. J. Brothers (1974). Mutant genes in the Mexican axolotl. *Science*, **184**, 1142–1147. *(119)*

Malinin, T., and J. D. Deck (1958). The effects of implantation of embryonic and tadpole tissues into adult frog limbs. I. Regeneration after amputation. *JEZ*, **139**, 307–327. *(39)*

Manner, H. W. (1953). The origin of the blastema and of new tissues in the regenerating forelimb of the adult *Triturus viridescens* (Rafinesque). *JEZ*, **122**, 229–257. *(11)*

Manner, H. W. (1955). The effect of cortisone acetate on the wound healing phase of *Triturus viridescens. Growth*, **19**, 169–175. *(64)*

Manson, J., R. Tassava, and M. Nishikawara (1976). Denervation effects on aspartate carbamyl transferase, thymidine kinase, and uridine kinase activities in newt regenerates. *DB*, **50**, 109–121. *(42)*

Marcucci, E. (1916). Capacità rigenerativa degli arti nelle larve de anuri e condizioni che ne determinano la perdita. *Arch. zool. ital. Napoli*, **8**, 89–117. *(58, 234)*

Mark, R. F. (1969). Matching muscles and motoneurones. *Brain Res.*, **14**, 245–254. *(35)*

Matagne-Dhoossche, F. (1964). Action cellulaire du myleran chez l'axolotl au cours de la régénération. *Arch. Biol.*, **75**, 93–106. *(106)*

Mazia, D. (1958). SH compounds in mitosis. *Exp. Cell Res.*, **14**, 486–494 *(105)*

Mescher, A. L., and D. Gospodarowicz (1979). Mitogenic effect of a growth factor derived from myelin on denervated regenerates of newt forelimbs. *JEZ*, **207**, 497–503. *(232)*

Mescher, A. L., and R. A. Tassava (1975). Denervation effects on DNA replication and mitosis during the initiation of limb regeneration in adult newts. *DB*, **44**, 187–197. *(43, 99)*

Mettetal, C. (1939). La régénération des membres chez la salamandre et le triton. *Arch. Anat. Histol. Emb.*, **28**, 1–214. *(196)*

Mettetal, C. (1952). Action du support sur la différenciation des segments proximeaux dans les régénérats de membre chez les amphibiens urodèles. *C.R. Acad. Sci.*, *Paris*, **D234**, 675–677. (*196*)

Michael, M. I., and F. K. Aziz (1976). Effect of sodium perchlorate on the restoration of the limb regenerative ability in a metamorphic stage of *Bufo regularis* Reuss. *Folia Biol. Krakow*, **24**, 309–315. (*61*, *107*)

Michael, M. I., and J. Faber (1961). The self-differentiation of the paddle-shaped limb regenerate, transplanted with normal and reversed proximo-distal orientation after removal of the digital plate (*Ambystoma mexicanum*). *Arch. Biol.*, **72**, 301–330. (*198*)

Michael, M. I., and J. Faber (1971). Morphogenesis of mesenchyme from regeneration blastemas in the absence of digit formation in *Ambystoma mexicanum*. *RA*, **168**, 174–180. (*146*, *200*)

Michael, M. I., and N. M. El Malkh (1969). Hind limb histogenesis and regeneration in larvae and metamorphic stages of the Egyptian toad, *Bufo regularis* Reuss. I. Transection at the thigh level. *Arch. Biol.*, **80**, 299–326. (*4*)

Michael, M. I., and I. A. Niazi (1972). Hind limb regeneration in tadpoles of *Bufo viridis* Laurenti, and cartilage formation from cells of non-chondrogenic origin in the thigh. *Acta Emb. exper.*, **1972**, 349–363. (*4*)

Michael, M. I., and A. J. Al Sammak (1970). Regeneration of limbs in adult *Rana ridibunda* Pallas. *Experientia*, **26**, 920–921. (*4*, *115*)

Milojevic, B. D. (1924). Beiträge zur Frage über die Determination der Regenerate. *RA*, **103**, 80–94. (*157*, *165*, *178*, *205*)

Milojevic, B. D., and N. Grbic (1925). La régénération et l'inversion de la polarité des extrémités chez les tritons adultes, à la suite d'une transplantation hétérotope. *C.R. Soc. Biol.*, **93**, 649–651. (*181*)

Mizell, M. (1968). Limb regeneration: induction in the newborn opossum. *Science*, **161**, 283–286. (*39*)

Monroy, A. (1941). Recerche sulle correnti elettriche derivabili dalla superficie del corpo di tritoni adulti normali e durante la rigenerazione degli arti e della coda. *Pubb. Staz. Zool. Napoli*, **18**, 265–281. (*111–12*)

Monroy, A. (1942). La rigenerazione bipolare in segmenti di arti isolati di *Triton crist.* *Arch. Ital. Anat. Emb.*, **48**, 123–142. (*181*)

Monroy, A. (1943). Ricerche sulla rigenerazione degli arti negli anfibi urodeli. Nota II. Risultati di rotazioni prossimo-distali della metà radiale dello zigopodio. *Arch. Zool. Ital.*, 31, 13–23.

Monroy, A. (1946). Nota III. Osservazioni su rigenerati formatisi su doppie superfici di sezione e considerazioni sui processi determinativi della rigerazione. *Arch. Zool. Ital.*, **31**, 151–172. (*189–90*)

Monroy, A., and F. Oddo (1943). Nota I. Il valore degli elementi muscolari nel determinismo della natura dei rigenerati. *Arch. Zool. Ital.*, **31**, 1–12. (*213*)

Moolenaar, W. H., S. W. de Laat, and P. T. van der Saag (1979). Serum triggers a sequence of rapid ionic conductance changes in quiescent neuroblastoma cells. *Nature, London*, **279**, 721–723. (*117*)

Morgan, T. H. (1901). *Regeneration*, Macmillan, London. (*195*, *229*)

Morzlock, F. V., and D. L. Stocum (1971). Patterns of RNA synthesis in regenerating limbs of the adult newt, *Triturus viridescens*. *DB*, **24**, 106–118. (*100*)

Morzlock, F. V., and D. L. Stocum (1972). Neural control of RNA synthesis in regenerating limbs of the adult newt *Triturus viridescens*. *RA*, **171**, 170–180. (*100*)

Muller, H. J. (1928). The production of mutations by X-rays. *Proc. Natl. Acad. Sci., Wash.*, **14**, 714–726. (*71*)

Munro, A., and S. Bright (1976). Products of the major histocompatibility complex and their relationship to the immune response. *Nature, London*, **264**, 145–152. (*123*)

Namenwirth, M. (1974). The inheritance of cell differentiation during limb regeneration in the axolotl. *DB*, **41**, 42–56. (*62*, *81*, *90*, *134–7*)

Nassonov, N. V. (1930). Die Regeneration der Axolotl Extremitäten nach Ligaturanlegung. *RA*, **121**, 639–657. *(186)*

Needham, A. E. (1941). Some experimental biological uses of the element Beryllium (Glucinum). *Proc. Zool. Soc. London*, **111A**, 59–85. *(107–10, 149)*

Needham, A. E. (1952). *Regeneration and Wound Healing*, Methuen, London. *(110)*

Needham, J. (1931). *Chemical Embryology* (3 vols.), Cambridge U.P., Cambridge. *(194)*

Needham, J. (1942). *Biochemistry and Morphogenesis*, Cambridge U.P., Cambridge. *(131)*

Newth, D. R. (1958a). On regeneration after amputation of abnormal structures. I. Defective amphibian tails. *JEEM*, **6**, 297–307. *(147, 216)*

Newth, D. R. (1958b). II. Supernumerary induced limbs. *JEEM*, **6**, 384–392. *(147, 216)*

Nicholas, J. S. (1924). Regulation of posture in the fore limb of *Amblystoma punctatum*. *JEZ*, **40**, 113–159. *(222)*

Nicoll, C. S., and P. Licht (1971). Evolutionary biology of prolactins and somatotrophins. *Gen. comp. Physiol.*, **17**, 490–507. *(56)*

Niwelinski, J. (1958). The effect of prolactin and somatotrophin on the regeneration of the forelimb of the newt *Triturus alpestris* Laur. *Folia Biol. Krakow*, **6**, 9–36. *(56)*

Normandin, D. K. (1960). Regeneration of Hydra from the endoderm. *Science*, **132**, 678. *(140)*

Oberheim, K. W., and W. Luther (1958). Versuche uber die Extremitäten regeneration von Salamanderlarven bei umgekehrter Polarität des Amputationsstumpfes. *RA*, **150**, 373–382. *(182–3)*

Oberpriller, J. (1967). A radioautographic analysis of the potency of blastemal cells in the adult newt, *Diemictylus viridescens*. *Growth*, **31**, 251–296. *(140)*

Oberpriller, J. C. (1968). The action of X-irradiation on the regeneration field of the forelimb of the adult newt, *Diemictylus viridescens*. *JEZ*, **168**, 403–422. *(81, 97, 190–1)*

Ochs, S. (1972). Fast transport of materials in mammalian nerve fibers. *Science*, **176**, 252–260. *(37)*

Ochs, S. (1974). Systems of material transport in nerve fibers (axoplasmic transport) related to nerve function and trophic control. *Ann. N.Y. Acad. Sci.*, **228**, 202–223. *(37, 41)*

Ogawa, Y. (1959). Development of muscle proteins in the regenerating limb of *Triturus pyrrhogaster* Boie. *Ann. Rept Natl. Inst. Genet. Japan*, **10**, 151–152. *(103)*

Ogawa, Y. (1962). Synthesis of skeletal muscle proteins in early embryos and regenerating tissue of chick and *triturus*. *Exp. Cell Res.*, **26**, 269–274. *(103)*

O'Steen, W. K., and B. E. Walker (1961). Radioautographic studies of regeneration in the common newt. II. Regeneration of the forelimb. *Anat. Rec.*, **139**, 547–555. *(15, 20)*

Overton, J. (1950). Mitotic stimulation of amphibian epidermis by underlying grafts of central nervous tissue. *JEZ*, **115**, 521–559. *(43)*

Parker, G. H. (1932). On the trophic impulse so-called, its rate and nature. *Am. Nat.*, **66**, 147–158. *(36)*

Parker, G. H., and V. L. Paine (1934). Progressive nerve degeneration and its rate in the lateral-line nerve of the catfish. *Am. J. Anat.*, **54**, 1–25. *(36, 41)*

Patrick, J. R. Briggs (1964). Fate of cartilage cells in limb regeneration in the axolotl (*Ambystoma mexicanum*). *Experientia*, **20**, 431–432. *(138)*

Pautou, M-P. and M. Kieny (1973). Interaction ecto-mésodermique dans l'établissement de la polarité dorso-ventrale du pied de l'embryon de poulet. *C.R. Acad. Sci., Paris*, **D227**, 1225–1228. *(228)*

Peadon, A. M. (1953). The effects of thiourea on limb regeneration in the tadpole. *Growth*, **17**, 21–44. *(61, 107)*

Peadon, A. M., and M. Singer (1965). A quantitative study of forelimb innervation in relation to regenerative capacity in the larval, land stage, and adult forms of *Triturus viridescens*. *JEZ*, **159**, 337–345. *(29, 31)*

256

Peadon, A. M., and M. Singer (1966). The blood vessels of the regenerating limb of the adult newt, *Triturus*. *JM*, **118**, 79–89. *(99)*

Philipeaux, J-M. (1866). Expériences démontrant que les membres de la salamandre aquatique (*Triton cristatus* L.) ne se régénèrent qu'à la condition qu'on laisse au moins sur place la partie basilaire de ces membres. *C.R. Acad. Sci., Paris*, **63**, 576–578. *(156)*

Philipeaux, J-M. (1867). Sur. la régénération des membres chez l'axolotl (*Siren pisciformis*). *C.R. Acad. Sci., Paris*, **64**, 1204–1205. *(2)*

Philipeaux, J-M. (1876). Les membres de la salamandre aquatique bien extirpés ne se régénèrent point. *C.R. Acad. Sci., Paris*, **82**, 1162–1163. *(156)*

Piatt, J. (1942). Transplantation of aneurogenic forelimbs in *Amblystoma punctatum*. *JEZ*, **91**, 79–101. *(44, 92, 154)*

Pietsch, P. (1961). Differentiation in regeneration I. The development of muscle and cartilage following deplantation of regenerating limb blastemata of *Amblystoma* larvae. *DB*, **3**, 255–264. *(159)*

Pietsch, P. (1962). The influence of spinal cord on differentiation of skeletal muscle in regenerating limb blastema of *Amblystoma* larvae *Anat. Rec.*, **142**, 169–177.

Pietsch, P., and A. K. Bruce (1965). The significance of post-irradiation growth in regenerating limb and tail blastemas of *Amblystoma* larvae. *J. cell. comp. Physiol.*, **66**, 243–249. *(86)*

Pietsch, P., and R. H. Webber (1965). Innervation and regeneration in orbitally transplanted limbs of *Amblystoma* larvae. *Anat. Rec.*, **152**, 439–450. *(45)*

Polezhaev, L. (1936). La valeur de la structure de l'organe et les capacités du blastème régénératif dans le processus de la détermination du régénérat. *Bull. Biol. France et Belgique*, **70**, 54–86. *(144, 152)*

Polezhaev, L. W. (1946a). Morphological data on regenerative capacity in tadpole limbs as restored by chemical agents. *C.R. (Doklady) Acad. Sci. URSS*, **54**, 281–284. *(40, 58)*

Polezhaev, L. W. (1946b). Further investigation on the regeneration of limbs in adult Anura. *C.R. (Doklady) Acad. Sci. URSS*, **54**, 461–464. *(58)*

Polezhaev, L. W. (1946c). Morphology of limb regenerates in adult Anura. *C.R. (Doklady) Acad. Sci. URSS*, **54**, 653–656. *(58)*

Polezhaev, L. W. (1946d). The loss and restoration of regenerative capacity in the limbs of tailless amphibia. *Biol. Rev.*, **21**, 141–147. *(40, 45, 58)*

Polezhaev, L. V. (1958). The restoration of the regenerative capacity in mammals (English summary pp. 232–233) *Folia Biol. Krakow*, **6**, 203–238. *(40)*

Polezhaev, L. V. (1960). Limb regeneration in axolotls following transplantation of destroyed tissues of axolotls and mammals. *DAN*, **131**, 1468–1471 (AIBS translation pp. 242–245). *(88)*

Polezhaev, L. V. (1972a). *Loss and Restoration of Regenerative Capacity in Tissues and Organs of Animals*, Harvard U.P., Cambridge, Mass.

Polezhaev, L. V. (1972b). *Organ Regeneration in Animals*, Thomas, Springfield, Ill. *(107, 230)*

Polezhaev, L. V. (1977). Restoration of X-ray repressed regeneration ability (in Russian). *Ispekhi Sovrem. Biol. (Akad. Nauk SSSR)*, **84**, 96–112 *(88, 94)*

Polezhaev, L. V., and N. I. Ermakova (1960). Restoration of regeneration capacity of the extremities of axolotls depressed by roentgen radiation. *DAN*, **131**, 209–212 (AIBS translation pp. 227–229). *(79, 129)*

Polezhaev, L. V., and W. N. Favorina (1935). Uber die Rolle des Epithels in den Anfänglichen Entwicklungsstadien einer Regenerationsanlage der Extremität beim Axolotl. *RA*, **133**, 701–727. *(149)*

Polezhaev, L. V., and G. I. Ginsburg (1939). Studies by the method of transplantation on the loss and restoration of the regenerative power in the tailess amphibian limbs. *C.R. (Doklady) Acad. Sci. URSS*, **23**, 733–737. *(58)*

Polezhaev, L. V., N. A. Teplits, and N. I. Ermakova (1961). Restoration of the regenerative property of extremities in axolotls inhibited by X-ray irradiation, by means of proteins, nucleic acids and lyophilised tissues. *DAN*, **138**, 477–480 (AIBS translation pp. 407–410). (*79, 130*)

Polezhaev, L. V., N. A. Teplits, and S. Y. Tuchkova (1962). The importance of proteins and nucleic acids in the recovery of regenerative power of axolotl limbs after its suppression by X-rays. *DAN*, **144**, 930–933 (AIBS translation pp. 540–542). (*79, 88*)

Polezhaev, L. V., N. A. Teplits, and S. Y. Tuchkova (1963). Factors governing recovery of the ability of axolotls to regenerate limbs after its suppression by X-rays. *DAN*, **150**, 694–697 (AIBS translation pp. 690–693). (*79*)

Polezhaev, L. V., N. A. Teplits, and S. Y. Tuchkova (1964). Restoration of the regenerating power of the limbs of the axolotl when suppressed by X-ray irradiation by means of nuclei acids. *DAN*, **159**, 682–685 (AIBS translation pp. 798–801). (*81, 88*)

Popiela, H. (1976). In vivo limb tissue development in the absence of nerves: a quantitative study. *Exp. Neurol.*, **53**, 214–226. (*48*)

Powell, J. A. (1969). Analysis of histogenesis and regenerative ability of denervated forelimb regenerates of *Triturus viridescens*. *JEZ*, **170**, 125–148. (*36*)

Prestige, M. C. (1965). Cell turnover in the spinal ganglia of *Xenopus laevis* tadpoles. *JEEM*, **13**, 63–72. (*234*)

Prestige, M. C. (1967a). The control of cell number in the lumbar spinal ganglia during the development of *Xenopus laevis* tadpoles. *JEEM*, **17**, 453–471. (*234*)

Prestige, M. C. (1967b). The control of cell number in the lumbar ventral horns during the development of *Xenopus laevis* tadpoles. *JEEM*, **18**, 359–387. (*234*)

Procaccini, D. J., and C. M. Doyle (1970). Streptomycin induced teratogenesis in developing and regenerating amphibians. *Oncology*, **24**, 378–387. (*107*)

Procaccini, D. J., and C. M. Doyle (1972). The inhibition of limb regeneration in adult *Diemictylus viridescens* treated with streptomycin. *Oncology*, **26**, 393–404. (*107*)

Puckett, W. O. (1936). The effects of X-radiation on limb development and regeneration in *Amblystoma*. *JM*, **59**, 173–213. (*73, 80, 85*)

Quastler, H., and F. G. Sherman (1959). Cell population kinetics in the intestinal epithelium of the mouse. *Exp. Cell Res.*, **17**, 420–438. (*17*)

Rahmani, T. M. Z. (1960). Conflit de potentialités morphogènes et duplicature. *RSZ*, **67**, 589–675. (*168, 173, 210*)

Rahmani, T., and V. Kiortsis (1961). Le rôle de la peau et des tissus profonds dans la régénération de la patte. *RSZ*, **68**, 91–102. (*81, 168*)

Repesh, L. A., and J. C. Oberpriller (1978). Scanning electron microscopy of epidermal cell migration on wound healing during limb regeneration in the adult newt, *Notophthalmus viridescens*. *Am. J. Anat.*, **151**, 539–559. (*148*)

Reyer, R. W. (1977). The amphibian eye: development and regeneration, in *Handbook of Sensory Physiology* (Ed. F. Crescitelli), vol. VII/5, pp. 309–390, Springer-Verlag, Berlin. (*229*)

Richards, C. M., and 4 others (1977). A scanning electron microscopic study of differentiation of the digital pad in regenerating digits of the Kenyan reed frog. *Hyperolius viridiflavus ferniquei. JM*, **153**, 387–395. (*4*)

Richardson, D. (1940). Thyroid and pituitary hormones in relation to regeneration. I. The effect of anterior pituitary hormone on regeneration of the hind leg in normal and thyroidectomized newts. *JEZ*, **83**, 407–429. (*60*)

Richardson, D. (1945). II. Regeneration of the hind leg of the newt, *Triturus viridescens* with different combinations of thyroid and pituitary hormones. *JEZ*, **100**, 417–429. (*55, 60*)

Roguski, H. (1961). Regeneration from implanted dissociated cells. III. Regenerative capacity of blastemal cells. *Folia Biol. Krakow*, **9**, 269–302. (*2, 81*)

Rose, F. C., H. Quastler, and S. M. Rose (1955). Regeneration of X-rayed salamander limbs provided with normal epidermis. *Science*, **122**, 1018–1019. (*79–83*)

Rose, F. C., and S. M. Rose (1965). The role of normal epidermis in recovery of regenerative ability in X-rayed limbs of *Triturus*. *Growth*, **29**, 361–393. (*16, 84–7, 95*)

Rose, F. C., and S. M. Rose (1967). Nerve penetration and regeneration after X-rayed limbs of *Triturus viridescens* were covered with un-X-rayed epidermis. *Growth*, **31**, 375–380. (*91*)

Rose, F. C., and S. M. Rose (1974). Regeneration of aneurogenic limbs of salamander larvae after X-irradiation. *Growth*, **38**, 97–108. (*78–80, 92–5*)

Rose, S. M. (1944). Methods of initiating limb regeneration in adult anura. *JEZ*, **95**, 149–170. (*39, 67*)

Rose, S. M. (1945). The effect of NaCl in stimulating regeneration of limbs of frogs. *JM*, **77**, 119–139. (*39, 67*)

Rose, S. M. (1948). Epidermal dedifferentiation during blastema formation in regenerating limbs of *Triturus viridescens*. *JEZ*, **108**, 337–361. (*11*)

Rose, S. M. (1962). Tissue-arc control of regeneration in the amphibian limb, in *Regeneration* (Ed. D. Rudnick), pp. 153–176, Ronald Press, New York. (*79, 91–4*)

Rose, S. M. (1964). Regeneration, in *The Physiology of Amphibia* (Ed. J. A. Moore), pp. 545–622, Academic Press, New York. (*69*)

Rose, S. M. (1970a). Differentiation during regeneration caused by migration of repressors in bioelectric fields. *Am. Zool.*, **10**, 91–99. (*91*)

Rose, S. M. (1970b). *Regeneration*, Appleton-Century-Crofts, New York. (*112, 163, 194*)

Rose, S. M., and F. C. Rose (1974). Electrical studies on normally regenerating, on X-rayed, and on denervated limb stumps of *Triturus*. *Growth*, **38**, 363–380. (*91, 113*)

Ruben, L. N. (1960). An immunological model of implant-induced supernumerary limb formation. *Am. Nat.*, **94**, 427–434. (*126*)

Ruben, L. N., and J. M. Stevens (1963). Post-embryonic induction in urodele limbs. *JM*, **112**, 279–301. (*126*)

Rzehak, K., and M. Singer (1966). Limb regeneration and nerve fiber number in *Rana sylvatica* and *Xenopus laevis*. *JEZ*, **162**, 15–22. (*32*)

Sato, N. L., and S. Inoue (1973). Effects of growth hormone and nutrient on limb regeneration in hypophysectomised adult newts. *JM*, **140**, 477–485. (*145*)

Saunders, J. W. (1948). The proximo-distal sequence of origin of the parts of the chick wing and the role of the ectoderm. *JEZ*, **108**, 363–403. (*149*)

Saunders, J. W. (1977). The experimental analysis of chick limb bud development, in *Vertebrate Limb and Somite Morphogenesis* (Eds. D. A. Ede *et al.*), pp. 1–24, Cambridge U.P., Cambridge. (*151, 163, 228*)

Saxen, L., and 3 others (1957). Quantitative investigation of the anterior pituitary–thyroid mechanism during frog metamorphosis. *Endocrinology*, **61**, 35–44. (*67*)

Saxena, S., and I. A. Niazi (1977). Effect of vitamin A excess on hind limb regeneration in tadpoles of the toad, *Bufo andersonii* (Boulanger). *Indian J. exp. Biol.*, **15**, 435–439. (*107*)

Scadding, S. R. (1977). Phylogenetic distribution of limb regeneration capacity in adult amphibia. *JEZ*, **202**, 57–67. (*2*)

Schaper, A. (1904). Experimentelle Untersuchungen uber den Einfluss der Radiumstrahlen und der Radiumemanation auf embryonale und regenerative Entwicklungsvorgänge. *Anat. Anz.*, **25**, 298–314, 326–337. (*71, 97*)

Schauble, M. K., and M. R. Nentwig (1974). Temperature and prolactin as control factors in newt forelimb regeneration. *JEZ*, **187**, 335–344. (*7, 56*)

Scheremetieva, E. A., and V. V. Brunst (1938). Preservation of the regeneration capacity in the middle part of the limb of newt and its simultaneous loss in the distal and proximal parts of the same limb. *Bull. Biol. Méd. exp. URSS*, **6**, 723–724. (*75, 80*)

Scheving, L. E., and J. J. Chiakulus (1965). Twenty-four-hour periodicity in the uptake of tritiated thymidine and its relation to mitotic rate in urodele larval epidermis. *Exp. Cell Res.*, **39**, 161–169. *(15)*

Schmidt, A. J., (1958a). Forelimb regeneration of thyroidectomized adult newts. I. Morphology. *JEZ*, **137**, 197–226. *(61, 233)*

Schmidt, A. J. (1958b). II. Histology. *JEZ*, **139**, 95–135. *(61)*

Schmidt, A. J. (1966). Electrophoretic separation of soluble proteins extracted from regenerating forelimbs of the adult newt, *Diemictylus viridescens*. *Anat. Rec.*, **154**, 417–418. *(101)*

Schmidt, A. J. (1968). *Cellular Biology of Vertebrate Regeneration and Repair*, U. Chicago Press, Chicago. *(61, 101, 112, 141, 230)*

Schotté, O. E. (1926a). Système nerveux et régénération chez le triton. Action globale des nerfs. *RSZ*, **33**, 1–211. *(23, 294)*

Schotté, O. E. (1926b). Hypophysectomie et régénération chez les batraciens urodèles. *C.R. Soc. phys. Hist. nat. Genève*, **43**, 67–72. *(54, 64–6)*

Schotté, O. E. (1926c). La régénération de la queue des Urodèles est liée à l'intégrité du teritoire caudal. *C.R. Soc. phys. Hist. nat. Genève*, **43**, 126–128. *(156)*

Schotté, O. E. (1939). The origin and morphogenetic potencies of regenerates. *Growth*, suppl. 1, 59–76.

Schotté, O. E. (1953). The role of the pituitary, of ACTH, and of some adreno-cortical extracts in the regeneration of limbs in adult *Triturus*. *Anat. Rec.*, **117**, 575–576. *(62)*

Schotté, O. E. (1959). The role of hormones in regeneration, in *Regeneration in Vertebrates* (Ed. C. S. Thornton), p. 81, U. Chicago Press, Chicago. *(63)*

Schotté, O. E. (1961). Systemic factors in initiation of regenerative processes in limbs of larval and adult amphibians, in *Synthesis of Molecular and Cellular Structures* (Ed. D. Rudnick), pp. 161–192, Ronald Press, New York. *(62–6)*

Schotté, O. E., and R. H. Bierman (1956). Effects of cortisone and allied adrenal steroids upon limb regeneration in hypophysectomized *Triturus viridescens*. *RSZ*, **63**, 353–375. *(63)*

Schotté, O. E., and E. G. Butler (1941). Morphological effects of denervation and amputation of limbs in urodele larvae. *JEZ*, **87**, 279–322. *(23, 29, 51)*

Schotté, O. E., and E. G. Butler (1944). Phases in regeneration of the urodele limb and their dependence upon the nervous system. *JEZ*, **97**, 95–121. *(23, 31, 214)*

Schotté, O. E., E. G. Butler, and R. T. Hood · (1941). Effects of transplanted blastemas on amputated nerveless limbs of urodele larvae. *Proc. Soc. exp. Biol. Med.*, **48**, 500–503. *(145, 208)*

Schotté, O. E., and J. L. Chamberlain (1955). Effects of ACTH upon limb regeneration in normal and in hypophysectomized *Triturus viridescens*. *RSZ*, **62**, 253–279. *(63)*

Schotté, O. E., and W. B. Christiansen (1957). Influence of Amphenone B—a cortical inhibitor—upon regeneration in normal and in hypophysectomized newts. *Anat. Rec.*, **128**, 619–620. *(63)*

Schotté, O. E., and A. Droin (1965). The competence of pituitaries and limb regeneration during metamorphosis of *Triturus (Diemictylus) viridescens*. *RSZ*, **72**, 205–223. *(65)*

Schotté, O. E., and A. B. Hall (1952). Effect of hypophysectomy upon phases of regeneration in progress (*Triturus viridescens*). *JEZ*, **121**, 521–559. *(56, 62)*

Schotté, O. E., and M. Harland (1943a). Effects of blastema transplantations on regeneration processes of limbs in *Amblystoma* larvae. *Anat. Rec.*, **87**, 165–180. *(208)*

Schotté, O. E., and M. Harland (1943b). Effects of denervation and amputation of hindlimbs in anuran tadpoles. *JEZ*, **93**, 453–493. *(23, 31)*

Schotté, O. E., and M. Harland (1943c). Amputation level and regeneration in limbs of late *Rana clamitans* tadpoles. *JM*, **73**, 329–363. *(58, 234)*

Schotté, O. E., and S. R. Hilfer (1957). Initiation of regeneration in regenerates after hypophysectomy in adult *Triturus viridescens*. *JM*, **101**, 25–55. (*56*)

Schotté, O. E., and D. A. B. Lindberg (1954). Effect of xenoplastic adrenal transplants upon limb regeneration in normal and hypophysectomized newts (*Triturus viridescens*). *Proc. Soc. exp. Biol. Med.*, **87**, 26–29. (*63*)

Schotté, O. E., and R. A. Liversage (1959). Effects of denervation and amputation upon the initiation of regeneration in regenerates of *Triturus*. *JM*, **105**, 495–528. (*31, 36, 147*)

Schotté, O. E., and A. Tallon (1960). The importance of autoplastically transplanted pituitaries for survival and for regeneration of adult *Triturus*. *Experientia*, **16**, 72–74. (*54, 65*)

Schotté, O. E., and W. W. Washburn (1954). The effects of thyroidectomy on the regeneration of the forelimb in *Triturus viridescens*. *Anat. Rec.*, **120**, 156. (*61*)

Schotté, O. E., and J. F. Wilber (1958). Effects of adrenal transplants upon forelimb regeneration in normal and in hypophysectomized adult frogs. *JEEM*, **6**, 247–269. (*67, 69*)

Schueing, M. R., and M. Singer (1957). The effects of microquantities of beryllium ion on the regenerating forelimb of the adult newt, *Triturus*. *JEZ*, **136**, 301–327. (*107, 110*)

Schultheis, H. (1979). Maintenance of growth- and thyroid-stimulating properties of ectopic pituitaries in the Mexican axolotl (*Ambystoma mexicanum*). *Gen. comp. Endocrinol.*, **38**, 75–82. (*233*)

Schwidefsky, G. (1935). Entwicklung und Determination der Extremitäten-Regenerate bei den Molchen. *RA*, **132**, 57–114. (*2, 7, 205–7, 222*)

Settles, H. E. (1970). Morphogenetic effects of limb skin during limb regeneration in the adult newt, *Triturus viridescens*. *Anat. Rec.*, **166**, 375. (*168*)

Shuraleff, N. C., and C. S. Thornton (1967). An analysis of distal dominance in the regenerating limb of the axolotl. *Experientia*, **23**, 747–748. (*204*)

Sicard, R. E. (1975a). The effects of hypophysectomy upon the endogenous levels of cyclic AMP during forelimb regeneration of adult newts (*Notophthalmus viridescens*). *RA*, **177**, 159–162. (*39, 102, 109*)

Sicard, R. E. (1975b). Cyclic AMP and neuroendocrine influence upon forelimb regeneration in the adult newt, *Notophthalmus viridescens*. *Oncology*, **32**, 190–195. (*39, 107–9*)

Siderova, O. F. (1951). Regeneration of irradiated axolotl legs after grafts of unspecific tissues (in Russian). *DAN*, **81**, 297–299. (*141*)

Sidman, R. L., and M. Singer (1951). Stimulation of forelimb regeneration in the newt, *Triturus viridescens*, by a sensory nerve supply isolated from the central nervous system. *Am. J. Physiol.*, **165**, 257–260. (*26*)

Sidman, R. L., and M. Singer (1960). Limb regeneration without innervation of the apical epidermis in the adult newt, *Triturus*. *JEZ*, **144**, 105–110. (*24, 92, 150*)

Simnett, J. D. (1964). Histocompatibility in the platanna, *Xenopus laevis laevis* (Daudin), following nuclear transplantation. *Exp. Cell. Res.*, **33**, 232–239. (*122*)

Simnett, J. D. (1965). The prolongation of homograft survival time in the platanna, *Xenopus laevis laevis* (Daudin), by exposure to low environmental temperature. *J. Cell. comp. Physiol.*, **65**, 293–298. (*122*)

Simpson, S. B. (1961). Induction of limb regeneration in the lizard, *Lygosoma laterale*, by augmentation of the nerve supply. *Proc. Soc. exp. Biol. Med.*, **107**, 108–111. (*39*)

Singer, M. (1942). The nervous system and regeneration of the forelimb of adult *Triturus*. I. The role of the sympathetics. *JEZ*, **90**, 377–399. (*23–7, 102*)

Singer, M. (1943). II. The role of the sensory supply. *JEZ*, **92**, 297–315. (*25, 36*)

Singer, M. (1945). III. The role of the motor supply. *JEZ*, **98**, 1–21. (*25*)

Singer, M. (1946a). IV. The stimulating action of a regenerated motor supply. *JEZ*, **101**, 221–239. (*26, 51*)

261

Singer, M. (1946b). V. The influence of number of nerve fibers, including a quantitative study of limb innervation. *JEZ*, **101**, 299–337. (*22, 25–7*)

Singer, M. (1947a). VI. A further study of the importance of nerve number, including quantitative measurements of limb innervation. *JEZ*, **104**, 223–249. (*27*)

Singer, M. (1947b). VII. The relation between number of nerve fibers and surface area of amputation. *JEZ*, **104**, 251–265. (*27*)

Singer, M. (1949). The invasion of the epidermis of the regenerating forelimb of the urodele, *Triturus*, by nerve fibers. *JEZ*, **111**, 189–209. (*148*)

Singer, M. (1952). The influence of the nerve in regeneration of the amphibian extremity. *Quart. Rev. Biol.*, **27**, 169–200. (*32*)

Singer, M. (1954). Induction of regeneration of the forelimb of the postmetamorphic frog by augmentation of the nerve supply. *JEZ*, **126**, 419–471. (*32, 39, 67*)

Singer, M. (1959). The acetylcholine content of the normal forelimb regenerate of the adult newt, *Triturus*. *DB*, **1**, 603–620. (*103*)

Singer, M. (1960). Nervous mechanisms in the regeneration of body parts in vertebrates, in *Developing Cell Systems* (Ed. D. Rudnick), pp. 115–133, Ronald Press, New York. (*37–40*)

Singer, M. (1961). Induction of regeneration of body parts in the lizard, *Anolis*. *Proc. Soc. exp. Biol. Med.*, **107**, 106–108. (*39*)

Singer, M. (1965). A theory of the trophic nervous control of amphibian limb regeneration, including a re-evaluation of quantitative nerve requirements, in *Regeneration in Animals and Related Problems* (Eds. V. Kiotsis and H. A. L. Trampusch), pp. 20–30, North-Holland, Amsterdam. (*32, 48*)

Singer, M. (1974). Neurotrophic control of limb regeneration in the newt. *Ann. N.Y. Acad. Sci.*, **228**, 308–321. (*37*)

Singer, M., and J. D. Caston (1972). Neurotrophic dependent of macromolecular synthesis in the early limb regenerate of the newt, *Triturus*. *JEEM*, **28**, 1–11. (*41–2*)

Singer, M., and L. Craven (1948). The growth and morphogenesis of the regenerating forelimb of adult *Triturus* following denervation at various stages of development. *JEZ*, **108**, 279–308. (*42, 147, 214*)

Singer, M., M. H. Davis, and E. S. Arkowitz (1960). Acetylcholinesterase activity in the regenerating forelimb of the adult newt, *Triturus*. *JEEM*, **8**, 98–111. (*103*)

Singer, M., and F. R. L. Egloff (1949). The nervous system and regeneration of the forelimb of adult *Triturus*. VIII. The effect of limited nerve quantities on regeneration. *JEZ*, **111**, 295–314. (*29, 31*)

Singer, M., D. Flinker, and R. L. Sidman (1956). Nerve destruction by colchicine resulting in suppression of limb regeneration in adult *Triturus*. *JEZ*, **131**, 267–299. (*38, 106*)

Singer, M., and J. Ilan (1977). Nerve-dependent regulation of absolute rates of protein synthesis in newt limb regenerates. *DB*, **57**, 174–187. (*101*)

Singer, M., and S. Inoue (1964). The nerve and the epidermal apical cap in regeneration of the forelimb of adult *Triturus*. *JEZ*, **155**, 105–116. (*150*)

Singer, M., R. P. Kamrin, and A. Ashbaugh (1957). The influence of denervation upon trauma-induced regenerates of the forelimb of the post-metamorphic frog. *JEZ*, **136**, 35–51. (*39, 69*)

Singer, M., C. E. Maier, and W. S. McNutt (1976). Neurotrophic activity of brain extracts in forelimb regeneration of the urodele, *Triturus*. *JEZ*, **196**, 131–150. (*37*)

Singer, M., and E. Mutterperl (1963). Nerve fiber requirements for regeneration in forelimb transplants of the newt *Triturus*. *DB*, **7**, 180–191. (*28–9, 45–7*)

Singer, M., E. K. Ray, and A. M. Peadon (1964). Regional growth differences in the early regenerate of the adult newt, *Triturus viridescens*, correlated with the position of the larger nerves. *Folia Biol. Krakow*, **12**, 347–362. (*187*)

Singer, M., K. Rzehak, and C. S. Maier (1967). The relation between the caliber of the axon and the trophic activity of nerves in limb regeneration. *JEZ*, **166**, 89–98. (*32*)

Singer, M., and M. M. Salpeter (1961). Regeneration in vertebrates: the role of the wound epithelium, in *Growth in Living Systems* (Ed. M. Zarrow), pp. 277–311, Basic Books, New York. *(149)*

Skowron, S., and H. Roguski (1958). Regeneration from implanted dissociated cells. I. Regenerative potentialities of limb- and tail-cells. *Folia Biol. Krakow*, **6**, 163–173. *(92, 125, 158)*

Skowron, S., K. Rzehak, and K. Maron (1963). The effect of blastemal tissue on regeneration. *Folia Biol. Krakow*, **11**, 259–267. *(108)*

Skowron, S., and J. Walknoswska (1963). The development of regeneration buds after transplantation. *Folia Biol. Krakow*, **11**, 421–442. *(108)*

Slack, J. M. W. (1976). Determination of polarity in the amphibian limb. *Nature, London*, **261**, 44–46. *(163, 216)*

Slack, J. M. W. (1977). Determination of anteroposterior polarity in the axolotl forelimb by an interaction between limb and flank rudiments. *JEEM*, **39**, 151–168. *(216, 227)*

Slack, J. M. W. (1980). A serial threshold theory of regeneration. *J. theoret. Biol.*, **82**, 105–140. *(204, 216)*

Slack, J. M. W., and S. Savage (1978a). Regeneration of reduplicated limbs in contravention of the complete circle rule. *Nature, London*, **271**, 760–761. *(216)*

Slack, J. M. W., and S. Savage (1978b). Regeneration of mirror symmetrical limbs in the axolotl. *Cell*, **14**, 1–8. *(216)*

Smith, A. R., and A. M. Crawley (1977). The pattern of cell division during growth of the blastema of regenerating newt forelimbs. *JEEM*, **37**, 33–48. *(145, 203)*

Smith, A. R., and 3 others (1974). A quantitative study of blastemal growth and bone regression during limb regeneration in *Triturus cristatus*. *JEEM*, **32**, 375–390. *(7, 78, 202)*

Smith, S. D. (1967). Induction of partial limb regeneration in *Rana pipiens* by galvanic stimulation. *Anat. Rec.*, **158**, 89–97. *(114)*

Smith, S. D. (1974). Effects of electrode placement on stimulation of adult frog limb regeneration. *Ann. N.Y. Acad. Sci.*, **238**, 500–507 *(114)*

Smith, S. D., and G. L. Crawford (1969). Initiation of regeneration in adult *Rana pipiens* limbs by injection of homologous liver nuclear RNP. *Oncology*, **23**, 299–307. *(107–9)*

Spallanzani, L. (1768). *Prodromo sa un Opera da Imprimersi sopra le Riproduzioni animali*. Modena. *(7)*

Steen, T. P. (1968). Stability of chondrocyte differentiation and contribution of muscle to cartilage during limb regeneration in the axolotl (*Siredon mexicanum*). *JEZ*, **167**, 49–78. *(15, 134, 138–9)*

Steen, T. P. (1970a). Cell differentiation during salamander limb regeneration, in *Regeneration of Striated Muscle and Myogenesis* (Eds. A. Mauro, S. A. Shafiq, and A. T. Milhorat), pp. 73–90, Excerpta Medica, Amsterdam. *(140)*

Steen, T. P. (1970b). Origin and differentiative capacities of cells in the blastema of the regenerating salamander limb. *Am. Zool.*, **10**, 119–132. *(139)*

Steen, T. P., and C. S. Thornton (1963). Tissue interaction in amputated aneurogenic limbs of *Ambystoma* larvae. *JEZ*, **154**, 207–221. *(46, 150)*

Sterling, K., and 4 others (1978). Mitochondrial thyroid hormone receptor: localization and physiological significance. *Science*, **201**, 1126–1129. *(236)*

Stinson, B. D. (1963). The response of X-irradiated limbs of adult urodeles to normal tissue grafts. I. Effects of autografts of sixty-day forearm regenerates. *JEZ*, **153**, 37–52. *(188)*

Stinson, B. D. (1964a). II. Effects of autografts of anterior or posterior halves of sixty-day forearm regenerates. *JEZ*, **155**, 1–24. *(81–2, 187–9)*

Stinson, B. D. (1964b). III Comparative effects of autografts of complete forearm regenerates and longitudinal half forearm regenerates. *JEZ*, **156**, 1–18. *(81, 187)*

Stinson, B. D. (1964c). IV. Comparative effects of autografts and homografts of complete forearm regenerates. *JEZ*, **157**, 159–178. (*81, 126*)

Stocum, D. L. (1968a). The urodele limb regneration blastema: a self-organizing system. I. Differentiation in vitro. *DB*, **18**, 441–456. (*110, 142, 198*)

Stocum, D. L. (1968b). II. Morphogenesis and differentiation of autografted whole and fractional blastemas. *DB*, **18**, 457–480. (*142, 147, 159, 198–9*)

Stocum, D. L. (1975a). Regulation after proximal or distal transposition of limb regeneration blastemas and determination of the proximal boundary of the regenerate. *DB*, **45**, 112–136. (*126, 179*)

Stocum, D. L. (1975b). Outgrowth and pattern formation during limb ontogeny and regeneration. *Differentiation*, **3**, 167–182. (*148, 200*)

Stocum, D. L. (1978a). Regeneration of symmetrical hindlimbs in larval salamanders. *Science*, **200**, 790–793. (*213*)

Stocum, D. L. (1978b). Organization of the morphogenetic field in regenerating amphibian limbs. *Am. Zool.*, **18**, 883–896. (*214*)

Stocum, D. L. (1979). Stages of forelimb regeneration in *Ambystoma maculatum*. *JEZ*, **209**, 395–416. (*6*)

Stocum, D. L., and G. E. Dearlove (1972). Epidermal-mesodermal interaction during morphogenesis of the limb regeneration blastema in larval salamanders. *JEZ*, **181**, 49–61. (*147, 151, 198*)

Stocum, D. L., and D. A. Melton (1977). Self-organizational capacity of distally transplanted regeneration blastemas in larval salamanders. *JEZ*, **201**, 451–461. (*179*)

Sturdee, A., and M. Connock (1975). The embryonic limb bud of the urodele: morphological studies of the apex. *Differentiation*, **3**, 43–49. (*149*)

Summerbell, D. (1979). The zone of polarizing activity: evidence for a rôle in normal chick limb morphogenesis. *JEEM*, **50**, 217–233. (*226*)

Swett, F. H. (1924). Regeneration after amputation of abnormal limbs in *Amblystoma*. *Anat. Rec.*, **27**, 273–287. (*147*)

Swett, F. H. (1927). Differentiation of the amphibian limb. *JEZ*, **47**, 385–435. (*163–4, 227*)

Swett, F. H. (1928). Further experiments on the mediolateral axis of the fore limb of *Amblystoma punctatum* (Linn.). *JEZ*, **50**, 51–70. (*163–4, 227*)

Swett, F. H. (1932). Reduplications in heteroplastic limb grafts. *JEZ*, **61**, 129–148. (*146, 162*)

Swett, F. H. (1937). Experiments upon delayed determination of the dorsoventral limb-axis in *Amblystoma punctatum* (Linn). *JEZ*, **75**, 143–153. (*227*)

Swett, F. H. (1938). Experiments upon the relationship of surrounding areas to polarisation of the dorsoventral limb-axis in *Amblystoma punctabum* (Linn.). *JEZ*, **78**, 81–100. (*227*)

Taban, C. (1949). Les fibres nerveuses et l'épithelium dans l'édification des régénérats de pattes (in situ ou induites) chez le triton. *Arch. Sci.*, **2**, 553–561. (*148*)

Taban, C., M. Cathieni, and M. Schorderet (1978). Cyclic AMP and noradrenaline sensitivity fluctuations in regenerating newt tissue. *Nature, London*, **271**, 470–472. (*102*)

Tank, P. W. (1977). The timing of morphogenetic events in the regenerating forelimb of the axolotl, *Ambystoma mexicanum*. *DB*, **57**, 15–32. (*175*)

Tank, P. W. (1978a). The occurrence of supernumerary limbs following blastemal transplantation in the regenerating forelimb of the axolotl, *Ambystoma mexicanum*. *DB*, **62**, 143–161. (*125, 222*)

Tank, P. W. (1978b). The failure of double-half forelimbs to undergo distal transformation following amputation in the axolotl, *Ambystoma mexicanum*. *JEZ*, **204**, 325–336. (*214–15*)

Tank, P. W. (1979). Positional information in the forelimb of the axolotl: experiments with double half-limbs. *DB*, **73**, 11–24. (*214–15*)

Tank, P. W., B. M. Carlson, and T. G. Connelly (1976). A staging system for forelimb regeneration in the axolotl, *Ambystoma mexicanum*. *JM*, **150**, 117–128. (*6*)

Tank, P. W., B. M. Carlson, T. G. Connelly (1977). A scanning electron microscopic comparison of the development of embryonic and regenerating limbs in the axolotl. *JEZ*, **201**, 417–429. (*149*)

Tank, P. W., and N. Holder (1978). The effect of healing time on the proximodistal organization of double-half forelimb regenerates in the axolotl, *Ambystoma mexicanum*. *DB*, **66**, 72–85. (*214–15*)

Tank, P. W., and N. Holder (1979). The distribution of cells in the upper forelimb of the axolotl, *Ambystoma mexicanum*. *JEZ*, **209**, 435–442. (*209*)

Tassava, R. A. (1969a). Hormonal and nutritional requirements for limb regeneration and survival of adult newts. *JEZ*, **170**, 33–53. (*54, 61–3*)

Tassava, R. A. (1969b). Survival and limb regeneration of hypophysectomized newts with pituitary xenografts from larval axolotls, *Ambystoma mexicanum*. *JEZ*, **171**, 451–458. (*54, 65*)

Tassava, R. A., F. J. Chlapowski, and C. S. Thornton (1968). Limb regeneration in *Ambystoma* larvae during and after treatment with adult pituitary hormones. *JEZ*, **167**, 157–164. (*65–6*)

Tassava, R. A., L. L. Bennett, and G. D. Zitnick (1974). DNA synthesis without mitosis in amputated denervated forelimbs of larval axolotls. *JEZ*, **190**, 111–116. (*20, 43*)

Tassava, R. A., and D. J. Garling (1979). Regenerative responses in larval axolotl limbs with skin grafts over the amputation surface. *JEZ*, **208**, 97–109. (*150*)

Tassava, R. A., and R. M. Lloyd (1977). Injury requirement for initiation of regeneration of newt limbs which have whole skin grafts. *Nature, London*, **268**, 49–50. (*150*)

Tassava, R. A., and W. D. McCullough (1978). Neural control of cell cycle events in regenerating salamander limbs. *Am. Zool.*, **18**, 843–854. (*18, 43*)

Tassava, R. A., and A. L. Mescher (1975). The role of injury, nerves, and the wound epidermis during initiation of amphibian limb regeneration. *Differentiation*, **4**, 23–24. (*43*)

Tassava, R. A., and A. L. Mescher (1976). Mitotic activity and nucleic acid precursor incorporation in denervated and innervated limb stumps of axolotl larvae. *JEZ*, **195**, 253–262. (*20, 43*)

Taurog, A. and 4 others (1974). The role of TRH in the neoteny of the Mexican axolotl (*Ambystoma mexicanum*). *Gen. comp. Endocrinol.*, **24**, 267–279. (*61*)

Taylor, A. C., and J. J. Kollros (1946). Stages in the normal development of *Rana pipiens* larvae. *Anat. Rec.*, **94**, 7–23. (*234*)

Taylor, R. E., and S. B. Barker (1965). Transepidermal potential difference: development in anuran larvae. *Science*, **148**, 1612–1613. (*114*)

Teplits, N. A. (1962). Changes in the nucleic acid content in the limb of the axolotl after irradiation with roentgen rays and restoration of the power of regeneration. *DAN*, **147**, 244–247 (AIBS translation pp. 1312–1315). (*88*)

Teplits, N. A. (1964a). Changes in nucleic acid content following X-irradiation suppression and restoration of the regenerative power of axolotl limbs. *DAN*, **156**, 1207–1209 (AIBS Biophysics pp. 94–96). (*100*)

Teplits, N. A. (1964b). Changes in nucleic acid content of axolotl limb tissues after X-irradiation and during normal regeneration. *DAN*, **159**, 442–445 (AIBS Biophysics pp. 169–172). (*100*)

Thornton, C. S. (1938a). The histogenesis of muscle in the regenerating forelimb of larval *Amblystoma punctatum*. *JM*, **62**, 17–47.

Thornton, C. S. (1938b). The histogenesis of the regenerating fore limb of larval *Amblystoma* after exarticulation of the humerus. *JM*, **62**, 219–241. (*85, 139, 147*).

Thornton, C. S. (1942). Studies on the origin of the regeneration blastema in *Triturus viridescens*. *JEZ*, **89**, 375–390. (*81, 92, 126, 159*)

Thornton, C. S. (1943). The effect of colchicine on limb regeneration in larval *Amblystoma. JEZ*, **92**, 281–295. (*106*)

Thornton, C. S. (1949). Beryllium inhibition of regeneration. I. Morphological effects of beryllium on amputated fore limbs of larval *Amblystoma. JM*, **84**, 459–494. (*107, 110*)

Thornton, C. S. (1950). II. Localization of the beryllium effect in amputated limbs of larval *Amblystoma JEZ*, **114**, 305–333. (*107, 110*)

Thornton, C. S. (1951). III. Histological effects of beryllium on the amputated fore limbs of *Amblystoma* larvae. *JEZ*, **118**, 467–493. (*107, 110*)

Thornton, C. S. (1953). Histological modifications in denervated injured fore limbs of *Amblystoma* larvae. *JEZ*, **122**, 119–149.

Thornton, C. S. (1954). The relation of epidermal innervation to limb regeneration in *Amblystoma* larvae. *JEZ*, **127**, 577–601. (*149*)

Thornton, C. S. (1956a). Epidermal modifications in regenerating and in non-regenerating limbs of anuran larvae. *JEZ*, **131**, 373–393. (*31, 149*)

Thornton, C. S. (1956b). The relation of epidermal innervation to the regeneration of limb deplants in *Amblystoma* larvae. *JEZ*, **133**, 281–299. (*31, 149*)

Thornton, C. S. (1957). The effect of apical cap removal on limb regeneration in *Amblystoma* larvae. *JEZ*, **134**, 357–381. (*150*)

Thornton, C. S. (1958) The inhibition of limb regeneration in urodele larvae by localized irradiation with ultraviolet light. *JEZ*, **137**, 153–179. (*97, 150*)

Thornton, C. S. (1960a). Regeneration of asensory limbs of *Amblystoma* larvae. *Copeia*, **1960**, 371–373. (*92, 150*)

Thornton, C. S. (1960b). Influence of an eccentric epidermal cap on limb regeneration in *Amblystoma* larvae. *DB*, **2**, 551–569. (*151*)

Thornton, C. S. (1962). Influence of head skin on limb regeneration in urodele amphibians. *JEZ*, **150**, 5–15. (*150*)

Thornton, C. S. (1965). Influence of the wound skin on blastemal cell aggregation, in *Regeneration in Animals and Related Problems* (Eds. V. Kiortsis and H. A. L. Trampusch), pp. 333–340, North-Holland, Amsterdam. (*47*)

Thornton, C. S. (1968). Amphibian limb regeneration. *Adv. Morphogenesis*, **7**, 205–249. (*114*)

Thornton, C. S. (1970). Amphibian limb regeneration and its relation to nerves. *Am. Zool.*, **10**, 113–118.

Thornton, C. S., and D. W. Kraemer (1951). The effect of injury on denervated unamputated fore limbs of *Amblystoma* larvae. *JEZ*, **117**, 415–439. (*149, 204*)

Thornton, C. S., and T. P. Steen (1962). Eccentric blastema formation in aneurogenic limbs of *Ambystoma* larvae following epidermal cap deviation. *DB*, **5**, 328–343. (*150*)

Thornton, C. S., and R. A. Tassava (1969). Regeneration and supernumerary limb formation under sparsely innervated conditions. *JM*, **127**, 225–232. (*29, 46–50, 126*)

Thornton, C. S., and M. T. Thornton (1965). The regeneration of accessory limb parts following epidermal cap transplantation in urodeles. *Experientia*, **21**, 146–148. (*147, 151*)

Thornton, C. S., and M. T. Thornton (1970). Recuperation of regeneration in denervated limbs of *Ambystoma* larvae. *JEZ*, **173**, 293–301. (*47–50*)

Todd, T. J. (1823). On the process of reproduction of the members of the aquatic salamander. *Quart. J. Lit. Sci. Arts*, **16**, 84–96. (*22*)

Tompkins, R. (1977). Grafting analysis of the periodic albino mutant of *Xenopus laevis. DB*, **57**, 460–464. (*121*)

Toole, B. P., and J. Gross (1971). The extracellular matrix of the regenerating newt limb: synthesis and removal of hyaluronate prior to differentiation. *DB*, **25**, 57–77. (*104*)

Tornier, G. (1906). Kampf der Gewebe im Regenerat bei Begünstigung der Hautregeneration. *RA*, **22**, 348–369. (*149*)

Torrey, T. W. (1934). The relation of taste buds to their nerve fibers. *J. comp. Neurol.*, **59**, 203–220 *(41)*

Torrey, T. W. (1936). The relation of nerves to degenerating taste buds. *J. comp. Neurol.*, **64**, 325–336. *(41)*

Trampusch, H. A. L. (1951). Regeneration inhibited by X-rays and its recovery. *Proc. K. Ned. Akad. Weten. Amsterdam*, **54C**, 373–385. *(80, 86, 95)*

Trampusch, H. A. L. (1958a). The action of X-rays on the morphogenetic field. I. Heterotopic grafts on irradiated limbs. *Proc. K. Ned. Akad. Weten. Amsterdam*, **61C**, 417–30. *(80, 95, 159)*

Trampusch, H. A. L. (1958b). II. Heterotopic skin on irradiated tails. *Proc. K. Ned. Akad. Weten. Amsterdam*, **61C**, 530–545. *(80, 95, 159)*

Trampusch, H. A. L. (1959). The effect of X-rays on regenerative capacity, in *Regeneration in Vertebrates* (Ed. C. S. Thornton), pp. 83–94, U. Chicago Press, Chicago. *(86, 95, 125)*

Trampusch, H. A. L. (1964). Nerves as morphogenetic mediators in regeneration. *Progr. Brain Res.*, **13**, 214–227. *(91)*

Trampusch, H. A. L. (1972). Cytodifferentiation. *Acta Morphol. Neerl.-Scand.*, **10**, 155–163. *(95, 159)*

Trampusch, H. A. L., and A. E. Harrebomée (1961). Embryonic induction in the adult organism. *Acta Morphol. Neerl.-Scand.*, **4**, 287–288. *(158)*

Trampusch, H. A. L., and A. E. Harrebomée (1961). Embryonic induction in the adult organism. *Acta Morphol. Neerl.-Scand.*, **4**, 287–288. *(158)*

Trampusch, H. A. L., and A. E. Harrebomée (1969). Dedifferentiation and the interconvertibility of the different cell-types in the amphibian extremity. *Acta Emb. exper.*, **1969**, 35–69. *(158)*

Tuchkova, S. Y. (1964a). Causes of the loss of regenerative power of axolotl limbs after X-irradiation. *DAN*, **158**, 1420–1423 (AIBS Biophysics pp. 159–162). *(84, 94)*

Tuchkova, S. Y. (1964b). Changes in the nucleic acid content after suppression of regeneration of the limbs in axolotls by X-ray irradiation. *DAN*, **159**, 215–218 (AIBS translation pp. 791–793). *(84, 94)*

Tuchkova, S. Y. (1966). Role of trauma in restoration of the regenerating power of the limbs in axolotls irradiated with X-rays. *DAN*, **168**, 473–476 (AIBS translation pp. 349–352). *(131)*

Tuchkova, S. Y. (1967a). Changes in the mitotic activity in tissues of axolotl limbs during recovery of powers of regeneration suppressed by roentgen irradiation. *DAN*, **172**, 222–224 (AIBS translation pp. 37–39). *(84–8)*

Tuchkova, S. Y. (1967b). Tissue resources active in restoring X-ray depressed limb regenerative capacity in the axolotl. *DAN*, **176**, 1181–1184 (AIBS translation pp. 591–593). *(84)*

Tuchkova, S. Y. (1969). Autoradiographic investigation on DNA synthesis in axolotl limbs during recovery of regenerative power suppressed by X-irradiation. *DAN*, **186**, 724–727 (AIES translation pp. 452–456). *(84–8)*

Tuchkova, S. Y. (1973a). Autoradiographic data on the participation of irradiated tissues in the regeneration of axolotl extremities suppressed by roentgen irradiation. *DAN*, **210**, 986–988 (AIBS translation pp. 202–204). *(88)*

Tuchkova, S. Y. (1973b). Morphogenetic potencies of regeneration blastema studied by methods of transplantation and autoradiography. *DAN*, **213**, 1217–1220 (AIBS translation pp. 490–493). *(140)*

Tuchkova, S. Y. (1976). On the participation of irradiated tissues in the formation of a limb regenerate in axolotls. *DAN*, **228**, 752–755 (AIBS translation pp. 243–245). *(89, 131)*

Tweedle, C. (1971). Transneuronal effects on amphibian limb regeneration. *JEZ*, **177**, 13–29. *(7–9, 35, 94)*

Umanski, E. (1937). Untersuchung des Regenerationsvorganges bei Amphibien mittels Ausschaltung der einzelnen Gewebe durch Röntgenbestrahlung. *Biol. Zhurn.*, **6**, 757–758 (German summary). *(62, 81, 95, 159)*

Umanski, E. (1938a). The regeneration potencies of axolotl skin studied by means of exclusion of the regeneration capacity of tissues through exposure to X-rays. *Bull. Biol. Méd. exp. URSS*, **6**, 141–145. *(81, 95, 159)*

Umanski, E. (1938b). A study of regeneration of the azolotl limb upon substitution of the inner tissues by dorsal musculature. *Bull. Biol. Méd. exp., URSS*, **6**, 383–386. *(159)*

Umanski, E. E., V. K. Tkatsch and V. P. Kondokotsev (1951). Dielectric anisotropy of the skin in axolotls (in Russian). *DAN*, **76**, 465–467. *(81, 117)*

Vale, W., G. Grant and R. Guilleimin (1973). Chemistry of the hypothalamic releasing factors, in *Frontiers of Neuroendocrinology* (Eds. W. Ganong and L. Martin), vol. 3, pp. 375–413, Oxford U.P., Oxford. *(61)*

Van Stone, J. M. (1955). The relationship between innervation and regenerative capacity in hind limbs of *Rana sylvatica*. *JM*, **97**, 345–392. *(31)*

Van Stone, J. M. (1964). The relationship of nerve number to regenerative capacity in the developing hind limb of *Rana sylvatica*. *JEZ*, **155**, 293–302. *(31)*

Van't Hof, J. (1965). Relationships between mitotic cycle duration, S-period duration and the average rate of DNA synthesis in the root meristem of several plants. *Exp. Cell. Res.*, **39**, 48–58. *(17)*

Varga, L., R. Wenner, and E. del Pozo (1973). Treatment of galacto-amenorrhea syndrome with Br-ergocryptine (CB 154): restoration of ovulatory function and fertility. *Am. J. Obstet. Gynec.*, **117**, 75–79. *(56)*

Vergroesen, A. J. (1958). Het aandeel van de zenuwen tijdens de regeneratie van bestraalde ledematen bij Amphibia. *Ned. Tidjschr. Geneesk.*, **102**, 1624. *(91)*

Verma, K. (1965). Regional differences in skin gland differentiation in *Rana pipiens*. *JM*, **117**, 73–85. *(236)*

Vethamany-Globus, S., and R. A. Liversage (1973a). In vitro studies of the influence of hormones on tail regeneration in adult *Diemictylus viridescens*. *JEEM*, **30**, 397–413. *(64, 111)*

Vethamany-Globus, S., and R. A. Liversage (1973b). The relationship between the anterior pituitary gland and the pancreas in tail regeneration of the adult newt. *JEEM*, **30**, 415–426. *(64)*

Vethamany-Globus, S., and R. A. Liversage (1973c). Effects of insulin insufficiency on forelimb and tail regeneration in adult *Diemictylus viridescens*. *JEEM*, **30**, 427–447. *(64)*

Vethamany-Globus, S., and 4 others (1977). Hormone control in regeneration: effects of somatostatin on appendage regeneration, blood glucose and liver glycogen in *Diemictylus viridescens*. *JEEM*, **40**, 115–124. *(56)*

Vethamany-Globus, S., M. Globus and B. Tomlinson (1978). Neural and hormonal stimulation of DNA and protein synthesis in cultured regeneration blastemata in the newt *Notophthalmus viridescens*. *DB*, **65**, 183–192. *(111)*

Vorontsova, M. A. (1929). Morphogenetische Analyse der Farbung bei weissen Axolotln. *RA*, **115**, 93–109. *(66)*

Vorontsova, M. A. (1937). Die Regenerationspotenzen der Muskulatur verschiedener Extremitätensegmente beim Axolotl. *Bull. Biol. Méd. exp. URSS*, **45**, 156. *(180)*

Vorontsova, M. A., and L. D. Liosner (1960). *Asexual Propagation and Regeneration*, Pergamon, London. *(229)*

Wake, D. B. (1976). On the correct scientific names of urodeles. *Differentiation*, **6**, 195. *(3)*

Wallace, B., and H. Wallace (1973). Participation of grafted nerves in amphibian limb regeneration. *JEEM*, **29**, 559–570. *(95, 143)*

Wallace, H. (1963). Nucleolar growth and fusion during cellular differentiation. *JM*, **112**, 261–278. *(121)*

Wallace, H. (1972). The components of regrowing nerves which support the regeneration of irradiated salamander limbs. *JEEM*, **28**, 419–435. *(80, 91, 95, 142)*

Wallace, H. (1978). Testing the clockface model of amphibian limb regeneration. *Experientia*, **34**, 1360–1361. *(221)*

Wallace, H. (1980). Regeneration of reversed aneurogenic arms of the axolotl. *JEEM*, **56**, 309–317. *(184–5)*

Wallace, H., and M. Maden (1976a). Irradiation inhibits the regeneration of aneurogenic limbs. *JEZ*, **195**, 353–358. *(80, 93)*

Wallace, H., and M. Maden (1976b). The cell cycle during amphibian limb regeneration. *J. Cell Sci.*, **20**, 539–547. *(16–19)*

Wallace, H., M. Maden, and B. M. Wallace (1974). Participation of cartilage grafts in amphibian limb regeneration. *JEEM*, **32**, 391–404. *(125, 136–7, 147)*

Wallace, H., and A. Watson (1979). Duplicated axolotl regenerates. *JEEM*, **49**, 243–258. *(210, 221–2)*

Wallace, H., S. Wessels, and H. Conn (1971). Radioresistance of nerves in amphibian limb regeneration. *RA*, **166**, 219–225. *(80, 91, 171)*

Waller, A. (1850). Experiments on the section of the glossopharyngeal and hypoglossal nerves of the frog, and observations of the alterations produced thereby in the structure of the primitive fibres. *Phil. Trans. R. Soc., London*, **140**, 423–429. *(35)*

Warren, E. A., and C. M. Bower (1939). The influence of normal and induced metamorphosis on hind limb regeneration in *Rana sylvatica*. *Anat. Rec.* (suppl.), **73**, 55–56. *(67)*

Weber, M. and K. Maron (1965). The effect of blastemal tissue on regeneration. II. Influence of blastema cell fractions on regenerative growth. *Folia Biol. Krakow*, **13**, 383–396. *(108)*

Weis, J. (1972). The effects of nerve growth factor on bullfrog tadpoles (*Rana catesbiana*) after limb amputation. *JEZ*, **180**, 385–391. *(39)*

Weis, J. S., and L. P. Bleier (1973). The effects of hydrocortisone on limb regeneration in the bullfrog tadpole, *Rana catesbiana*. *JEEM*, **29**, 65–71. *(67)*

Weis, J. S., and P. Weis (1970). The effect of nerve growth factor on limb regeneration in *Ambystoma*. *JEZ*, **174**, 73–78. *(39)*

Weiss, P. (1924). Regeneration aus doppelten Extremitätenquerschitt (an *Triton cristatus*). *Anz. Akad. Wiss. Wein, Math-naturw. Kl.*, **61**, 45–47. *(189)*

Weiss, P. (1925). Abhängikeit der Regeneration entwickelter Amphibienextremitäten vom Nervensystem. *Arch. f. mik. Anat.*, 104, 317–358. *(23)*

Weiss, P. (1926). Ganzregenerate aus Halbem Extremitätenquerschnitt. *RA*, **107**, 1–53. *(146, 186)*

Weiss, P. (1927). Potenzprufung am Regenerationsblastem. I. Extremitätenbildung aus Schwanzblastem in Extremitätenfeld bei Triton. *RA*,, **111**, 317–340. *(157)*

Weiss, P. (1939). *Principles of Development*, Holt, New York. *(12, 139, 195)*

Weiss, P. (1973). Differentiation and its three facets: facts, terms, and meaning. *Differentiation*, **1**, 3–10. *(12)*

Weiss, P., and H. B. Hiscoe (1948). Experiments on the mechanism of nerve growth. *JEZ*, **107**, 315–395 *(35)*

Wertz, R. L., and D. J. Donaldson (1979). Effects of X-rays on nerve-dependent (limb) and nerve-independent (jaw) regeneration in the adult newt, *Notophthalmus viridescens*. *JEEM*, **53**, 315–325. *(81, 94)*

Wertz, R. L., and D. J. Donaldson (1980). Early events following amputation of adult newt limbs given regeneration-inhibitory doses of X irradiation. *DB*, **74**, 434–445. *(84, 94)*

Wertz, R. L., D. J. Donaldson and J. M. Mason (1976). X-ray induced inhibition of DNA synthesis and mitosis in internal tissues during the initiation of limb regeneration in the adult newt. *JEZ*, **198**, 253–259. *(81, 84)*

Wey-Schué, M. (1953). Étude histologique de l'origine et de la migration des cellules de régénération chez *Triturus cristatus*. *Archs Zool. exp. gén.*, **91**, 5–16. (*78*)

Whimster, I. W. (1965). An experimental approach to the problem of spottiness. *Brit. J. Derm.*, **77**, 397–420. (*226*)

Whimster, I. W. (1971). The group behaviour of pigment cells. *Trans. St. Johns Hospital Derm. Soc.*, **57**, 57–86. (*226*)

Whimster, I. W. (1979). The focal differentiation of pigment cells. *JEZ*, **208**, 153–159. (*226*)

Wilde, C. E. (1950). Studies on the organogenesis in vitro of the urodele limb bud. *JM*, **86**, 73–113. (*110*)

Wilder, I. W. (1924). The relation of growth to metamorphosis in *Eurycea bislineata* (Green). *JEZ*, **40**, 1–112. (*77*)

Wilkerson, J. A. (1963). The role of growth hormone in regeneration of the forelimb of the hypophysectomized newt. *JEZ*, **154**, 223–230. (*54*)

Williams, D. D. (1959). Limb regeneration in the salamander, *Triturus viridescens*, after large initial and during prolonged smaller injections of cortisone acetate. *Endocrinology*, **64**, 292–296. (*64*)

Wolff, E., and M. Wey-Schué (1952). Démonstration expérimentale de la migration des cellules de régénération dans la régénération des membres de *Triton cristatus*. *C.R. Soc. Biol.*, **146**, 113–117. (*77, 80*)

Wolpert, L. (1969). Positional information and the spatial pattern of cellular differentiation. *J. theoret. Biol.*, **25**, 1–47. (*195, 202*)

Wolpert, L. (1971). Positional information and pattern formation. *Current Topics in Dev. Biol.*, **6**, 183–224 (*195*)

Wolsky, A. (1974). The effects of chemicals with gene-inhibiting activity on regeneration, in *Neoplasia and Cell Differentiation* (Ed. G. V. Sherbet), pp. 153–188, Karger, Basel. (*104–5, 144*)

Wolsky, A., and N. Van Doi (1965). The effect of actinomycin on regeneration processes in amphibians. *Trans. N.Y. Acad. Sci.*, **27**, 882–892. (*107*)

Wolsky, A., and M. I. Wolsky (1969). Nucleic acid control of regeneration in amphibian larvae. *Ann. Emb. Morph.*, Suppl. **1**, 248–249. (*107*)

Wolsky, M. de I., and L. Fogarty (1962). Effect of regeneration blastema autotransplantation on subsequent regeneration in *Triturus (Diemictylus) viridescens*. *Nature, London*, **195**, 621–622. (*108*)

Wright, M. R., and M. E. Plumb (1960). Regeneration in hypophysectomized anuran tadpoles. *Proc. Soc. exp. Biol. Med.*, **103**, 672–674. (*67*)

Yamada, T., and D. S. McDevitt (1974). Direct evidence for transformation of differentiated iris epithelial cells into lens cells. *DB*, **38**, 104–118. (*140*)

Yamada, T., M. E. Roesel, and J. J. Beauchamp (1975). Cell cycle parameters in dedifferentiating iris epithelial cells. *JEEM*, **34**, 497–510. (*17–18*)

Yankow, M. (1965). Effect of growth inhibitors and their combinations on the tail regeneration of amphibia. *Trans. N.Y. Acad. Sci.*, **27**, 878–881. (*106*)

Yntema, C. L. (1959a). Regeneration in sparsely innervated and aneurogenic forelimbs of *Amblystoma* larvae. *JEZ*, **140**, 101–123. (*31, 46, 92*)

Yntema, C. L. (1959b). Blastema formation in sparsely innervated and aneurogenic forelimbs of *Amblystoma* larvae. *JEZ*, **142**, 423–439. (*46*)

Yntema, C. L. (1962). Duplication and innervation in anterior extremities of *Amblystoma* larvae. *JEZ*, **149**, 127–145. (*46*)

Young, J. Z. (1942). The functional repair of nervous tissue. *Physiol. Rev.*, **22**, 318–374. (*35*)

Zalik, S. E., and T. Yamada (1967). The cell cycle during lens regeneration. *JEZ*, **165**, 385–393. (*18*)

Zelena, J. (1964). Development, degeneration and regeneration of receptor organs. *Prog. Brain Res.*, **13**, 175–213. (*36*)

Zwilling, E. (1956). Interaction between limb bud ectoderm and mesoderm in the chick embryo. IV. Experiments with a wingless mutant. *JEZ*, **132**, 241–253. (*119*)

Zwilling, E. (1974). Effects of contact between mutant (wingless) limb buds and those of genetically normal chick embryos: confirmation of a hypothesis. *DB*, **39**, 37–48. (*119*)

Zwilling, E., and L. A. Hansborough (1956). Interaction between limb bud ectoderm and mesoderm in the chick embryo. III. Experiments with polydactylous limbs. *JEZ*, **132**, 219–239. (*119*)

Subject Index

274

Micronucleus, 86, 105
Mitotic index, 10–11, 19–20, 87
Modulation, 12, 132–4, 141–4, 153
Morphallaxis, 195
Morphogenetic field, 94–7, 157–62,
    187–90, 194–5, 224–5
  mutants, 119
Muscle grafts, 135, 159, 168, 173–7
  homogenates, 128–30
Myleran, 106–8
Myoblast, 103, 153–4
Myosin, 103–4

Necturus maculosus, 2
Neoblast, 12, 140
Nerve, 22–52
  collateral, 31
  motor, 22–6, 35, 225
  regrowth, 22–6, 46–51
  sensory, 22–6, 35, 226
  structure, 33–5
  sympathetic, 23–7, 102
Nerve dependency, 22–5, 31–3, 41–8,
    101
  quantitative, 25–33, 48
Nerve deviation, 32–5, 39, 156, 176–8,
    225–7
Nerve extract, 38, 41
Nerve fibre count, 25–9, 31–2
  density, 27–9, 31–2, 45
Nerve graft, 91, 95, 140–3, 180
Nerve growth factor, 39–40
Nervous control, 22–52
  irradiation, 91–4, 97
  mechanism, 33–43
Neuronal chromatolysis, 9, 35
  degeneration, 234–5
  integrity, 35–6
Neurotransmitter, 41, 102
Neurotrophic factor, 37–51, 57, 93, 102,
    154, 232–4
  theory, 36–51
Neutrons, 97
Nitrogen mustard, 106–8
Norepinephrine, 41, 102
Notch, 6–8
Notophthalmus viridescens, 2–3
  allografts, 122, 126
  blastema culture, 64, 111
  blastemal grafts, 165–7, 178, 208–10
  blastemal physiology, 9–19, 101–11
  denervation, 22–9, 42–3, 111–13, 150
  double limb stump, 97, 189–91
  extirpated skeleton, 139

half blastema, 146
half limb stump, 186–8, 212–13
hypophysectomy, 53–7, 62–6
irradiation, 80–94, 113, 134, 187
larvae, 29, 65–6
nerve fibres 25–30
neurotrophic factor, 37–41
rate of regeneration, 6–8, 60
reversed arm, 182
stump current, 115
supernumerary arms, 28, 45, 126,
    165–6, 208–10
surface potential, 112–13
thyroidectomy, 60–1
thyroxine, 59–62, 233
tissue grafts, 89, 134, 140, 152, 157,
    212–13
transplanted arm, 45
Nuclear transplantation, 121, 144
Nucleolus, 89–90, 100, 121, 133–8
Nucleus, 72–3, 83–6,

Organ specificity, 156–60
Osteoclast, 14

Palette, 6–8
Pancreas, 64
Parabiosis, 46
Parathyroid hormone, 61
Perchlorate, 61, 107–9, 236
Phosphatidyl choline, 41
Pituitary, 53–8, 65–7
  extirpation, 53–7, 60–7
  grafts, 54–7, 65–6, 122
Planarian, 140, 194
Pleurodeles waltl, 2–3
  allografts, 124
  blastemal grafts, 218–21
  defective regenerates, 60
  rate of regeneration, 59–60
  surface potential, 111, 114
  supernumerary arms, 59, 176–7
  tissue grafts, 169–73, 176–8
Polar confrontation, 217–18
Polar coordinates, 211
Polarity, 194, 201–5, 217–18
Polarizing zone, 146, 163, 218, 226–8
Polydactyly, 105, 119–20
Positional information, 195, 202, 211, 224
Potential difference, 111–15
Progress zone, 202–4
Prolactin, 53–7, 61, 67
Protein Synthesis, 38–42, 100–4, 107–11
  inhibitors, 104–8